The Handbook of Environmental Chemistry

Volume 3 Anthropogenic Compounds
Part X

O. Hutzinger
Editor-in-Chief

Springer

Berlin
Heidelberg
New York
Hong Kong
London
Milan
Paris
Tokyo

Synthetic Musk Fragrances in the Environment

Volume Editor: Gerhard G. Rimkus

With contributions by
G. G. Rimkus · C. Sommer · H.-D. Eschke · C. Fooken
P. E. G. Leonards · J. de Boer · R. Kallenborn
R. Gatermann · W. Butte · T. Heberer · S. Jürgensen
H. Fromme · J. Hajšlová · L. Šetková · S. Biselli
L. K. Sydnes · H. Hühnerfuss · D. R. Dietrich
B. C. Hitzfeld · S. Schwarz · V. Berding · M. Matthies
H. Brunn · N. Bitsch · J. Amberg-Müller · P. Slanina
F. Balk · D. Salvito · H. Blok

Springer

Volume Editor

Dr. Gerhard G. Rimkus
Official Food and Veterinary
Institute Schleswig-Holstein
Department of Residue
and Contamination Analysis
P.O. Box 2743
24517 Neumünster
Germany

Present address:
European Commission
DG SANCO (Health and
Consumer Protection)
Food and Veterinary Office (FVO)
Grange, Dunsany
Co. Meath, Ireland
E-mail: Gerhard.Rimkus@cec.eu.int

ISSN 1433-6847
ISBN 3-540-43706-1
DOI 10.1007/b14173
Springer-Verlag Berlin Heidelberg New York

Library of Congress Cataloging-in-Publication Data
The Natural environment and the biogeochemical cycles / with
contributions by P. Craig ... [et al.].
v. <A–F > : ill. ; 25 cm. – (The Handbook of environmental chemistry :
v. 1) Includes bibliographical references and indexes.
ISBN 0-387-09688-4 (U.S). – ISBN 3-540-55255-3 (pt. F : Berlin). –
ISBN 0-387-55255-3 (pt. F : New York)
1. Biogeochemical cycles. 2. Environmental chemistry.
I. Craig, P. J., 1944– . II. Series.
QD31. H335 vol. 1 [QH344] 628.5 s

Springer-Verlag Berlin Heidelberg New York
is a part of Springer Science+Business Media GmbH
springeronline.com

Printed in Germany

Production Editor: Christiane Messerschmidt, Rheinau
Cover Design: E. Kirchner, Springer-Verlag
Typesetting: Fotosatz-Service Köhler GmbH, Würzburg
Printed on acid-free paper 52/3020 – 5 4 3 2 1 0

The Handbook of Environmental Chemistry is also Available Electronically

For all customers who have a subscription to The Handbook of Environmental Chemistry, we offer the electronic version via SpringerLink free of charge. Please contact your librarian who can receive a password for free access to the full articles by registering at:

http://www.springerlink.com

If you do not have a subscription, you can still view the tables of contents of the volumes and the abstract of each article by going to the SpringerLink Homepage, clicking on "Browse by Online Libraries", then "Chemical Sciences", and finally choose The Handbook of Environmental Chemistry.

You will find information about the

- Editorial Bord
- Aims and Scope
- Instructions for Authors
- Sample Contribution

at http://www.springeronline.com using the search function.

Preface

Environmental Chemistry is a relatively young science. Interest in this subject, however, is growing very rapidly and, although no agreement has been reached as yet about the exact content and limits of this interdisciplinary discipline, there appears to be increasing interest in seeing environmental topics which are based on chemistry embodied in this subject. One of the first objectives of Environmental Chemistry must be the study of the environment and of natural chemical processes which occur in the environment. A major purpose of this series on Environmental Chemistry, therefore, is to present a reasonably uniform view of various aspects of the chemistry of the environment and chemical reactions occurring in the environment.

The industrial activities of man have given a new dimension to Environmental Chemistry. We have now synthesized and described over five million chemical compounds and chemical industry produces about hundred and fifty million tons of synthetic chemicals annually. We ship billions of tons of oil per year and through mining operations and other geophysical modifications, large quantities of inorganic and organic materials are released from their natural deposits. Cities and metropolitan areas of up to 15 million inhabitants produce large quantities of waste in relatively small and confined areas. Much of the chemical products and waste products of modern society are released into the environment either during production, storage, transport, use or ultimate disposal. These released materials participate in natural cycles and reactions and frequently lead to interference and disturbance of natural systems.

Environmental Chemistry is concerned with reactions in the environment. It is about distribution and equilibria between environmental compartments. It is about reactions, pathways, thermodynamics and kinetics. An important purpose of this Handbook, is to aid understanding of the basic distribution and chemical reaction processes which occur in the environment.

Laws regulating toxic substances in various countries are designed to assess and control risk of chemicals to man and his environment. Science can contribute in two areas to this assessment; firstly in the area of toxicology and secondly in the area of chemical exposure. The available concentration ("environmental exposure concentration") depends on the fate of chemical compounds in the environment and thus their distribution and reaction behaviour in the environment. One very important contribution of Environmental Chemistry to the above mentioned toxic substances laws is to develop laboratory test methods, or mathematical correlations and models that predict the environmental fate of new chemical compounds. The third purpose of this Handbook is to help in the basic understanding and development of such test methods and models.

The last explicit purpose of the Handbook is to present, in concise form, the most important properties relating to environmental chemistry and hazard assessment for the most important series of chemical compounds.

At the moment three volumes of the Handbook are planned. Volume 1 deals with the natural environment and the biogeochemical cycles therein, including some background information such as energetics and ecology. Volume 2 is concerned with reactions and processes in the environment and deals with physical factors such as transport and adsorption, and chemical, photochemical and biochemical reactions in the environment, as well as some aspects of pharmacokinetics and metabolism within organisms. Volume 3 deals with anthropogenic compounds, their chemical backgrounds, production methods and information about their use, their environmental behaviour, analytical methodology and some important aspects of their toxic effects. The material for volume 1, 2 and 3 was each more than could easily be fitted into a single volume, and for this reason, as well as for the purpose of rapid publication of available manuscripts, all three volumes were divided in the parts A and B. Part A of all three volumes is now being published and the second part of each of these volumes should appear about six months thereafter. Publisher and editor hope to keep materials of the volumes one to three up to date and to extend coverage in the subject areas by publishing further parts in the future. Plans also exist for volumes dealing with different subject matter such as analysis, chemical technology and toxicology, and readers are encouraged to offer suggestions and advice as to future editions of "The Handbook of Environmental Chemistry".

Most chapters in the Handbook are written to a fairly advanced level and should be of interest to the graduate student and practising scientist. I also hope that the subject matter treated will be of interest to people outside chemistry and to scientists in industry as well as government and regulatory bodies. It would be very satisfying for me to see the books used as a basis for developing graduate courses in Environmental Chemistry.

Due to the breadth of the subject matter, it was not easy to edit this Handbook. Specialists had to be found in quite different areas of science who were willing to contribute a chapter within the prescribed schedule. It is with great satisfaction that I thank all 52 authors from 8 countries for their understanding and for devoting their time to this effort. Special thanks are due to Dr. F. Boschke of Springer for his advice and discussions throughout all stages of preparation of the Handbook. Mrs. A. Heinrich of Springer has significantly contributed to the technical development of the book through her conscientious and efficient work. Finally I like to thank my family, students and colleagues for being so patient with me during several critical phases of preparation for the Handbook, and to some colleagues and the secretaries for technical help.

I consider it a privilege to see my chosen subject grow. My interest in Environmental Chemistry dates back to my early college days in Vienna. I received significant impulses during my postdoctoral period at the University of California and my interest slowly developed during my time with the National Research

Council of Canada, before I could devote my full time of Environmental Chemistry, here in Amsterdam. I hope this Handbook may help deepen the interest of other scientists in this subject.

Amsterdam, May 1980 *O. Hutzinger*

Twentyone years have now passed since the appearance of the first volumes of the Handbook. Although the basic concept has remained the same changes and adjustments were necessary.

Some years ago publishers and editors agreed to expand the Handbook by two new open-end volume series: Air Pollution and Water Pollution. These broad topics could not be fitted easily into the headings of the first three volumes. All five volume series are integrated through the choice of topics and by a system of cross referencing.

The outline of the Handbook is thus as follows:

1. The Natural Environment and the Biochemical Cycles,
2. Reaction and Processes,
3. Anthropogenic Compounds,
4. Air Pollution,
5. Water Pollution.

Rapid developments in Environmental Chemistry and the increasing breadth of the subject matter covered made it necessary to establish volume-editors. Each subject is now supervised by specialists in their respective fields.

A recent development is the accessibility of all new volumes of the Handbook from 1990 onwards, available via the Springer Homepage http://www.springer. de or http://Link.springer.de/series/hec/ or http://Link.springerny.com/ series/hec/.

During the last 5 to 10 years there was a growing tendency to include subject matters of societal relevance into a broad view of Environmental Chemistry. Topics include LCA (Life Cycle Analysis), Environmental Management, Sustainable Development and others. Whilst these topics are of great importance for the development and acceptance of Environmental Chemistry Publishers and Editors have decided to keep the Handbook essentially a source of information on "hard sciences".

With books in press and in preparation we have now well over 40 volumes available. Authors, volume-editors and editor-in-chief are rewarded by the broad acceptance of the "Handbook" in the scientific community.

Bayreuth, July 2001 *Otto Hutzinger*

Contents

Foreword

More than 100 years ago a nitroaromatic compound with a typical and strong musk odor, a so-called nitro musk, was found by chance in the laboratory. This discocery was the spark that started a fast growing worldwide industry producing and distributing a variety of synthetic musk fragrances with different chemical structures. These compounds are used as substitutes for natural musk, not only in all kind of cosmetics, but also in household and industrial products. In 1981, two nitro musks were found for the first time in fish, mussels, and water samples taken from the area of Tokyo, Japan. These analysis data provided the first evidence that synthetic musk fragrances had reached the aquatic environment and the food chain. However, it was only the findings of nitro and polycyclic musks in European fish and humans at the beginning of the 1990s which initiated a broad discussion and many research activities on this new group of environmental pollutants.

In several aspects, the synthetic musk fragrances are similar to the "classical" organic environmental pollutants like DDT, PCBs etc. and they are typical xenobiotics. Due to their lipophilicity and persistence, they are stored as well in the lipid of aquatic organisms as in sediment and sewage sludge. Because of the worldwide production and usage of these chemicals, there is a ubiquitous distribution. From an analytical point of view, the nitro musks can be screened sensitively by the same method (GC-ECD) as the organochlorine pollutants. However, there are also several differences worth mentioning. To date there has been a continuous production and use of synthetic musks. Consequently, there has been a constant input into the environment. In contrast, the classical pollutants including the POPs are mostly not in use anymore (in particular in western countries) and in general, a decrease in their environmental concentrations can be observed. The synthetic musk levels in human samples such as human milk and adipose tissue are mainly caused not by the uptake of polluted food but by absorption through the skin. This explains why the levels in human samples in general do not correlate with the age. Skin absorption of lipophilic substances from cosmetics is a new aspect for both researchers and the cosmetic industry. In this case, chemicals used daily in personal care and household products, effectively operate as environmental pollutants. Therefore, the high levels of contamination which have been observed in the aquatic environment are in general not found at industrial or agricultural sites, but in highly populated regions, especially in the sewage treatment plants. Several synthetic musk compounds are therefore specific and sensitive indicator substances for sewage contamination. The environmental impact of synthetic musks is an example of the life cycle assessment of chemicals and demonstrates the necessity to assess the fate of a chemical after its usage.

From the regulatory point of view several nitro musks are now forbidden from use in cosmetics by European legislation. Other nitro musks are regulated by maximum authorized concentrations in finished cosmetic products. The polycyclic musks are under discussion in scientific committees advising the European Commission. Up till now, there have been no regulations for the use of synthetic musks in non-cosmetic products.

This monograph represents the first comprehensive survey of the subject, compiled by the relevant research groups working in this field. This state-of-the-art review systematically summarizes all data, results and discussion topics of the last 10 years. Also many as yet unpublished data complete this survey. Naturally, this report does not represent a final assessment on the issue, and it also highlights the gaps in scientific knowledge where more research and monitoring efforts are needed. For example more information is needed on the metabolism of these compounds in humans and in the aquatic environment, including elucidation of the toxicity of the resulting metabolites. The parameters and factors influencing musk concentrations in human samples remain to be established. In particular, more knowledge is needed about the mechanism of skin absorption and the role of indoor pollution and air-borne transport of synthetic musks. Comprehensive long-term monitoring programs will be necessary to study in detail temporal contamination trends in environmental and human samples. With classical musk fragrances being substituted to an increasing extent by the so-called macrocyclic compounds, data on their environmental behavior and toxicity need to be generated, given the lack of published information on these substances. In general, the monograph highlights the fact that more work is urgently needed in the fields of toxicology and risk assessment in order to comprehensively evaluate the toxic risk to both consumer and environment from these substances.

I am honored to be able to acknowledge the efforts of all of the colleagues involved in preparing this monograph. Time is indeed precious, and free time is all too rare in the scientific world. Given the busy schedules of all concerned, it should be mentioned that most chapters were complied in the private time of the authors. Therefore I am very grateful that we have been able to complete this book in spite of several adverse conditions. In particular, I would like to thank Prof. Dr. O. Hutzinger for inviting me to compile this book and to Springer-Verlag for their encouragement and long-lasting patience with editor and authors. Last but not least, I would like to thank my family for their understanding and support during the many evenings and weekends taken to complete this project.

Finally I hope and wish that the present monograph will inspire further research projects and monitoring studies in order to further our understanding of this important subject.

Navan (Ireland), December 2003 Gerhard G. Rimkus

The Handbook of Environmental Chemistry Vol. 3, Part X (2004): 1–16
DOI 10.1007/b14130

The Role of Musk and Musk Compounds in the Fragrance Industry

Cornelia Sommer

Official Food and Veterinary Institute (LVUA) Schleswig-Holstein, Eckernförder Straße 421, 24107 Kiel, Germany
E-mail: Cornelia.Sommer@lvua-sh.de

Abstract An overview of the role of musk and musk compounds in the fragrance industry is given. Discovery and syntheses of representatives occurring naturally in animals and plants as well as of artificial substances possessing musk-like odor properties are reviewed. Examples of the three major classes – nitro musks, polycyclic musks, and macrocyclic musks – are covered. The importance of these compounds as fragrance ingredients of cosmetics and detergents is shown. The impact of environmental and toxicological data on the actual use and ongoing developments of this important class of fragrances are described.

Keywords Musk · Musk deer · Nitro musks · Polycyclic musks · Macrocyclic musks

1 Introduction

Musk is a gland secretion produced by the male musk deer (*Moschus moschiferus* L.) which has been used as fragrance material for centuries [1, 2]. In addition, the term "musk" also refers to a diverse spectrum of chemically defined substances which are quite different in their chemical structures but exhibit a common, distinct, and typical flavor. These musk compounds comprise representatives occurring naturally in animals and plants as well as artificial substances possessing musk-like odor properties [3,.4].

The use of musk flavor has a long history dating back to ancient times. Until the end of the nineteenth century the popular fragrance was only obtained from natural sources. Nowadays synthesized compounds are almost exclusively used [5]. They can be divided into three major classes: aromatic nitro musks, polycyclic substances, and macrocyclic musk compounds [6].

Representatives of the first two groups are broadly applied in industry [7]. They are components of fragrance compositions, which are added to cosmetics (e.g., perfumes, soaps, and creams) and to detergents. The detection of nitro musks in fish and human matrices (milk, fat) initiated a public debate on the use of these compounds. Later, the polycyclic musk compounds, which were increasingly used to replace the nitro musks, were also detected in environmental and human samples [8–10]. Therefore, macrocyclic musk compounds are expected to be of increasing importance in the future [5].

2 Natural Musk

An animal secretion called "musk" is the carrier of the natural musk aroma. It is produced by the male musk deer (*Moschus moschiferus* L.) in a gland situated in the prenuptial region between the abdomen and the genitals [1, 11–13]. The musk deer (Fig. 1) belongs to the family Moschidae and reaches approximately the size of the central European roe deer. It lives in upper regions of Eastern Asia, e.g., India, Tibet, China, Siberia, and Mongolia [3, 11–15].

Fig. 1 Musk deer (*Moschus moschiferus* L.) [11]

In order to get access to the natural musk, the animal must be killed to remove the gland, also called musk pod (Fig. 2). The fully developed pods (50–70 g) contain about 40% musk [11]. Upon drying, the reddish-brown paste turns into a black, granular material (musk grain) which is used for alcoholic solutions. The aroma of the tincture, which is described for example as animal-like, earthy, and woody, becomes more intensive during storage. Only after considerable dilution does the obtained extract exhibit a pleasant odor [1, 2, 14]. No other natural product possesses such a complex aroma associated with many often contradictory descriptions [16]. The commercially used products are differentiated according to their provenance. The best qualities, called Tonkin musk, originate from Tibet and China [1, 14].

Discovery and use of musk date back to ancient China and pre-historic India. In these societies musk was of extraordinary cultural importance and was also used as a universal drug [15]. The crusaders eventually brought musk from the Orient to Europe. There it was also used as drug as well as ingredient of perfumes. It was highly appreciated due to its properties to enhance, harmonize, and round off perfume compositions [1, 15]. Comparable to ancient times, musk is still today one of the most expensive natural products [15]. In 1998 the value of 1 g of musk ranged from 30 to 50 US $. Thus, its price was higher than that of gold (10 US $ g^{-1}) [17].

Owing to the limited availability, the high price, and attempts to save the musk animals, the fragrance industry increasingly replaces natural musk by chemically synthesized musk compounds [1, 15].

Trade of musk from Afghanistan, Bhutan, India, Myanmar, Nepal, and Pakistan has been forbidden since 1979 by the Convention on International Trade in Endangered Species of Wild Fauna and Flora (CITES) and the import of musk from other countries is restricted by control of documents. Despite these regulations,

Fig. 2 Musk pods [62]

musk animals are still an endangered species. The main reason is the use of musk in traditional Asian medicine. About 500–1000 kg musk per year are used in China for production of drugs, resulting in the death of about 100,000 animals [17].

The main sources of musk used by the fragrance industry today are China, Arabia, and Russia [17]. In the European Union the trade of musk from China and Russia has been forbidden since 1999 [18]. In recent years, France has been the only European country using natural musk (annual amount in the kg range) [17].

3 History of Compounds with Musk Odor

Owing to the limited availability and the high price of natural musk, there were early attempts to find replacements. First indications date from 1759, when the chemist Markgraf detected products with musk-like odors in the course of the nitration of amber oil. Although these results were of no immediate practical importance, they stimulated and influenced future investigations [15]. In 1890, several years before the isolation and structural elucidation of the natural carrier of the musk aroma, Baur succeeded in synthesizing the first chemically defined substance with musk odor by nitration of *meta-tert*-butyl-toluene [3, 5, 15]. 2-(1,1-Dimethylethyl)-4-methyl-1,3,5-trinitro-benzene (Fig. 3) was patented and commercialized as "Musc Baur" [19]. Later, other members of this class of compounds, called nitro musks, were synthesized and gained considerable commercial importance.

In contrast to the development of synthetic musk compounds, the first major success of research activities on the natural musk constituents was only reported in 1906 [3]. Walbaum isolated a ketone, which he named muscone, as the major odor-contributing constituent of the secretion from the musk gland [20]. In 1915 Sack isolated another ketone with musk odor from the secretion of an animal called civet cat (*Viverra civetta* L.), which he named civetone [21]. In 1926 Ruzicka et al. eventually succeeded in characterizing muscone as 3-methylcyclopentadecanone and civetone as cycloheptadecen-1-one and confirmed their structures by synthesis [22–25]. This was the discovery of a new class of compounds, the macrocyclics [26]. One year later Kerschbaum detected additional macrocyclic lactones in angelica root oil and in ambrette seed oil [3, 27]. In 1928 Stoll and Ruzicka synthesized these compounds and identified them as cyclohexadecenolide (e.g., Ambrettolide) in ambrette seed oil and as cyclopentadecanolide (e.g., Exal-

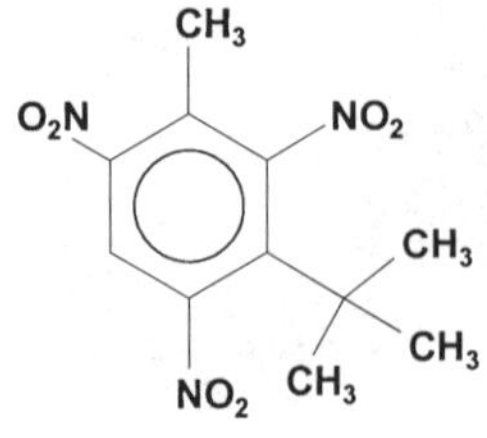

Fig. 3 Chemical structure of 2-(1,1-dimethylethyl)-4-methyl-1,3,5-trinitro-benzene ("Musc Baur")

tolide) in angelica root oil [3, 26]. In 1942 Stevens and Erickson identified cyclopentadecanone (e.g., Exaltone) and cycloheptadecanone (e.g., Dihydrocivetone) obtained from the American musk rat (*Ondatra zibethica* L.) [26].

The importance of these macrocyclic fragrance compounds of animal and plant origin stimulated the development of improved syntheses meeting the demands of industrial applications. However, the yields and the prices did not fulfill the expectations [15, 26]. Therefore, there was a search for compounds which could be synthesized more easily. This was achieved in the 1950s by the synthesis of the so-called polycyclic musk compounds, another nitro-free group of musks [3, 5]. In 1951 the synthesis of 6-acetyl-1,1,2,3,3,5-hexamethyldihydroindene (AHDI) (e.g., Phantolide) was described. Starting from this first industrially important member of this class of musks a broad spectrum of polycyclic musk compounds has been developed [6].

4
Synthetic Musk Compounds

Musk compounds traditionally belong to the most important substances used in the fragrance industry [28]. On one hand this is due to their odor properties which can be divided into types such as animal-like, flowery, and fruity. On the other hand, they are appreciated because of their abilities to improve the fixation of compounds and to round off fragrance compositions [3, 29]. Increased fixation improves the effectiveness of fragrances by slowing down the release of volatiles, thus contributing to a defined and stable quality over an extended period [2]. They are also known to bind fragrances to fabrics. Therefore, they are added as perfumery ingredients not only to cosmetics but also to detergents [30].

Synthetic musks comprise a broad spectrum of different substances. Commercially, only nitro derivatives, polycyclic, and macrocyclic compounds are of importance [4]. For many years the nitro musks dominated the market. Since 1983 their share has decreased continuously by 5% per year. In 1987 the total amount (7000 tonnes) of musk compounds produced worldwide comprised 61% polycyclic, 35% nitro musks, and 3–4% macrocyclic compounds [28].

4.1
Nitro Musk Compounds

The era of nitro musk compounds began with the discovery of the so-called "Musc Baur" by Baur at the end of the nineteenth century [19]. In the following years, other aromatic nitro compounds were synthesized, which gained considerable importance as replacements for natural musk. These artificial substances exhibit musk-like odors although they are structurally very different from the naturally occurring musk compounds [3, 5, 6, 15]. The best known nitro musks (musk ketone, musk xylene, musk ambrette, musk tibetene, musk moskene) are listed in Table 1. They are two- or threefold nitrated benzene derivatives with additional alkyl, keto, or methoxy groups. Musk moskene, synthesized in 1932 and identified as a dinitroindane derivative in 1955, can be seen as intermediate between nitro musks and the nitro-free indane substances (polycyclic musks) [5, 6].

Table 1 Commercially important nitro musks

CAS Name CAS No.	Trivial name	Molecular formula	Chemical structure
1-(1,1-Dimethylethyl)-3,5-dimethyl-2,4,6-trinitrobenzene 81–15–2	Musk xylene, Musk xylol	$C_{12}H_{15}N_3O_6$	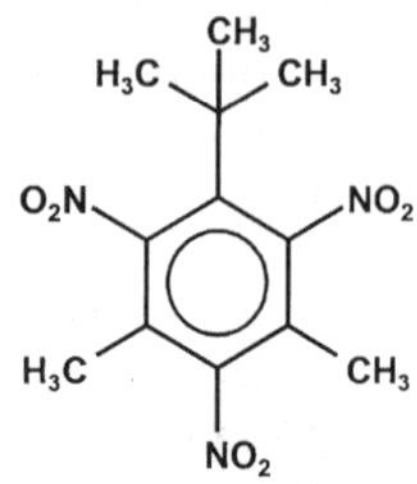
1-[4-(1,1-Dimethyl-ethyl)-2,6-dimethyl-3,5-dinitrophenyl]-ethanone 81–14–1	Musk ketone	$C_{14}H_{18}N_2O_5$	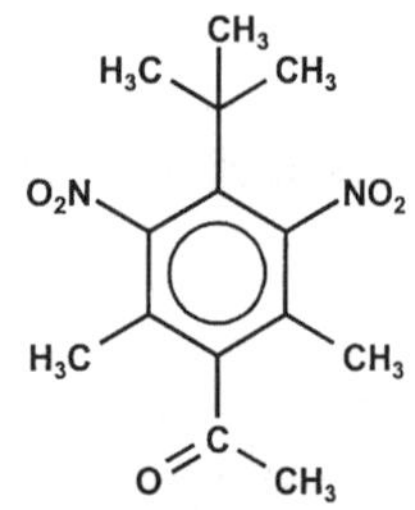
1-(1,1-Dimethylethyl)-2-methoxy-4-methyl-3,5-dinitro-benzene 83–66–9	Musk ambrette	$C_{12}H_{16}N_2O_5$	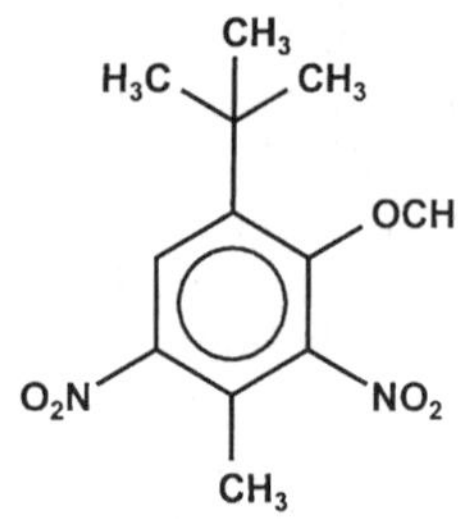
1-(1,1-Dimethylethyl)-3,4,5-trimethyl-2,6-dinitrobenzene 145–39–1	Musk tibetene	$C_{13}H_{18}N_2O_4$	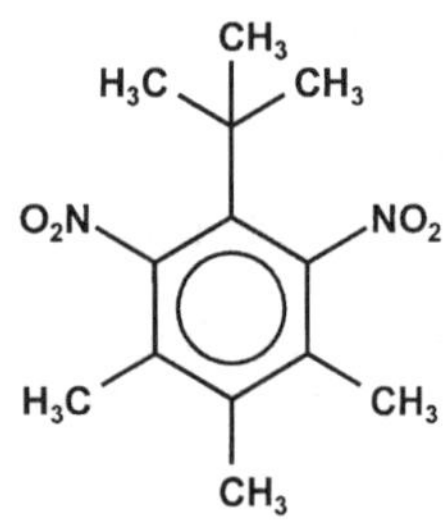
2,3-Dihydro-1,1,3,3,5-pentamethyl-4,6-dinitro-1*H*-indene 116–66–5	Musk moskene	$C_{14}H_{18}N_2O_4$	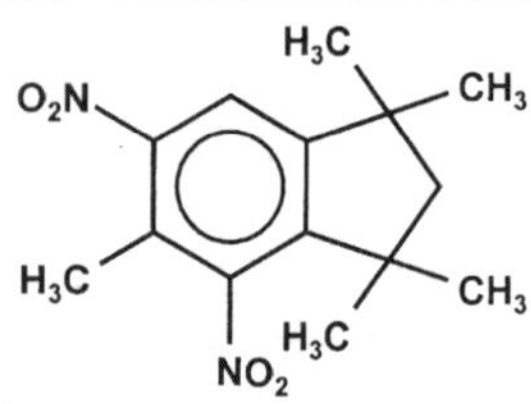

Comparable to the other musk substances, nitro musks are appreciated because of their odors, their role in fixation, and their versatile technological applicabilities [7, 15]. For many years they were the musk compounds produced in highest amounts, especially because of their low prices [6]. However, starting from 1983 the production rate decreased mainly because of reports on photoallergic reactions elicited by musk ambrette [28]. In 1981 musk xylene and musk ketone were detected for the first time in fish and water in Japan; the presence of both compounds in these samples was explained by their potential for bioaccumulation in aquatic systems [31, 32]. In 1983 musk xylene was also detected in fish in the USA. However, a final interpretation of these results was not possible due to potential laboratory contamination [33].

In 1993 the detection of $\mu g\ kg^{-1}$ (on wet weight basis) amounts musk xylene, musk ketone, and musk ambrette in fish initiated a broad public debate on the use of nitro musk compounds. Subsequent investigations of samples from humans (milk, fat) revealed the presence of musk xylene and musk ketone and in a few samples of musk ambrette and musk moskene [34–36].

In order to locate potential sources of contamination, the content of nitro musk compounds in low-priced cosmetics and detergents marketed in Germany was surveyed in 1992. It was found that 55% of the investigated cosmetics (perfumes, shaving lotions, shower gels, shampoos, creams) and 41.5% of the detergents contained nitro musks. There were significant differences in the amounts detected, e.g., musk ketone concentrations in cosmetics ranged from 4.0 to 2200 $mg\ kg^{-1}$. Musk ketone dominated in cosmetics; musk xylene was the main representative in detergents (Fig. 4). Musk ambrette could only be found in one cosmetic product [37]. This is in agreement with results reported by the Food and Drug Administration (FDA) [38, 39]. It reflects the voluntary compliance of the

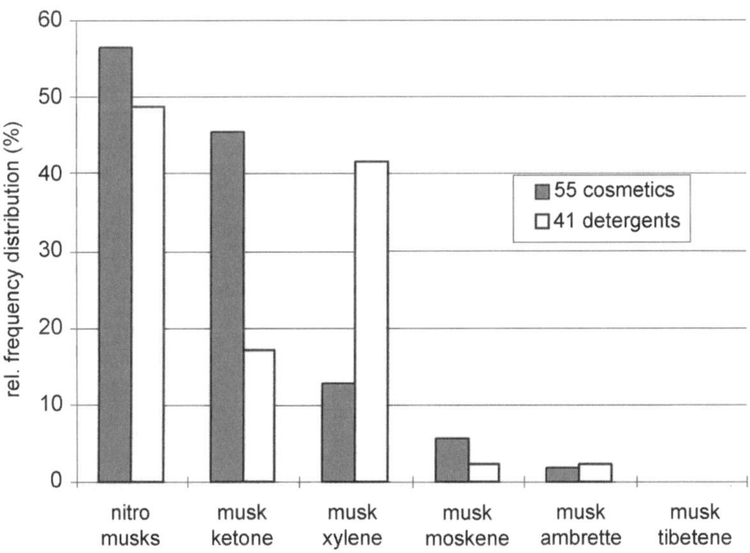

Fig. 4 Frequency distribution of nitro musks in cosmetics and detergents in 1992 [37]

fragrance industry with the 1985 recommendation of the International Fragrance Association (IFRA) not to use musk ambrette in any fragranced products coming into contact with the skin [37–39].

In 1993 the public discussion on nitro musks resulted in a recommendation of the German Cosmetic, Toiletry, Perfumery and Detergent Association (IKW) not to use musk xylene for the production of cosmetics, detergents, and other household products. The decision was based on the bioaccumulation of this compound and its potential carcinogenicity [5, 40].

In 1995 the strong photo-allergenicity of musk ambrette resulted in a prohibition of this compound in the production of cosmetics in the European Union [41]. Since 1998 musk moskene and musk tibetene are also included in the list of compounds which according to directive 76/768/EEC are not allowed to be used in cosmetics [42]. Recently, the Scientific Committee on Cosmetic Products and Non-Food Products (SCCNFP) of the EU Commission recommended the implementation of limits for the use of musk xylene and musk ketone in cosmetics [43, 44].

In Switzerland the prohibition of musk ambrette and limits for the other nitro musks were already implemented by 1995. The maximum concentrations of nitro musks are 50 mg kg^{-1} in deodorants and skin care products, 200 mg kg^{-1} in aqueous-alcoholic products, and 500 mg kg^{-1} in Eaux de Cologne and Eaux de Toilette. Shampoos and perfumes must be free of nitro musks [45, 46].

The intensive debate on nitro musks is also reflected in the commercial use of this group of musk compounds [47, 48]. In 1996 investigations of low-price cosmetics and detergents (mainly produced in Germany) revealed only 7 (12.5%) out of a total of 56 cosmetics to contain musk ketone, xylene, and tibetene. In the 33 detergents no nitro musks could be detected. A comparison with data obtained in 1992 showed that almost all producers of cosmetics (Fig. 5) and detergents in Germany had stopped using nitro musks. On the other hand, in 1995 the investigation of a spectrum of 42 high-priced, exclusive cosmetics mainly produced in France demonstrated the use of nitro musks in more than 50% of the products (Fig. 5) [49].

As shown in Table 2, there has been a significant decline in the usage of nitro musks by the European fragrance industry between 1992 and 1998 [50]. Worldwide the proportion of nitro musks (related to the total production of musk compounds) decreased from 35% in 1987 to about 12% in 1996 [6].

4.2 Polycyclic Musk Compounds

The polycyclic musk compounds were not discovered until the 1950s [5]. They are nitro-free substances, which can be divided into indane derivatives, tetraline derivatives, tricyclic compounds, and coumarin derivatives [6, 29]. The most important representatives are listed in Table 3. Analogous to the nitro musk compounds they are artificial compounds which do not occur in nature and have no chemical relationship to the natural musk compounds. Their use began after the synthesis of 6-acetyl-1,1,2,3,3,5-hexamethyl-dihydroindene (AHDI) (e.g., Phantolide) in 1951 [6]. They are appreciated not only because of their attractive odor properties but also because their synthesis is cheaper than that of the macrocyclic

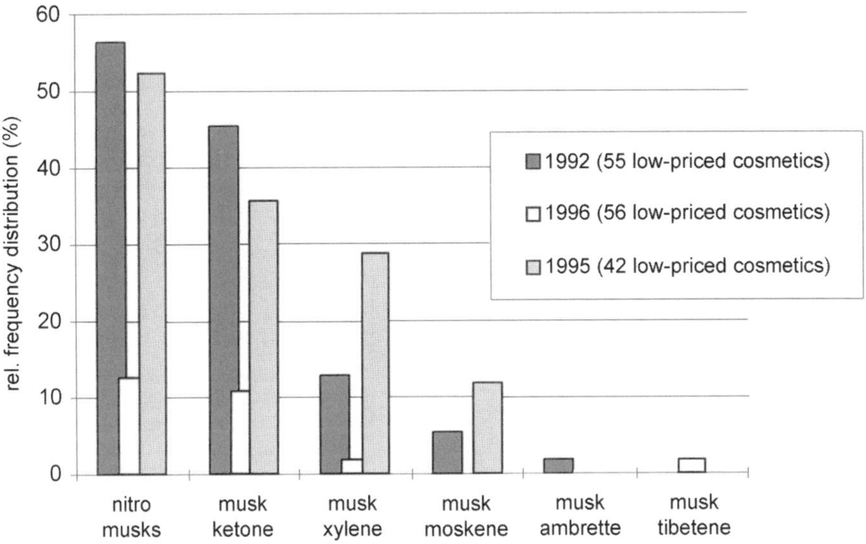

Fig. 5 Frequency distribution of nitro musks in cosmetics [37, 49]

Table 2 Industrial use of musk xylene and other nitro musks in Europe (in tonnes) [50]

Year	Musk xylene	Musk ketone	Musk moskene	Musk tibetene
1992	174	124		
1995	110	61	5	0.8
1998	86	40		

compounds, another group of nitro-free musks. Compared to the nitro musk compounds, they are superior in terms of resistance to light and alkali and in their abilities to bind to fabrics [3, 5, 6, 15]. Accordingly, they are mainly used in cosmetics and detergents. The most important representatives of this class of musks are 7-acetyl-1,1,3,4,4,6-hexamethyltetrahydronaphthalene (AHTN) and 1,3,4,6,7,8-hexahydro-4,6,6,7,8,8-hexamethylcyclopenta[*g*]-2-benzopyrane (HHCB) followed by 4-acetyl-1,1-dimethyl-6-*tert*-butyldihydroindene (ADBI) and 5-acetyl-1,1,2,6-tetramethyl-3-isopropyl-dihydrindene (ATII) [5, 51]. HHCB was used in higher amounts than AHTN in the early 1970s, due to more advanced production procedures and lower price. Since the 1980s these parameters have been comparable for both compounds [15]. 1500 tonnes AHTN and 3800 tonnes HHCB are used per year in the USA and in Europe [51]. These production volumes amount to about 95% of the commercially used polycyclic musk compounds [52]. In contrast, 7-acetyl-1,1,4,4-tetramethyl-6-ethyltetrahydro-naphthalene (ATTN) (Table 3) is only of historical importance. Owing to its neurotoxic properties, production and use have been terminated as from the beginning of the 1980s [5].

The decrease of the production rate of nitro musks was paralleled by an increase for the polycyclic compounds. A market share of 61% in 1987 corre-

Table 3 Commercially important polycyclic musks

CAS name CAS no.	Trade name(s)	Chemical name (abbreviation)	Molecular formula	Chemical structure
1,3,4,6,7,8-Hexahydro-4,6,6,7,8,8-hexamethyl-cyclopenta[g]-2-benzopyrane 1222–05–5	Galaxolide Abbalide Pearlide	1,3,4,6,7,8-*H*exahydro-4,6,6,7,8,8-*h*examethyl*c*yclopenta[*g*]-2-*b*enzo-pyrane (HHCB)	$C_{18}H_{26}O$	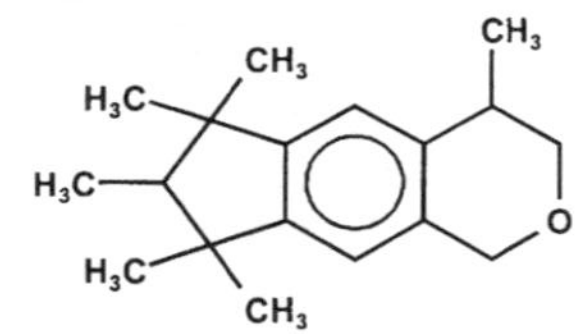
1-(5,6,7,8-Tetrahydro-3,5,5,6,8,8-hexamethyl-2-naphthalenyl)-ethanone 1506–02–1	Tonalide, Fixolide	7-*A*cetyl-1,1,3,4,4,6-*h*examethyl-*t*etrahydro*n*aphthalene (AHTN)	$C_{18}H_{26}O$	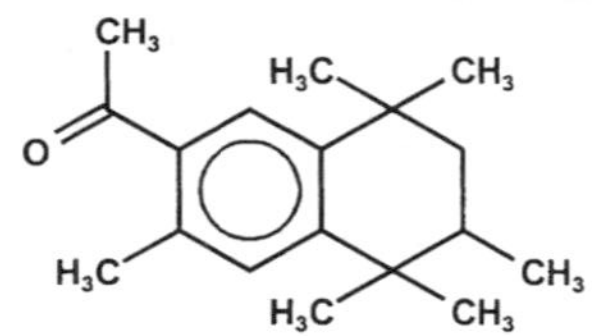
1-[6-(1,1-Dimethylethyl)-2,3-dihydro-1,1-dimethyl-1*H*-inden-4-yl]-ethanone 13171–00–1	Celestolide, Crysolide	4-*A*cetyl-1,1-*d*imethyl-6-*tert.* *b*utyldihydro*i*ndene (ADBI)	$C_{17}H_{24}O$	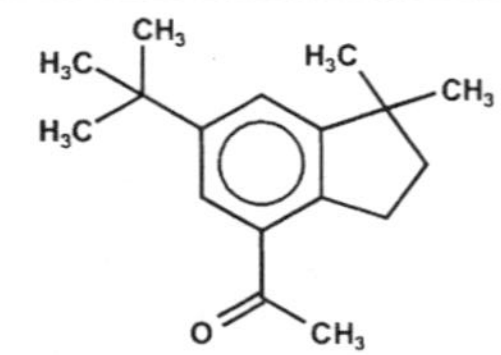
1-(2,3-Dihydro-1,1,2,3,3,6-hexamethyl-1*H*-inden-5-yl)-ethanone 15323–35–0	Phantolide	6-*A*cetyl-1,1,2,3,3,5-*h*examethyl-dihydro*i*ndene (AHDI)	$C_{17}H_{24}O$	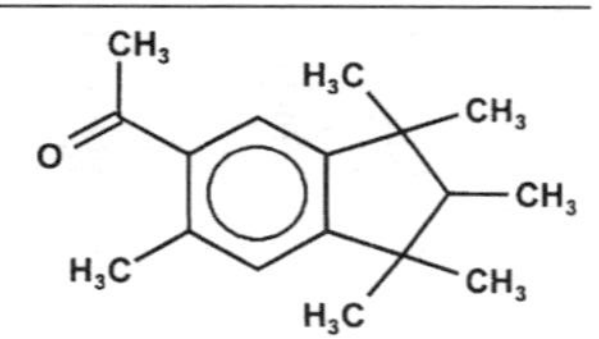

Table 3 (continued)

CAS name CAS no.	Trade name(s)	Chemical name (abbreviation)	Molecular formula	Chemical structure
1,2,3,5,6,7-Hexahydro-1,1,2,3,3-pentamethyl-4H-inden-4-one 33704–61–9	Cashmeran	6,7-*D*ihydro-1,1,2,3,3-*p*enta*m*ethyl-4(5*H*)*i*ndanone (DPMI)	$C_{14}H_{22}O$	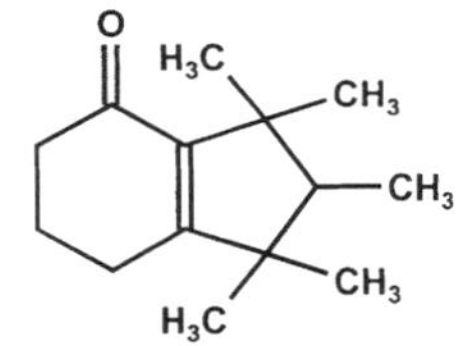
1-[2,3-Dihydro-1,1,2,6-tetra-methyl-3-(1-methyl-ethyl)-1H-inden-5-yl]-ethanone 68140–48–7	Traseolide	5-*A*cetyl-1,1,2,6-*t*etramethyl-3-*i*sopropyl-dihydro*i*ndene (ATII)	$C_{18}H_{26}O$	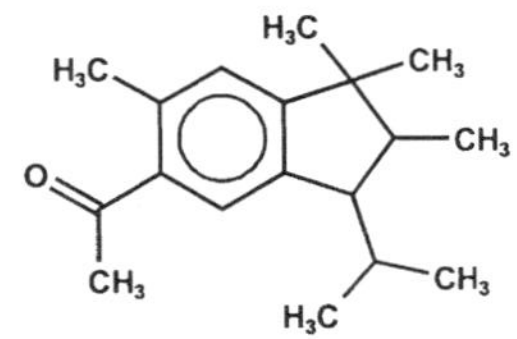
1-(3-Ethyl-5,6,7,8-tetrahydro-5,5,8,8-tetramethyl-2-naphthalenyl)-ethanone 88–29–9	Versalide	7-*A*cetyl-1,1,4,4-*t*etramethyl-6-ethyl*t*etrahydro*n*aphthalene (ATTN)	$C_{18}H_{26}O$	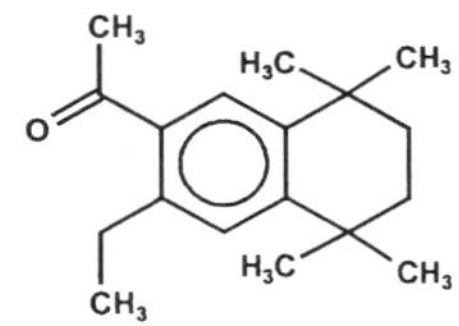

sponding to an amount of about 4300 tonnes per year increased to 70% in 1996 corresponding to 5600 tonnes per year [5, 28]. This development was mainly due to the role of HHCB and AHTN as replacements for the nitro musks [5, 6, 53].

An investigation of cosmetics and detergents in 1994/95 revealed HHCB and AHTN to be the mainly used polycyclic musks. The concentration of HHCB, e.g., in cosmetics ranged from 0.5 to 500 mg kg^{-1} and of AHTN from 1.1 to 520 mg kg^{-1}. Other representatives of this group play only a minor role [10].

The first report on the presence of polycyclic musks in fish and water dates back to 1994 [8]. One year later the compounds were also found in samples from humans (milk, fat) [9]. HHCB and AHTN were analyzed in highest amounts. The values were higher than those determined for the nitro musk compounds [8, 9].

Meanwhile many producers of cosmetics and detergents stopped using polycyclic musk compounds [5]. The effect on the overall use of these compounds in Europe is shown in Table 4 [50]. In the meantime, the polycyclic musk compounds are also being evaluated by the SCCNFP of the EU Commission [7, 54, 55]. A decision of the EU Commission on the regulatory status of HHCB and AHTN is expected [7].

4.3 Macrocyclic Musk Compounds

The development of the macrocyclic musk compounds began in 1926 with the structural characterization of muscone and civetone by Ruzicka and others [15, 22, 24–26]. They demonstrated the compounds to be cyclic macromolecules, the existence of which had been considered impossible according to the so-called "Baeyer's strain theory" [3, 15, 56].

After this breakthrough additional macrocyclic compounds exhibiting musk-like odors were isolated from natural materials, their structures were elucidated, and syntheses were developed [15, 16, 26]. The natural macrocyclic musk compounds turned out to be ketones (animal sources) and lactones (plant materials) [5, 15]. They are 15- or 17-membered ring systems. The type of odor is influenced by the ring size. Starting from 14 ring atoms, a weak musk scent is perceived. Compounds with 15–16 ring atoms exhibit strong musk odor [26].

Owing to their outstanding properties (stability to light and alkaline conditions, fixation, and high quality odors), macrocyclic musk compounds are of high value for the fragrance industry. Accordingly, there have been many attempts to improve syntheses of naturally occurring macrocyclic musks for industrial application and to develop new, more easily accessible members of this class [3, 15, 26]. The synthesized macrocyclic compounds can be divided into ketones, diketones, lactones, oxalactones (ether lactones), dilactones, ketolactones, and esters. Some of the most prominent examples are listed in Table 5.

In addition to the naturally occurring representatives, a wide array of other substances not being found in nature has been synthesized [26, 57–59]. One of the most important compounds of this group (production of about 300 tonnes per year) is the dilactone ethylene brassylate [59]. Ethylene brassylate is an inexpensive musk compound because of its easy synthesis and the low costs of the starting materials [5, 60]. Another inexpensive macrocyclic musk compound is Habanolide, the unsaturated version of Exaltolide [61].

Table 4 Industrial use of polycyclic musks in Europe (in tonnes) [50]

Year	HHCB	AHTN	ADBI	AHDI	ATII
1992	2400	885			
1995	1482	585	34	50	40
1998	1473	385	18	19	2

Table 5 Commercially important macrocyclic musks

CAS name CAS no.	Trade name(s)	Chemical name	Molecular formula	Chemical structure
9-Cyclohepta-decen-1-one 542–46–1	Civettone Civetone	*cis*-9-Cyclohepta-decenone, 10-Ketocyclo-heptadecene	$C_{17}H_{30}O$	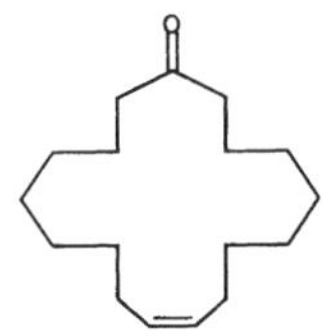
3-Methyl-cyclopenta-decanone 541–91–3	Muscone	3-Methylcyclo-pentadecanone	$C_{16}H_{30}O$	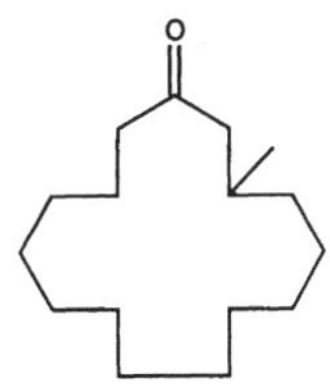
Oxacyclo-heptadec-8-en-2-one 123–69–3	Ambrettolide	7-Hexadecen-16-olide, 16-Hydroxy-7-hexadecenoic-acidlactone, Cyclohexade-cenolide	$C_{16}H_{28}O_2$	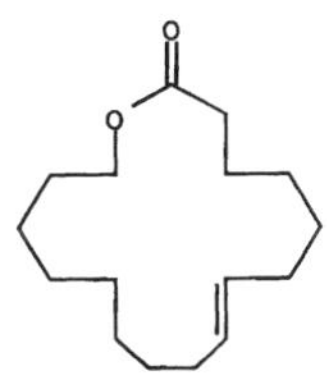
Oxacyclohexa-decan-2-one 106–02–5	Exaltolide, Muskalactone, Pentalide, Thibetolide	15-Pentade-canolide	$C_{15}H_{28}O_2$	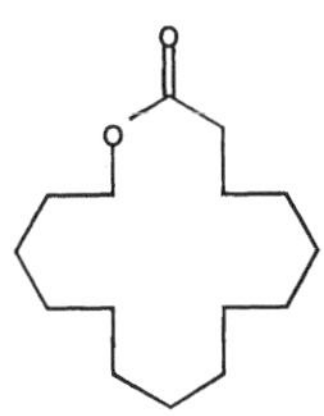

Table 5 (continued)

CAS name CAS no.	Trade name(s)	Chemical name	Molecular formula	Chemical structure
Cyclopenta-decanone 502–72–7	Exaltone, Normuscone	Cyclopenta-decanone	$C_{15}H_{28}O$	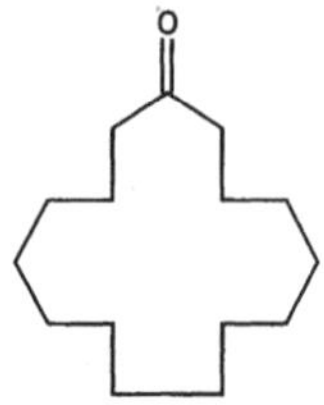
Cyclohepta-decanone 3661–77–6	Dihydro-civettone, Dihydro-civetone	Cyclohepta-decanone	$C_{17}H_{32}O$	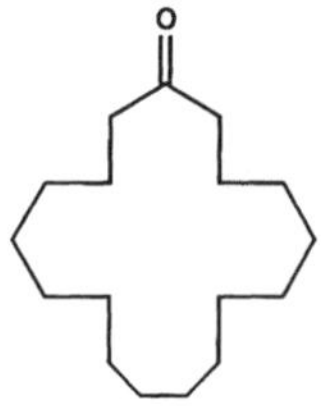
Oxacyclohexa-decen-2-one 34902–57–3	Habanolide, Globalide	Oxacyclohexa-decen-2-one	$C_{15}H_{26}O_2$	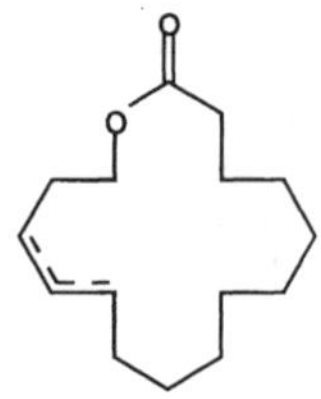
1,4-Dioxa-cyclohepta-decane-5,17-dione 105–95–3	Musk T, Musk NN, Astratone, Musk MC-5	Ethylene brassylate, Ethylene-1, 13-tride-canedioate	$C_{15}H_{26}O_4$	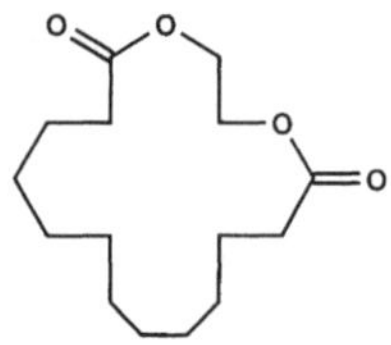
1,4-Dioxacy-clohexadecane-5,16-dione 54982–83–1	Musk MC-4, Musk C14	Ethylenedode-candioate	$C_{14}H_{24}O_4$	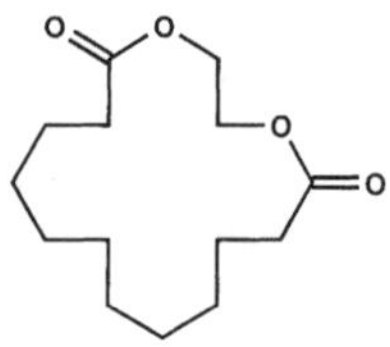
1,6-Dioxacy-cloheptadecan-7-one 6707–60–4	Musk 781, Cervolide	12-Oxahexa-decanolide, 12-Oxa-1,16-hexadecanolide	$C_{15}H_{28}O_3$	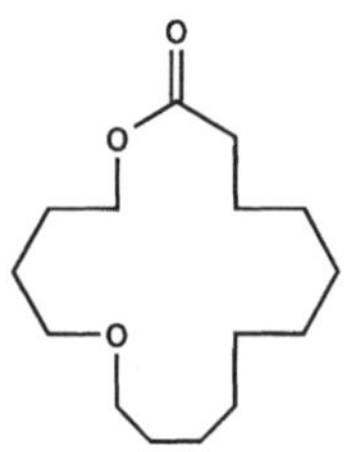

The synthesis of macrocyclic musk compounds is difficult and in many cases a multi-step procedure. Due to the relatively high production costs, their economical importance is still limited. In 1996 they comprised about 5% of the total amount (8000 tonnes) of musk compounds [5]. In contrast to the nitro musks and the polycyclic musk compounds which are offered for 10–30 DM kg^{-1} and 20–60 DM kg^{-1}, respectively, the price for the macrocyclic representatives ranges from 50 to 5000 DM kg^{-1}. Macrocyclic musks are expected to be of increasing importance in the future, especially because many of them are naturally occurring and even the artificial representatives (e.g., ethylene brassylate) closely resemble the natural counterparts [5]. In addition, the progress in synthetic chemistry contributes to declining prices and will stimulate increased use of this type of musks [60].

5 Perspectives

Due to critical public debates on the use of nitro musks and polycyclic musk compounds and the resulting regulatory limitations, the fragrance industry has put increasing emphasis on the development of macrocyclic and other musk odorants. A promising new class are the so-called linear musks. The first representative, a cyclopentenyl ester, was synthesized in 1975 and is being marketed as Cyclomusk. In 1990 another example (Helvetolide) of this class of compounds was discovered [61]. The future will show to what degree these new compounds will replace the "traditional" synthetic musk substances used so far to supply the fragrance industry with the desired musk odor.

6 References

1. Falbe J, Regitz M (1991) Römpp Chemie Lexikon, 9th edn, vol 4. Georg Thieme, Stuttgart, p 2858
2. Fey H, Otte I (1985) Wörterbuch der Kosmetik, 2nd edn. Wissenschaftliche Verlagsgesellschaft, Stuttgart, pp 93, 172
3. Mignat S (1960) Dragoco Rep 2:27
4. Wagner M (1998) SÖFW-J 124:554
5. Gebauer H, Bouter T (1997) Euro Cosmet 1:30
6. Rebmann A, Wauschkuhn C, Waizenegger W (1997) Dtsch Lebensm-Rundsch 93:251
7. Muermann HE (1999) Parfüm Kosmet 80:12
8. Eschke HD, Traud J, Dibowski HJ (1994) UWSF-Z Umweltchem Ökotox 6:183
9. Eschke HD, Dibowski HJ, Traud J (1995) Dtsch Lebensm-Rundsch 91:375
10. Eschke HD, Dibowski HJ, Traud J (1995) UWSF-Z Umweltchem Ökotox 7:131
11. Kaester A (1995) Lehrbuch der Speziellen Zoologie, vol 2 Wirbeltiere, Gustav Fischer, Spektrum Akademischer Verlag, Heidelberg Berlin, p 1022
12. Remane A, Storch V, Welsch U (1997) Systematische Zoologie, 5th edn. Gustav Fischer, Stuttgart, p 738
13. Petzsch H (1992) Urania Tierreich Säugetiere, Urania, Leipzig, p 443
14. Nowak GA (1990) Die kosmetischen Präparate, 4th edn, vol 1. Die Parfümerie, Verlag für chem Industrie, Augsburg, p 261
15. Pilz W (1997) SEPAWA Conference Proceedings, p 43
16. Kastner D (1999) SEPAWA Conference Proceedings, p 218

17. Homes V (1999) On the scent: conserving musk deer – the uses of musk and Europe's role in its trade, TRAFFIC Europe, Brussels, ISBN 90-9012795-X
18. Commission Regulation (EC) No 1968/1999 (1999) Off J Europ Comm L244:22
19. Baur A (1891) Ber Dtsch Ges 24:2832
20. Walbaum H (1906) J Prakt Chem 73:488
21. Sack E (1915) Chem-Ztg 39:538
22. Ruzicka L (1926) Helv Chim Acta 9:230
23. Ruzicka L, Stoll M, Schinz H (1926) Helv Chim Acta 9:249
24. Ruzicka L (1926) Helv Chim Acta 9:715
25. Ruzicka L (1926) Helv Chim Acta 9:1008
26. Berends W (1966) SEPAWA Dokumenta 1:1
27. Kerschbaum M (1927) Ber Dtsch Ges 60:902
28. Barbetta L, Trowbridge T, Eldib IA (1988) Perfum Flavor 13:60
29. Boelens H (1967) Naarden Nachrichten 18:5
30. Middleton J (1999) IFSCC Mag 2:36
31. Yamagishi T, Miyazaki T, Horii S, Kaneko S (1981) Bull Environ Contam Toxicol 26:656
32. Yamagishi T, Miyazaki T, Horii S, Akiyamak K (1983) Arch Environ Contam Toxicol 12:83
33. Yurawecz MP, Puma BJ (1983) J Assoc Off Anal Chem 66:241
34. Rimkus G, Wolf M (1993) Dtsch Lebensm-Rundsch 89:171
35. Rimkus G, Wolf M (1993) Dtsch Lebensm-Rundsch 89:103
36. Hahn J (1993) Dtsch Lebensm-Rundsch 89:175
37. Sommer C (1993) Dtsch Lebensm-Rundsch 89:108
38. Wisneski HS, Havery DC (1996) Cosmet Toiletries 111:73
39. Wisneski HS (2001) J Assoc Off Anal Chem 84:376
40. IKW Rechtssammlung, Industrieverband Körperpflege- und Waschmittel e. V., Frankfurt am Main, Recommendation 35/2 and 56/2
41. Council Directive 95/34/EEC (1995) Off J Europ Comm L167:19
42. Council Directive 98/62/EEC (1998) Off J Europ Comm L253:20
43. Opinion of the Scientific Committee on cosmetic products and non-food products intended for consumers (SCCNFP) concerning musk ketone. SCCNFP/0162/99
44. Opinion of the Scientific Committee on cosmetic products and non-food products intended for consumers (SCCNFP) concerning musk xylene. SCCNFP/0163/99
45. Noser J, Sutter A, Auckenthaler A (2000) Mitt Lebensm Hyg 91:102
46. Eidg Department des Innern (1998) Verordnung über kosmetische Mittel, Eidg. Drucksachen- und Materialverwaltung, Bern
47. Eymann W, Roux B, Zehringer M (1999) Mitt Lebensm Hyg 90:318
48. Klemm U (2000) Mitt Lebensm Hyg 91:464
49. Sommer C (1997) Parfüm Kosmet 78:22
50. Grundschober F (2000) International Fragrance Association (IFRA). Personal communication
51. Ford RA (1998) Dtsch Lebensm-Rundsch 94:268
52. Balk F, Ford RA (1999) Toxicol Letters 111:57
53. Rimkus G, Brunn H (1996) Ernährungs-Umschau 43:442
54. Opinion of the Scientific Committee on cosmetic products and non-food products intended for consumers (SCCNFP) concerning 6-Acetyl-1,1,2,4,4,7-hexamethyltetraline (AHTN), SCCNFP/0372/00
55. Opinion of the Scientific Committee on cosmetic products and non-food products intended for consumers (SCCNFP) concerning hexahydro-hexamethyl-cyclopenta(*g*)-2-benzopyran (HHCB). SCCNFP/0403/00
56. Wood T (1975) Givaudanian 6:6
57. Warty VS, Balasubramanian (1974) Bombay Technol 24:3
58. Anonis DP (1992) Perfum Flavor 17:23
59. Warwel S, Bachem H, Deckers A, Döring N, Kätker H, Rose E (1989) SÖFW-J 115:538
60. Williams AS (1999) Synthesis 10:1707
61. Kraft P, Bajgrowicz JA, Denis C, Frater G (2000) Angew Chem 112:3107
62. STERN Jahrbuch 1978, 1st edn. 1979 Gruner und Jahr, Hamburg, Germany, p 180

The Handbook of Environmental Chemistry Vol. 3, Part X (2004): 17–28
DOI 10.1007/b14131

Synthetic Musks in Different Water Matrices

Hans-Dietrich Eschke

Parsevalstrasse 32 d, 45470 Mülheim a. d. Ruhr, Germany
E-mail: hes@ruhrverband.de

Abstract Nitro musks and polycyclic musks are used as synthetic musk compounds in almost all scented consumer products, such as perfumes, cosmetics, and certain cleaning agents. After application they are dumped via waste water treatment plants into the aquatic environment. In this chapter all data so far published on synthetic musks in surface, waste, and drinking water are presented and discussed. Furthermore, a brief description is given of special aspects of the analysis of nitro and polycyclic musks in water matrices.

The list of contaminants in all sorts of water is topped by the polycyclic musk compounds HHCB and AHTN, which are present in the order of magnitude of micrograms per liter and whose concentrations in all samples analyzed exceeded those of the nitro musks musk xylene and musk ketone. The highest concentrations of synthetic musks were found in waste water and surface water near the tributaries of sewage treatment plants. The published data suggest an ubiquitous distribution of these chemicals in the aquatic environment.

Keywords Synthetic musk compounds · Surface water · Waste water · Drinking water

1 Introduction

The high costs of natural musk and the technical limitations connected with its use prompted early efforts to find synthetic substitutes. Meanwhile, this development has brought about an industry of world-wide reach, which provides different fragrance components used in perfumery and meets the rising demand for basic materials. Among the economically most important classes of synthetic musk substitutes are the nitro musk compounds and the polycyclic musk fragrances. Both classes present ideal technical properties and can be synthesized at comparatively low costs. Due to this positive constellation synthetic musks are used as basic materials in a great number of fragrance formulae and as additives in a lot of products such as cleaning agents and detergents, soaps, and other odor-improving cosmetics.

The use of cleaning agents, for example for personal hygiene or textile and room cleaning, implies that considerable quantities of synthetic musk compounds end up together with surfactants and dirt particles in municipal waste water and, consequently, in the sewerage. Part of the substances are in the course of the subsequent sewage treatment retained in the sewage sludge by adsorption and/or partially by degradation (see Chap. 4). Residual synthetic musks leave the sewage treatment plant in the clarified waste water, thus contaminating the surface waters (usually rivers and brooks).

The entry of synthetic musks via the waste water path is to be regarded as the major contamination path of the aquatic environment. The first reports on the presence of nitro musk components in the aquatic environment go back to 1981 [1, 2]. Yamagashi et al. recognized during their analyses of different samples from the River Tama and adjacent sewage treatment plants a relationship between the use of synthetic musk fragrances and their occurrence in the waters. Fish and shellfish from this river contained the nitro musk compounds musk xylene and musk ketone. The range of concentrations in the water and biota samples showed that these compounds due to their lipophilic character and their obvious persistence possess a considerable bioaccumulation potential. The BCF values derived from the data measured for musk xylene and musk ketone were 4.100 and 1.100, respectively.

In the early 1990s these observations were confirmed by numerous studies by Rimkus and Wolf and other authors [3–6] for European waters, also extending the scope by detection of nitro musks in biota, human fat and milk. In addition, studies carried out in 1994/95 [7, 8] showed polycyclic musks which are widely used as fragrances in a great variety of consumer products ending up in the aquatic environment as well. Fish from the River Ruhr (Germany) and from polishing ponds contained considerable quantities of polycyclic musks, correlated to their fat content and the synthetic musk concentrations found in water. HHCB and AHTN, detected preferentially in fish, exceeded the nitro musk levels recorded so far.

Thus synthetic musks have gained attention as potential environmental contaminants. In the following section these compounds will be described as to relevant aspects in fresh and waste waters (including analytical aspects), with particular consideration of the water matrix.

2
Analysis of Synthetic Musks in Water Matrices

Polycyclic and nitro musk compounds are relatively low-molecular substances with molecular weights ranging between 200 and 300. They are sufficiently volatile and readily vaporizable without decomposition, thus complying with essential requirements of the gas chromatographic (GC) technique. A combination of gas chromatography with mass spectrometric detection (GC/MS) is commonly used for the determination of polycyclic musks [8–10]. The EI mass spectra of polycyclic musks are adequately structured and show typical mass fragments, which are well suited for the identification and quantification of the substances. On the other hand, nitro musk compounds with their specific molecular structural elements are ideally determined by specific, highly sensitive detectors (ECD, NPD) [1, 3, 4]. Musk xylene and musk ketone in water samples are detectable and quantifiable by ECD with a limit of detection of 1 ng l^{-1}.

The preparation of water samples is characterized by extraction and ensuing clean-up steps depending upon the matrix. Usually the total water sample including the suspended particulate matter (SPM) content is extracted. If the fragrances are to be determined in SPM the solids are separated for example by centrifugation (see Chap. 4).

The increased contamination risk which this group of substances entails because of their ample use and their physical and chemical properties is discussed particularly in connection with the extraction and clean-up operations [11].

For analyte extraction both the classical liquid-liquid and the solid-phase extraction (SPE) are used. Water quantity varies between one and several liters, the exact volume depending upon the concentrations expected and the required limits of detection. Bester et al. [12], analyzing sea water from the German Bight, for instance used 100 l of water for extraction with *n*-pentane and could improve the limits of detection for HHCB and AHTN to 0.04 and 0.03 ng l^{-1}, respectively. For the liquid-liquid extraction use is made of, apart from *n*-pentane, *n*-hexane and dichloromethane [1, 4, 13–15]. The solvent extracts are usually dried with sodium sulfate prior to further processing (clean-up).

A number of authors prefer SPE for separating and enriching the synthetic musk fragrances from less contaminated surface waters [3, 4, 10].

The clean-up steps preceding the determination of synthetic musks are similar to those taken for the analytical preparation of other lipophilic residues, such as PCBs or certain pesticides. Accordingly, the liquid-liquid or the solid-phase extraction is followed by column-chromatographic steps on the basis of various adsorbents like silica gel or florisil. This is especially true of waste waters with a complex matrix. Surface water samples need as a rule not be cleaned up. *n*-Pentane and *n*-hexane extracts are only dried with sodium sulfate and partially evaporated prior to GC.

Several authors describe the recently developed solid-phase microextraction (SPME) technique for the screening of water samples [16–19]. This technique permits an efficient and fast preparation of surface and waste water samples on the basis of fibers coated with solid-phase material, which are introduced via the headspace of the sample or directly into the water phase for the purpose of enriching the analyte and finally are thermally desorbed in the GC injector. The solid-phase materials tested and used with success were, apart from polydimethylsiloxane (PDMS) [17] and polydimethylsiloxane-divinylbenzol (PDMS-DVB) [19] of various film thickness, polyacrylate and carboxen fibers [16, 19]. Among the fibers investigated, PDMS-DVB fibers obviously provided the best reproducible determination under non-equilibrium conditions based on an internal standard quantification method [19]. Another simplification of the extraction procedure is the use of Empore® disks [20] and semipermeable membrane devices (SPMDs) [21]. With the help of C_{18} Speed Disks with graded prefilter, relatively large volumes of solids containing waters, for example influents of waste water treatment plants, were extracted [22].

After separation by GC, nitro musk compounds are preferentially measured with highly sensitive detectors such as ECD or NPD. These detectors are very useful for applications exclusively related to nitro musks. In these cases the mass spectrometry is only used for confirmation of the results [1, 3, 4]. Methods for the simultaneous determination of synthetic musks are based on the GC/MS technique. However, GC/MS has the disadvantage of a lower sensitivity of the EI/MS detection for nitro musks, which is compensated in quadrupole devices by selecting the multiple ion detection (MID) [13] or the selected ion monitoring (SIM) mode [3]. For the trace analysis of polycyclic musk compounds, including nitro musk compounds, a wide range of mass spectrometric systems is used, such as mass-selective detectors [3, 17, 23], ion trap instruments [7–9, 24], high-resolution mass spectrometers, and hybrid systems [12].

The use of sophisticated ion trap devices in GC/MS/MS experiments has brought about a decisive reduction of the chemical noise and thus an improvement in sensitivity and selectivity [9, 25]. Combination of these techniques with the single ion storage (SIS) mode of the ion trap systems also implies a high sensitivity in the determination of all synthetic musk fragrances in one chromatographic run [26].

To improve the quantitative results of the analytical methods used, some authors make use of internal standards, such as d_6-musk xylene, d_7-musk ketone [22], d_3-AHTN [22, 26], d_8-naphthalene [12], and aldrine [17].

There are also reports about the application of the HPLC technique for the determination of HHCB, in spite of its well-known lower separation and detection efficiency [27]. The limits of determination of this method after detection with the UV or fluorescence detector (5 $\mu g\ L^{-1}$ and 1.5 $mg\ L^{-1}$, respectively), however, make this technique suitable only for higher concentrations, for instance in waste water.

3
Synthetic Musks in Waste Water

A large part of the synthetic musk compounds is used as an additive to products intended for application in water. These are mainly cleaning agents and detergents and a number of articles for personal hygiene, which after their use are released at least partly into the waste water. Another part of these substances is retained in the washing or on the body, especially because synthetic musks are much less volatile than other fragrances of perfumery oils. Due to their higher adherence they tend to remain longer on textile fibers and on the skin, where they produce a prolonged scent.

After their application, certainly the major part of the musk compounds is carried by municipal waste waters via the sewage system into municipal treatment plants. Therefore, it is not surprising that the highest concentrations of synthetic musks are measured in untreated waste water and in the influents of waste water treatment plants. The waste water concentrations so far reported (Tables 1

Table 1 Polycyclic and nitro musks in waste water influent samples (concentrations in ng L^{-1})

Origin	Samples (n)[a]	HHCB	AHTN	ADBI	AHMI	ATII	DPMI	MX	MK	Reference
Germany (1993)	1	–[b]	–	–	–	–	–	53	–	[3]
Germany (1994)	7	500–2900 ($\bar{x}$[c]=1460)	800–4400 ($\bar{x}$=2240)	40–140 ($\bar{x}$=80)	–	–	–	90–1700	570–2400	[4, 7]
Germany (1996)	1	–	–	–	–	–	–	150	550	[30]
Germany (1998)	6	200–6000	–	–	–	–	–	–	–	[28]
U.S. (1997)	1[d]	9,810	10,000	–	–	–	–	339	488	[22]
	3[e]	$\bar{x}$=13,700	$\bar{x}$=10,700	–	–	–	–	$\bar{x}$=376	$\bar{x}$=569	[22]

[a] Number of samples.
[b] No data.
[c] $\bar{x}$, mean.
[d] Three-day composite sample of a trickling filter wastewater treatment plant, based on plant flow.
[e] Three daily composite samples of a activated sludge wastewater treatment plant, based on plant flow.

Table 2 Polycyclic and nitro musks in waste water effluent samples (concentrations in ng L^{-1})

Origin	Samples (n)[a]	HHCB	AHTN	ADBI	AHMI	ATII	DPMI	MX	MK	Reference
Japan (1981)	3	–[b]	–	–	–	–	–	25–36 ($\bar{x}^c$=32)	140–410 ($\bar{x}$=270)	[1]
Germany (1993)	1	–	–	–	–	–	–	22	–	[3]
Sweden (1993/1994)	3	1000–6000	–	–	–	–	–	–	1000–5000	[29]
Germany (1994)	7	600–2000 ($\bar{x}$=1090)	800–2400 ($\bar{x}$=1400)	40–70 ($\bar{x}$=60)	–	–	–	30–310	220–1300	[4, 7]
The Netherlands (1997)	3	170–290 ($\bar{x}$=230)	110–420 ($\bar{x}$=230)	–	–	–	–	–	–	[20]
Switzerland (1997)	3	–	–	–	–	–	–	<25	100–150	[18]
	8	1900–3900	1400–2800	55–140	90–120	30–60	<500	–	–	
Germany (1996)	1	–	–	–	–	–	–	10	6	[30]
Switzerland (1998)	17	1100–5600	500–2400	<100–160	–	–	–	–	–	[23]
Germany (1998)	8	160–1500	–	–	–	–	–	–	–	[28]
Germany (1999)	2	2500; 5700	530; 610	<50; 60	120	<59; 60	<50; 60	<50; 160	<50	[36]
U.S. (1997)	1[d]	1630	1660	–	–	–	–	31	96	[22]
	3[e]	$\bar{x}$=1170	$\bar{x}$=1180	–	–	–	–	$\bar{x}$=5	$\bar{x}$=99	[22]

[a] Number of samples.
[b] No data.
[c] $\bar{x}$, mean.
[d] Three-day composite sample of a trickling filter wastewater treatment plant, based on plant flow.
[e] Three daily composite samples of a activated sludge wastewater treatment plant, based on plant flow.

and 2) show the highest values for polycyclic musks. The values for HHCB and AHTN are in the order of magnitude of several micrograms per liter. The other substances of this group are found in by far lower concentrations of up to 0.1 μg L^{-1}. The main representatives of the nitro musk group are musk xylene and musk ketone, whose concentrations mostly remain below 1 μg L^{-1}. Musk ambrette, musk moskene, and musk tibetene or DPMI were found in none of the studies.

In treated waste waters the musk compounds appear in lower concentrations. The removal of fragrance materials during waste water treatment process is based on adsorption to solids and degradation of musk fragrances into polar metabolites. The degree of removal of these substances in treatment plants varies from one study to another. In treated water measurements show a reduction of 82–95% and 50–81% for MX and MK, respectively, as opposed to 34–87% and 60–86% for the polycyclic musks HHCB and AHTN, respectively [4, 22]. These relatively high variations are on one hand a reflection of the wide range of concentrations reported and on the other hand a lack of representative data in the studies published so far.

In a comprehensive study influent, primary and final effluent samples from U.S. sewage water treatment plants (activated sludge and trickling filter plants) were taken to estimate the total amounts of fragrance material removed during waste water treatment [22]. To prevent over- or underestimation it seems necessary to collect samples hourly so as to consider the effect of the plant flow, especially because of the high diurnal fluctuation of the fragrance concentrations observed in the influent samples. The concentrations in the final effluent did not show such an extent of variation and remained relatively constant.

Nevertheless, more detailed studies are urgently needed to elucidate the behavior of synthetic musks in sewage water treatment plants. In many sewage water samples the AHTN concentrations were higher or equal to the HHCB values, which is in contrast to the higher application rates of HHCB [31]. This discrepancy shows that the results available at present seem to be inadequate to explain the qualitative and quantitative processes to which the synthetic musks are subjected during sewage water treatment.

4 Synthetic Musks in Surface Water

The treated waste water leaving the municipal sewage plants seems to be the main source of synthetic musk contamination in the aquatic environment. The highest concentrations in river systems are often those analyzed downstream the tributaries of such plants. The synthetic musk levels diminish as the dilution rate increases with the distance from the source. This is the uncontested result of longitudinal analyses of the Rivers Ruhr [4, 8] and Tama [32]. Thus, the basic load of synthetic musks in waters depends essentially upon its waste water share, which may be up to 90%, as found for example of the small brook Wuhle in Berlin, Germany [17]. Synthetic musk fragrances, especially HHCB and AHTN, therefore seem to be suited as organic indicators for the load of municipal sewage in water systems [21].

Table 3 Polycyclic and nitro musks in surface water samples (concentrations in ng L^{-1})

Water sample (origin)	Samples (n)[a]	HHCB	AHTN	ADBI	AHMI	ATII	DPMI	MX	MK	Reference
River water (River Tama, Japan, 1981)	8	–[b]	–	–	–	–	–	3.4–7.9 (x̄[c]=4.1)	<0.05–19 (x̄=9.9)	[1]
River water (River Lauchert, 1993)	17	–	–	–	–	–	–	<1–39	–	[3]
River water (River Elbe, Brunsbüttel, 1993)	1	–	–	–	–	–	–	0.8	4.6	[38]
River water (River Tamagawa, Japan, 1994)	5	0.7–100 median: 80	–	–	–	–	–	–	–	[32]
River water (River Ruhr, 1994)	23	<30–500	<30–300	(x̄=<30)	–	–	–	LOD[d]–30 (x̄=10)	LOD-30 (x̄=30)	[4, 7]
River water (River Elbe, Stade,1995)	1	95	67	–	–	–	–	–	–	[12]
River water (River Elbe, Torgau, 1995)	2	57; 92	62; 116	<80; nd[e]	–	–	–	–	–	[10]
River water (River Glatt, Switzerland, 1995)	1	136	75	3.2	–	–	–	0.62	8.3	[13]
River water (River Rhine, The Netherlands, 1994–1996)	32	10–220 median: 60	10–130 median: 50	–	–	–	–	–	–	[34]
River water (River Meuse, The Netherlands, 1994–1996)	35	10–260 median: 80	10–400 median: 70	–	–	–	–	–	–	[34]
River water (River Ruhr, 1995/1996)	48/18	median: 275	median: 100	–	–	–	–	Six samples >LOD median: 1–3	Median: 10	[33]
Rivers, lakes and canals in Berlin (1996)	30	20–12,470 (x̄=2492)	30–6780 (x̄=1344)	10–520 (x̄=104)	–	–	–	1 sample >LOD (180)	LOD-390 (x̄=80)	[17]
Magdeburg, 1996/1997)		(x̄=114)	(x̄=66)	(x̄=4)					(x̄=6)	

Table 3 (continued)

Water sample (origin)	Samples (n)[a]	HHCB	AHTN	ADBI	AHMI	ATII	DPMI	MX	MK	Reference
River water after centrifugation (River Elbe, Magdeburg, 1996/1997)	31	36–126 ($\bar{x}$=89)	24–71 ($\bar{x}$=50)	2–5 ($\bar{x}$=3)	–	–	–	–	2–11 ($\bar{x}$=5)	[35]
River water (River Elbe, Schmilka to Grauerort, 1996–1997)	30	29–410 median: 116	30–250 median: 85	–	–	–	–	–	–	[36]
River water (River Mulde, Saale 1996–1997)	11	150–610 median: 380	140–330 median: 245	–	–	–	–	–	–	[36]
River water (River Rhine, Switzerland, 1997)	1	25	30	<10	<10	<10	<500	–	–	[18]
River water (River Ergolz, Birs, Bils, Digterbach, Switzerland, 1997)	6	10–260	10–200	<10	<10	<10	<500	–	–	[18]
River and lake water (River Aare, Saane, Emme, Kiesen, Diesbach, Bielersee, Thunersee, Switzerland, 1997)	20	<25–244	<25–95	<10	–	–	–	–	–	[23]
River water (River Isar, Eisbach, Seebach, Ilm, Naab, 1998)	14	60–330	–	–	–	–	–	–	–	[28]
Bank filtrate (River Elbe, Torgau, 1995)	1	<30	<30	–	–	–	–	–	–	[10]
Barrage water (Barrage Rappbode, 1995)	1	<30	<30	nd	–	–	–	–	–	[10]
Sea water (North Sea, German Bight, 1990)	6	0.09–0.88 ($\bar{x}$=0.33)	0.09–0.94 ($\bar{x}$=0.37)	–	–	–	–	–	–	[12]
Sea water (North Sea, 1993)	33	–	–	–	–	–	–	<0.02–0.17 median: 0.03	<0.02–0.06 median: 0.06	[38]
Sea water (North Sea, German Bight, 1995)	6	0.15–4.8 ($\bar{x}$=1.03)	0.08–2.6 ($\bar{x}$=0.58)	–	–	–	–	–	–	[12]

[a] Number of samples; [b] No data; [c] $\bar{x}$, mean; [d] LOD, limit of determination; [e] Not detectable.

Apart from the variation of the load itself, no seasonal variability in sewage plant discharges are to be recognized [26, 35]. Therefore the concentrations in surface waters also show no significant seasonal differences. The level of synthetic musks essentially depends on the runoff of rivers because of changes in dilution rate, which is influenced for instance by "clean" tributaries or rain events.

Over the last years the data base for synthetic musk compounds in surface waters has grown considerably and now permits a more representative picture of the pollution situation. However, a major part of the samples analyzed stems from European waters. Samples were taken from rivers, brooks, lakes and canals, but also from the North Sea. Table 3 gives an overview of the concentrations measured in surface waters.

The relations between the various compounds found in waste water are observed at least approximately also in surface waters. The predominant load components in surface waters are again HHCB and AHTN. The highest concentrations ever published for surface or sewage water samples so far were measured in various rivers, lakes, and canals in Berlin consisting exclusively of sewage effluents (about 12 $\mu g\ L^{-1}$ HHCB, about 6.8 $\mu g\ L^{-1}$ AHTN) [17].

The mean concentrations of the polycyclic musks HHCB and AHTN measured in waters with a lower share of waste water, however, were within the range of 100–400 ng L^{-1} and 50–240 ng L^{-1}, respectively. The concentrations of ADBI like those of musk xylene and musk ketone are considerably lower, often lying near or below the limit of determination.

The HHCB and AHTN values in the River Elbe were the highest among all lipophilic organic pollutants analyzed, such as polychlorinated biphenyls (PCBs) and polycyclic aromatic hydrocarbons (PAHs) [35]. The situation in marine waters is similar. The HHCB and AHTN concentrations of samples taken in the German Bight in the North Sea lay in accordance with the high dilution factor in the lower nanogram per liter range, but were still in the order of magnitude of other organic contaminants analyzed, such as α-HCH and γ-HCH, nitro benzene, certain thiophosphates, and methylbenzothiazol [12].

5
Synthetic Musks in Drinking Water

The regulations and directives of the EU and the national legislation of the Member States set high standards for the quality of drinking water in terms of chemical and biological, but also organoleptic properties. These latter concern the flavor and odor as criteria for a high-quality drinking water. Studies on the threshold concentrations for the odor of fragrance substances in drinking water also included HHCB and AHTN [37]. The relatively low threshold concentrations of 80 ng L^{-1} HHCB and 40 ng L^{-1} AHTN were slightly exceeded in surface waters, which might affect odor if these waters were to be used for the processing of drinking water. The detection of polycyclic musks in drinking water partially processed from surface water confirms this hypothesis [37]. In this context studies should be mentioned dealing with the experimental determination of the distribution pattern of the fragrances in the water-suspended particulate matter system of the River Elbe. These studies suggest that higher levels of HHCB,

Table 4 Polycyclic and nitro musks in drinking water samples (concentrations in ng L^{-1})

Origin	Samples (n)[a]	HHCB	AHTN	ADBI	AHMI	ATII	DPMI	MX	MK	Reference
Germany (1996)	2	<30	<30	nd[b]	–[c]	–	–	–	–	[10]
Switzerland (1997)	10	–	–	–	–	–	–	<25	<25	[18]
	9	<LOD[d]	<LOD	<LOD	<LOD	<LOD	<LOD	–	–	[18]
Germany (1998)	5	<30	–	–	–	–	–	–	–	[28]

[a] Number of samples.
[b] Not detectable.
[c] No data.
[d] LOD, limit of determination.

AHTN, and ADBI seem to be dissolved in the water phase than would be expected for these nonpolar and lipophilic compounds [35].

An important step in the processing of drinking water from surface waters in urban areas is the artificial recharge of groundwater by abstracting water from rivers and leading it into slow-rate sand filters, which obviously eliminate synthetic musks. So, for instance, HHCB and AHTN were no longer detected in the bank filtrate of the River Elbe [10]. In another study at a groundwater enrichment plant, however, the amount of polycyclic musks was reduced by no more than 20–60% by sand filtration as compared with the river water used [18]. On the other hand, the few drinking water analyses made so far yielded no measurable quantities of synthetic musk compounds (Table 4). These sometimes contradictory results point to a number of unresolved problems in the environmental behavior of synthetic musks. In addition, it should be mentioned that in this extremely low concentration range several analytical and blank problems may occur which must be critically assessed.

6 Conclusions

Synthetic musks are significant contaminants in waste water polluted waters and have obviously found a wide distribution in the aquatic environment. Especially the polycyclic musks HHCB and AHTN can be detected in almost all sorts of waters except drinking water. As compared to other lipophilic organic traces, these chemicals belong to the dominating pollution components. Consequently, it is above all HHCB which seems appropriate as a sensitive organic indicator for contamination of waters by municipal waste water.

Conclusions on temporal trends of the musk load are made only in a few publications, suggesting that the rise in the production and sales figures for HHCB and AHTN is paralleled by increasing concentrations, as has been demonstrated for instance in the River Meuse [34].

Due to the lack of representative data so far, more comprehensive studies are needed to understand in depth the behavior and the removal of synthetic musks during waste water treatment.

7 References

1. Yamagishi T, Miyazaki T, Horii S, Akiyama K (1983) Arch Environ Contam Toxicol 12:89
2. Yamagishi T, Miyazaki T, Horii S, Kaneko S (1981) Bull Environ Contam Toxicol 26:656
3. Hahn J (1993) Dtsch Lebensm-Rundsch 89:175
4. Eschke H-D, Traud J, Dibowski H-J (1994) Vom Wasser 83:373
5. Rimkus G, Wolf M (1993) Dtsch Lebensm-Rundsch 89:103
6. Rimkus G, Wolf M (1993) Dtsch Lebensm-Rundsch 89:171
7. Eschke H-D, Traud J, Dibowski H-J (1994) UWSF – Z Umweltchem Ökotox 6:183
8. Eschke H-D, Dibowski H-J, Traud J (1995) UWSF – Z Umweltchem Ökotox 7:131
9. Eschke H-D, Dibowski H-J, Traud J (1995) Dtsch Lebensm-Rundsch 91:375
10. Lagois U (1996) Wasser Abwasser 137:154
11. Rimkus G, Wolf M (1996) Chemosphere 33:2033
12. Bester K, Hühnerfuss H, Lange W, Rimkus GG, Theobald N (1998) Water Res 32:1857
13. Müller S, Schmid P, Schlatter C (1996) Chemosphere 33:17
14. van Dijk A (1996) Report to RIFM, RCC Umweltchemie AG, Project 364 825, cited in [31]
15. van Dijk A (1996) Report to RIFM, RCC Umweltchemie AG, Project 381418, cited in [31]
16. Mang N (1996) Diploma thesis, University of Essen, Essen, Germany
17. Heberer T, Gramer S, Stan H-J (1999) Acta Hydrochim Hydrobiol 27:150
18. Noser J, Sutter A, Auckenthaler A (2000) Mitt Gebiete Lebensm Hyg 91:102
19. Winkler M, Headley JV, Peru KM (2000) J Chromatogr A 903:203
20. Verbruggen E (1997) Letter of RITOX, Utrecht, The Netherlands, cited in [31]
21. Gatermann R (1998) PhD thesis, University of Hamburg, Hamburg, Germany
22. Simonich SL, Begley WM, Debaere G, Eckhoff WS (2000) Environ Sci Technol 34:959
23. Bundesamt für Umwelt, Wald und Landschaft (BUWAL) Switzerland (1998) unpublished results
24. Fromme H, Otto T, Pilz K (2001) Water Res 35:121
25. Eschke H-D (1996), Eighteenth International Symposium on Capillary Chromatography, vol. III, p 1573
26. Eschke H-D (2000) unpublished results
27. Schüssler W, Nitschke L (1998) Fresenius J Anal Chem 361:220
28. Nitschke L (2000) unpublished results
29. Paxeus N (1996) Water Res 30:1115
30. Gatermann R, Hühnerfuss H, Rimkus G, Atar A, Kettrup A (1998) Chemosphere 36:2535
31. van de Plassche EJ, Balk F (1997) Environmental risk assessment of the polycyclic musks AHTN and HHCB according to the EU-TGD. RIVM Rep No 601503008. RIVM, Bilthoven, The Netherlands
32. Yun S-J, Teraguchi T, Zhu X-M, Iwashima K (1994) J Environ Chem 4:325
33. Eschke H-D (1996) Ruhrwassergüte. Ruhrverband, Essen, Germany, p 86
34. Breukel RMA, Balk F (1996), RIZA Werkdocument 96.197x. RIZA, The Netherlands, cited in [31]
35. Winkler M, Kopf G, Hauptvogel C, Neu T (1998) Chemosphere 37:1139
36. Wiegel S, Harms H, Stachel B, Reincke H (2000) Synthetische Moschus-Duftstoffe in der Elbe. Arbeitsgemeinschaft für die Reinhaltung der Elbe, Hamburg, Germany
37. Schlett C (1995) Ruhrwassergüte. Ruhrverband, Essen, Germany, p 111
38. Gatermann R, Hühnerfuss H, Rimkus G, Wolf M, Franke S (1995) Mar Pollut Bull 30:221

The Handbook of Environmental Chemistry Vol. 3, Part X (2004): 29–47
DOI 10.1007/b14129

Synthetic Musks in Suspended Particulate Matter (SPM), Sediment, and Sewage Sludge

Cornelia Fooken

Im Münchfeld 21, 55122 Mainz, Germany
E-mail: cornelia.fooken@netart-net.de

Abstract Nitro musk and polycyclic musk fragrances accumulate in solid matters of the aquatic environment. They are analyzed by extraction with an organic solvent and subsequent measurement by GC/MS. Among the polycyclic musks AHTN and HHCB are the dominating contaminants; DPMI, ADBI, AHDI, and ATII are partly found, but at significantly lower concentrations. Within the nitro musk compounds only musk xylene and musk ketone are detected in a part of the samples with concentration levels below those of the polycyclic musks.

The suspended particulate matter (SPM) samples were usually obtained by flow-through centrifuges. In many surface waters, the AHTN and HHCB concentrations showed median values of 40–500 μg kg^{-1} dry matter (dm) in SPM. The polycyclic musk content in SPM collected by sedimentation chambers was only approximately 15% of that in the centrifuge-type SPM. In general, sediment samples were about one order of magnitude less contaminated by musk fragrances than SPM samples. Mean AHTN and HHCB concentrations of 5000–28,000 μg kg^{-1} dm were usually observed in sewage sludge. Sewage sludge from industrial waste water showed lower musk levels than that from domestic sewage. In sewage treatment plants, metabolism of nitro musks was observed; some of the transformation products were identified.

Partition coefficients were determined for several synthetic musks. Obviously the adsorption of musk compounds to solid matter is of minor importance only: about 90% of AHTN or HHCB and 97% of musk xylene or musk ketone are dissolved in the water phase.

Keywords Polycyclic musks · Nitro musks · Suspended particulate matter · Sediment · Sewage sludge · Partition coefficient

1 Introduction

In previous years, synthetic musk compounds have been investigated in many surface or sewage waters, whereas less data were reported for the corresponding solid matter matrices [1]. Meanwhile, the results of musk contents in solid matters have increased. This review presents the results of studies analyzing nitro and/or polycyclic musks in suspended particulate matter (SPM), sediment, and sewage sludge.

Besides a survey of the analytical methods, the musk concentrations in all investigated solid matter samples are compiled. Furthermore the partition equilibrium of musk compounds between water and solid matter phase is described. As far as possible, all available literature data including unpublished results have been considered in this review. The results of the investigations are summarized and critically discussed.

2 Analytical Methods

The analytical methods for the determination of musk compounds in solid matter samples follow in general the same procedure: extraction with an organic solvent, clean-up, and subsequent measurement by GC/MS. The methods of the studies which are presented here are described in the following sections.

2.1 Analysis of Synthetic Musks in SPM or Sediment

In most of the studies, the solid matter samples were freeze-dried before analysis. Only in [2, 3] were the original wet samples used. The extraction solvents applied were quite different: acetone together with other solvents in subsequent extraction steps [2, 3], dichloromethane [4], toluene [5–7], cyclohexane [8, 9], or hexane/acetone [10, 11]. Also different extraction techniques were applied: Soxhlet extraction was the most frequently used method [4–7, 10, 11]; in addition, cold stirring [2, 3], ultrasonic extraction [8], or steam-distillation/solvent extraction [9] were used. In the latter case, aqueous suspensions of the samples were extracted. The clean-up was often performed by chromatography on silica gel [2–7]. No clean-up step was used in [8]. Sometimes pyrogenic copper was applied for desulfuration [4, 9]. Separation and detection were carried out with GC/MS, in general in the selected ion monitoring (SIM) mode. The detection limits of the different methods varied from 0.5 $\mu g\ kg^{-1}$ dry matter (dm) to approximately 50 $\mu g\ kg^{-1}$ dm.

2.2 Analysis of Synthetic Musks in Sewage Sludge

The wet original samples were used for the extractions. Different extraction solvents were applied: acetone followed by other solvents [2, 3, 12], dichloromethane [11, 13], or hexane [14]. The samples were stirred or agitated at ambient temperature. The clean-up was usually performed by gel permeation chromatography and chromatography on silica gel. Separation and detection were carried out in the same way as already described. The resulting detection limits showed a range mostly comparable to the one mentioned above (or partly higher values up to 1000 $\mu g\ kg^{-1}$ dm [11, 13]).

2.3 Note on the HHCB Standard

The analytical standard of HHCB generally is not chemically pure but has only technical purity. In the past the standard material obtained by Promochem (Wesel, Germany) contained approximately 50% diethylphthalate according to the supplier's declaration. In many of the studies described in this chapter, the HHCB reference material was obtained by Promochem. Unfortunately, the HHCB impurity was not known and taken into account by all laboratories. As a consequence, the HHCB concentrations reported in some studies are too high. Especially the HHCB results in [5–7, 14] have to be corrected by the factor 0.5. In this review, the originally reported HHCB data are presented without correction.

Recently, the purity of the HHCB reference material supplied by Promochem has improved. Measurements carried out by Promochem gave purities of about 70%. Furthermore, there are some indications that the present HHCB standard does not contain diethylphthalate any longer.

3
Results and Discussion

3.1
Synthetic Musk Compounds in Suspended Particulate Matter

3.1.1
Musk Xylene, Musk Ketone, AHTN, and HHCB in SPM

The results of the studies analyzing these four musk compounds in SPM are summarized in Table 1. The table contains information about the sample origin (river and year), the number of samples (n), and the musk content in SPM; the concentrations are given as range (minimum to maximum) and median values. The samples were mainly taken in the years 1996 to 1999. With the exception of [5], the SPM samples were obtained by flow-through centrifuges. The SPM in [5] was collected in sedimentation chambers yielding another type of SPM whose results are described separately (see Sect. 3.1.2).

SPM samples were taken in 11 Hessian rivers (e.g., Fulda, Lahn, Main) once a year [2, 3]. Musk xylene was mostly not found and musk ketone was partly detected with a median of 5 $\mu g\ kg^{-1}$ dm. AHTN or HHCB showed concentrations of 50–1130 $\mu g\ kg^{-1}$ dm with a median of about 250 $\mu g\ kg^{-1}$ dm. Additionally, brooks having a waste water content of 100% during periods of low water discharge were also investigated. The musk levels in these small waters were one order of magnitude higher; the polycyclic musks AHTN and HHCB reached maximum values of about 13,000 $\mu g\ kg^{-1}$ dm in SPM. The medians were 6 $\mu g\ kg^{-1}$ dm musk xylene, 88 $\mu g\ kg^{-1}$ dm musk ketone, and about 2800 $\mu g\ kg^{-1}$ dm AHTN or HHCB.

In the river Elbe at Magdeburg, SPM samples were taken weekly or biweekly and investigated for musk fragrances [4]. The concentration of musk xylene was always below 2 $\mu g\ kg^{-1}$ dm and that of musk ketone was 4–22 $\mu g\ kg^{-1}$ dm. The AHTN or HHCB levels came to 150–770 $\mu g\ kg^{-1}$ dm with a median of about 450 $\mu g\ kg^{-1}$ dm. Over the sampling period of a year no seasonal variability could be seen. High weekly alterations of the SPM contamination were observed.

Musk xylene was determined in rivers of Baden-Württemberg over a long time [15]. The SPM samples were taken monthly at six sampling places in the rivers Rhine, Neckar and Danube. Between 1993 and 1998, the concentration of this nitro musk compound decreased. In 1994, at each sampling station musk xylene was detected 3–7 times per year, whereas in 1997/98 this frequency was only 0–2 times per year (detection limit: 2 $\mu g\ kg^{-1}$ dm). The maximum levels decreased from 17 to 6 $\mu g\ kg^{-1}$ dm during this period.

At the same sampling places in Baden-Württemberg, many musk compounds were analyzed monthly in a recent study [8]. The concentration of musk xylene (musk ketone) was always (mostly) below 10 $\mu g\ kg^{-1}$ dm. The content of AHTN or HHCB in SPM was <5 to 190 $\mu g\ kg^{-1}$ dm with a median of about 40 $\mu g\ kg^{-1}$ dm. The levels especially of AHTN showed a significant decrease from 1998 to 1999.

SPM measurements in the rivers Rhine and Meuse at the Dutch-German and Dutch-Belgian border (1994–1996) are reported in [10]. The concentrations of

Table 1 Musk xylene, musk ketone, AHTN, and HHCB in suspended particulate matter (SPM)[a] [μg kg^{-1} dry matter]

Region	River	Year	n[b]		Musk xylene	Musk ketone	AHTN	HHCB	Reference
Germany, Hessen	11 rivers	1996–98	34	Range	<1–14	<1–46	53–857	45–1134	[2, 3]
				Median	<2	5	237	266	
Germany, Hessen	Schwarzbach, Rodau and other brooks	1996–98	21	Range	<3–46	31–408	544–12,666	897–13,722	[2, 3]
				Median	6	88	2948	2600	
Germany, Saxony-Anhalt	Elbe	1996–97	31	Range	<2	4–22	194–770	148–736	[4]
				Median		7	458	442	
Germany, Baden-Württemberg	Rhine, Neckar, Danube	1993–98	307	Range	<2–17	–	–	–	[15]
				Median	<2	–	–	–	
Germany, Baden-Württemberg	Rhine, Neckar+ Danube	1998–99+	98+	Range	<10	<10–12	<5–199	<5–180	[8]
		1997	11	Median		<10	41	35	
The Netherlands, Dutch-German border	Rhine	1994–96	11	Range	<50	<50	100–540	<50–160	[10]
				Median			290	70	
The Netherlands, Dutch-Belgian border	Meuse	1994–96	11	Range	<50–80	<50	60–1200	<50–580	[10]
				Median	<50		830	240	
The Netherlands	Main surface waters	1997–98	24	Range	–	–	<100–1700	<100–1800	[11]
				Median	–	–	120	100	
Germany, from Czech border to Hamburg[c]	Elbe	1998–99	82	Range	<0.5–5	<0.5–5	4–130	3–240	[5]
				Median	<0.5	0.6	32	59	
Germany, Saxony-Anhalt[c]	Mulde, Saale, Schwarze Elster	1998–99	36	Range	<0.5–5	<0.5–8	10–230	29–400	[5]
				Median	0.6	1.4	69	133	

[a] SPM usually obtained by flow-through centrifuges.
[b] Number of samples.
[c] SPM obtained by sedimentation chambers.
–: no data.

both polycyclic musks were <50 to 1200 μg kg^{-1} dm. In the Rhine, a median of 290 μg AHTN kg^{-1} dm and of 70 μg HHCB kg^{-1} dm was found. The musk levels in the Meuse were higher (median: 830 μg AHTN kg^{-1} dm, 240 μg HHCB kg^{-1} dm).

SPM samples were taken in the Netherlands from the main waters at seven places [11]. The AHTN or HHCB contents were <100 to 1800 μg kg^{-1} dm with a median of about 110 μg kg^{-1} dm.

Comprehensively, in many German and Dutch surface waters the AHTN and HHCB concentrations in SPM showed median values between 100 and 500 μg kg^{-1} dm. The corresponding levels in the investigated rivers of Baden-Württemberg were distinctly lower and that in Hessian brooks one order of magnitude higher. The contamination of SPM by nitro musks is generally much smaller: musk xylene was mostly not detected, and musk ketone had average values of 1–10 μg kg^{-1} dm.

3.1.2
Musk Xylene, Musk Ketone, AHTN, and HHCB in Sedimenting SPM

Extensive measurements of musk compounds in SPM were carried out in [5]. The SPM samples were obtained by sedimentation chambers made of Plexiglas (the sediment traps are described in [16]). The surface water runs into the traps and its flow velocity is slowed down to about 1 cm s^{-1} making sedimentation possible. With a separation efficiency of 20–30%, the SPM was collected within four weeks yielding monthly mixed samples. In each case, two of these samples were mixed and subsequently analyzed.

The SPM samples were taken from seven stations in the river Elbe between the Czech-German border and the Elbe estuary and furthermore from three tributaries in the vicinity of their mouth into the river Elbe. As already stated above (see Sect. 2.3), the HHCB data as given in [5] and Table 1 have to be corrected by the factor 0.5.

Musk xylene and musk ketone were detected in the majority of all SPM samples; their concentrations ranged from <0.5 to 5 μg kg^{-1} dm. In 1998 (1999), the following results for AHTN were obtained (after correction, HHCB showed values very similar to that of AHTN): at many SPM sampling places on the river Elbe the annual mean values were about 60 (40) μg kg^{-1} dm, whereas in Hamburg the levels decreased to one-third due to the tide influence (the sand originating from sea water reduces the total organic carbon (TOC) and consequently the musk content in the SPM). In the three tributaries Schwarze Elster, Saale, and Mulde, the musk contamination was higher than in the Elbe (annual mean values of 80–120 (50–80) μg AHTN kg^{-1} dm). A decrease of approximately 30% was observed for the polycyclic musks in SPM from 1998 to 1999. There was a distinct seasonal variability with minimum levels in summer and maximum values in winter. These alterations were explained by metabolism processes being influenced by the different temperatures [5].

The musk contamination level obviously depends on the SPM type: the concentrations in the samples collected in sedimentation chambers are much lower than in the samples obtained by flow-through centrifuges (see Table 1). At the sampling place Elbe at Magdeburg, results for both SPM types are available (centrifuge, 1996/97 [4] and sedimentation trap, 1998 [5, 17]). The comparison of the mean AHTN or HHCB levels shows that the musk content in the trap-type SPM is only 14–17% of that in the centrifuge-type SPM. It is interesting that the analogous comparison for strongly adsorbing compounds like HCB, *p,p'*-DDT, or benzo(*a*)pyrene leads to deviating results, namely almost identical concentrations in both SPM types [4, 17]. In the river Ruhr, musk contents were also measured in the two different SPM types confirming the results of the Elbe samples [18]. Again, the trap-type SPM contained only 12–13% AHTN or HHCB as compared with the centrifuge-type SPM.

There is no simple explanation for the observed phenomenon. The mean TOC content of the two SPM types in the Elbe at Magdeburg was 9.5% for the centrifuge-type [19] and 6.4% for the trap-type SPM [17]. In general, the portion of the <63-μm fraction amounted to 95–100% in the centrifuge-type SPM [2] and to 82% in the trap-type SPM (mean in the Elbe at Magdeburg [17]). Thus, the relatively small differences regarding the organic carbon and the fine particle fraction only partly explain the clearly different musk levels in both SPM types. Therefore further investigations of this problem are necessary.

3.1.3
Other Nitro Musks and Polycyclic Musks in SPM

In Table 2 the results of studies on further musk fragrances in SPM are shown. The three nitro musk compounds musk ambrette, musk tibetene, and musk moskene as well as the polycyclic musk DPMI were not detected in any SPM sample. In the river Elbe at Magdeburg, ADBI was measured in more than half of the SPM samples with a median of 16 μg kg^{-1} dm [4]. In the trap-type SPM from the Elbe and its tributaries, the levels of ADBI, AHDI, and ATII were <0.5–17 μg kg^{-1} dm with a median of 1–3 μg kg^{-1} dm [5]. ATII was also detected in a part of the SPM samples from rivers in Baden-Württemberg (maximum: 38 μg kg^{-1} dm), whereas ADBI and AHDI always were below 5 μg kg^{-1} dm [8].

3.2
Synthetic Musk Compounds in Sediment

3.2.1
Musk Xylene, Musk Ketone, AHTN, and HHCB in Sediment

In Table 3, information about musk measurements in sediment samples is given. Most of the samples date from the years 1996–1999. The average musk concentrations are given as median or mean values. The HHCB values as presented in [5–7] and Table 3 must again be corrected by the factor 0.5 (see Sect. 2.3).

In [6, 7] a total of 54 sediment samples from big and smaller rivers in Lower Saxony/Germany were analyzed. In samples from the Elbe catchment area, the AHTN concentrations ranged from <0.5 to 144 μg kg^{-1} dm with a median of 8 μg kg^{-1} dm. Levels above 50 μg AHTN kg^{-1} dm were only found in three sediment samples; the highest musk contamination was observed in the river Este behind a sewage treatment plant.

Table 2 Musk ambrette (MA), musk tibetene (MT), musk moskene (MM), DPMI, ADBI, AHDI, and ATII in different solid matters [μg kg⁻¹ dry matter]

River or region[a]	Year	n[b]		MA	MT	MM	DPMI	ADBI	AHDI	ATII	Reference
Suspended particulate matter (SPM)[c]											
Hessen, rivers+brooks	1996–98	34+21	Range	<1–<20	–	–	–	–	–	–	[2, 3]
			Median	<4	–	–	–	–	–	–	
Elbe (Magdeburg)	1996–97	31	Range	–	–	–	–	<4–43	–	–	[4]
			Median	–	–	–	–	16	–	–	
Elbe+Mulde, Saale, Schwarze Elster[d]	1998–99	82+36	Range	<0.5	<0.5	<0.5	<0.5	<0.5–5	<0.5–15	<0.5–17	[5]
			Median					1	3	3	
Rhine, Neckar, Danube	1997–99	109	Range	<20	<10	<10	<5	<5	<5	<5–38	[8]
			Median							<5	
Sediment											
Ems, Weser, Leine etc. +Elbe and 15 tributaries	1996+ 1997–98	8+46	Range	<0.1	<0.1	<0.1	<0.5	<0.5–3	<0.5–6	<0.5–18	[6, 7]
			Median					<0.5	<0.5	0.8	
Saale	1999	4	Range	<0.5	<0.5	<0.5	<0.5	<0.5–6	1–24	2–26	[5]
			Mean					2	10	9	
Rhine, Neckar	1999	39	Range	<20	<10	<10	<5	<5	<5	<5–22	[8]
			Median							<5	
Spree, Dahme, Havel, Teltowkanal	1996–97	19; 20; 20[e]	Range	–	–	–	–	<4–68	<4–93	<4–220	[9]
			Median	–	–	–	–	<4; 8; 5	<4; 9; 31	<4; 21; 100	
Impounded lake Baldeney	1994	6	Range	–	<2	<2	–	–	–	–	[20]
			Median	–			–	–	–	–	
Sewage sludge											
Germany, Hessen	1996–98	27	Range	<9–<34	–	–	–	–	–	–	[2, 3]
			Median	<21	–	–	–	–	–	–	
Germany	–	2	Values	–	–	–	–	120; 290	–	–	[12]
			Mean	–	–	–	–	200	–	–	
Switzerland	1998	12	Range	<1	<1	<1	38–332	41–330	65–843	–	[14]
			Median				68	113	204	–	

[a] For details: see Tables 1, 3, and 4; [b] Number of samples; [c] SPM usually obtained by flow-through centrifuges; [d] SPM obtained by sedimentation chambers; [e] Samples from areas with low; moderate; high input of sewage water (range: all samples; median: data within the different areas); –: no data.

Table 3 Musk xylene, musk ketone, AHTN, and HHCB in sediment [$\mu g\ kg^{-1}$ dry matter]

Region	River (or lake)	Year	n[a]		Musk xylene	Musk ketone	AHTN	HHCB	Reference
Germany, Lower Saxony	Ems, Weser, Elbe, Leine, Oker	1996	8	Range Median	0.4–2 0.7	0.2–4 0.9	<0.5–4 <0.5	<0.5–54 7	[6]
Germany, Lower Saxony	Elbe and 15 tributaries in the Elbe catchment area	1997–98	46	Range Median	<0.1–2 0.1	<0.1–5 0.2	<0.5–144 8	<0.5–249 15	[7]
Germany, Saxony-Anhalt	Saale	1999	4	Range Mean	<0.5–0.6 not calculated[b]	<0.5–1.3 not calculated[b]	13–380 118	45–720 279	[5]
Germany, Baden-Württemberg	Rhine, Neckar	1999	39	Range Median	<10	<10–13 <10	<5–180 7	<5–110 6	[8]
Germany, Berlin	Spree, Dahme, Havel, Teltowkanal	1996–97	19; 20; 20[c]	Range Median	– –	– –	<20–2600 20; 240; 930	<30–2200 <30; 230; 910	[9]
Germany, Northrhine-Westfalia	Impounded lake Baldeney	1994	6	Range Median	<2	3–5 4	– –	– –	[20]
Germany, Schleswig-Holstein	Polishing pond of a sewage treatment plant	1997	2	Values Mean	3; 4 4	20; 35 28	420; 780 600	420; 890 655	[22]

[a] Number of samples.
[b] More than 50% of the values below detection limit.
[c] Samples from areas with low; moderate; high input of sewage water (range: all samples; median: data within the different areas).
–: no data.

Sediment samples from the river Saale near Halle showed AHTN levels varying between 13 and 380 μg kg^{-1} dm [5]. In Baden-Württemberg, sediment samples were taken from 39 different sampling places in the rivers Rhine and Neckar [8]. The median concentration of AHTN or HHCB was about 7 μg kg^{-1} dm. Only in two samples did the polycyclic musk levels exceed 30 μg kg^{-1} dm (values of 110 and 180 μg kg^{-1} dm).

Many sediment samples from several surface waters in Berlin were analyzed [9]. Besides the rivers Spree, Dahme, Havel, and the Teltowkanal, samples from their lake-like broadenings (e.g., Wannsee) were also taken. The surface waters contained different portions of municipal treated sewage effluents relative to the total water content. The polycyclic musk levels found in Berlin's river and lake sediments differed by two orders of magnitude (range of AHTN or HHCB: <20–2600 μg kg^{-1} dm). There was a relationship between the musk content in the sediment and the waste water portion in the corresponding surface water. The median concentrations of AHTN or HHCB in sediments from areas with a low, moderate, and high input of sewage water were about <30, 235, and 920 μg kg^{-1} dm, respectively.

In 1994, sediment from the impounded lake Baldeney was investigated [20]. The concentration of musk xylene was below 2 μg kg^{-1} dm and that of musk ketone 3–5 μg kg^{-1} dm. For sediment from the river Ruhr, indicative AHTN or HHCB concentrations of 150–300 μg kg^{-1} dm are reported [21]. In sediment samples from a polishing pond of a sewage treatment plant, relatively high mean levels of polycyclic musks were measured (AHTN: 600 μg kg^{-1} dm, HHCB: 655 μg kg^{-1} dm) [22].

Summarizing, the AHTN and HHCB levels in sediments were (with few exceptions) between the detection limit and 40 μg kg^{-1} dm in the investigated rivers of the Elbe catchment area and in Baden-Württemberg. Sediment samples from the rivers Saale and Ruhr showed average contents of 100–300 μg kg^{-1} dm. The highest values were observed in sediments of waters influenced by sewage (about 900 μg kg^{-1} dm HHCB in a polishing pond, 2600 μg kg^{-1} dm AHTN in a surface water of Berlin).

Musk xylene was not detected in the sediments of [8, 20] and was found in the other studies in the range of 1–4 μg kg^{-1} dm. Musk ketone was partly measured in sediment samples with values up to 5 μg kg^{-1} dm [5–7, 20], up to 13 μg kg^{-1} dm [8] or in the range of 20–35 μg kg^{-1} dm in a polishing pond [22].

In general, the sediment samples were less contaminated by musk compounds than the SPM samples. On the basis of the medians, the difference was often about one order of magnitude. There are several reasons for this observation: the smaller fine particle fraction and the lower TOC content in sediment compared to SPM and possible metabolism processes in the sediment. The TOC contents given in a few studies clearly differ between SPM and sediment samples. In SPM the TOC range usually was 6–13% [19], 4–12% [2], or 4–9% [8], whereas in sediments TOC values of 1–6% [7] and 2–5% [8] were obtained.

3.2.2
Other Nitro Musks and Polycyclic Musks in Sediment

In Table 2, the contents of further musk compounds in sediments are compiled. Like the results in SPM, musk ambrette, musk tibetene, and musk moskene as well as DPMI were detected in no sample. In sediments from many rivers in Lower Saxony/Germany and in the river Saale, ADBI, AHDI, and ATII were found with maximum levels of 3–26 $\mu g\ kg^{-1}$ dm [5–7]. ATII was measured in a few sediment samples from the rivers Rhine and Neckar with values up to 22 $\mu g\ kg^{-1}$ dm [8]. In the rivers and lakes of Berlin the synthetic musk contents were partly higher: in waters with a high input of sewage water maximum concentrations of 68 $\mu g\ kg^{-1}$ dm ADBI, 93 $\mu g\ kg^{-1}$ dm AHDI, and 220 $\mu g\ kg^{-1}$ dm ATII were determined in sediment [9].

3.3
Synthetic Musk Compounds in Sewage Sludge

3.3.1
Musk Xylene, Musk Ketone, AHTN, and HHCB in Sewage Sludge

In Table 4, musk contents in sewage sludge from municipal sewage treatment plants (STPs) are listed. The samples are described with regard to their origin (region and year), their type, and their quantity (number of investigated STPs and total sample number). The average musk concentrations are given as median or mean values. Most of the samples date from the years 1996 to 1998.

Samples of digested sewage sludge were taken from seven great Hessian STPs once a year [2, 3]. Musk xylene and musk ketone were not detected in any sample. The AHTN or HHCB concentrations were about 6000–22,000 $\mu g\ kg^{-1}$ dm with a median of about 13,000–14,000 $\mu g\ kg^{-1}$ dm. Together with unpublished data from recent measurements it is possible to study the temporal trend over five successive years (1996–2000). A distinct and continuous decrease of the polycyclic musk levels in sewage sludge can be noticed in the data of each of the seven STPs.

The two sludge samples analyzed in [12] showed similar results (mean concentration of the main polycyclic musks: about 8500 $\mu g\ kg^{-1}$ dm). In the same report, sewer slime was also investigated; it was taken from industrial or domestic sewer systems. Musk xylene and musk ketone were found in half of the samples with values up to 200 and about 1800 $\mu g\ kg^{-1}$ dm, respectively. The levels of AHTN and HHCB varied between about 100 and 37,000 $\mu g\ kg^{-1}$ dm.

In [21] concentrations of 42,000–45,000 $\mu g\ kg^{-1}$ dm are reported for polycyclic musks in sewage sludge of one STP.

In two Dutch studies, the musk contents in different sludge types (primary, activated, and digested) were determined [11, 13]. The polycyclic musk concentrations were not significantly different between these sludge types [23]. The mean or median levels were about 5000–16,000 $\mu g\ kg^{-1}$ dm AHTN and about 10,000–28,000 $\mu g\ kg^{-1}$ dm HHCB (values for all sludge samples in both studies).

The musk contamination of Swiss digested sewage sludge samples was investigated in [14]. Musk xylene was detected in only one sample; musk ketone was

Table 4 Musk xylene, musk ketone, AHTN, and HHCB in sewage sludge [$\mu g\ kg^{-1}$ dry matter]

Region	Sludge type	Year	n[a]		Musk xylene	Musk ketone	AHTN	HHCB	Reference
Germany, Hessen (7 STPs[b])	Digested	1996–98	21	Range	<3–<23	<3–<27	5750–20,110	6700–21,930	[2, 3]
				Median	<5	<9	13,030	14,080	
Germany (2 STPs)	–	–	2	Values	<5	<10; 60	4000; 12,600	4300; 13,400	[12]
				Mean		33	8300	8870	
Germany	Sewer slime	–	17; 2[c]	Range	<5–200	<10–1,780	130–36,700	80–21,800	[12]
				Mean	28; 95	120; 250	2080; 23,100	1430; 15,500	
Germany, Northrhine-Westfalia (1 STP)	–	1995	1	Value	–	–	42,000	45,000	[21]
The Netherlands (6 STPs)	Primary	1997	11	Range	–	–	3300–14,000	5400–27,000	[13]
				Mean	–	–	8300	13,900	
	Activated	1997	12	Range	–	–	2300–34,000	4400–63,000	[13]
				Mean	–	–	16,000	27,900	
	Digested+ thickened+ composted	1997	8+2+3	Range	–	–	4900–22,000	9000–31,000	[13]
				Median	–	–	16,000	23,000	
The Netherlands (3 STPs)	Primary	1997–98	8	Range	–	–	3700–11,700	6000–17,000	[11]
				Median	–	–	8200	13,500	
	Activated	1997–98	7	Range	–	–	290–13,500	4800–21,000	[11]
				Median	–	–	5300	9700	
	Digested	1997–98	2	Values	–	–	11,000; 13,000	19,000; 21,000	[11]
				Mean	–	–	12,000	20,000	
Switzerland (12 STPs)	Digested	1998	12	Range	<1–32	<1–7	741–4161	2293–12,157	[14]
				Median	<1	2	1321	3896	

[a] Number of samples.
[b] Sewage treatment plants.
[c] Samples from industrial; domestic sewer systems (range: all samples; mean: data within the different sewer systems).
–: no data.

found in half of the 12 samples with a median of 2 μg kg^{-1} dm. For AHTN (HHCB) concentrations between about 700 and 4200 (about 2300–12,000) μg kg^{-1} dm were measured. The HHCB values as given in [14] and Table 4 must probably be corrected by the factor 0.5 (see Sect. 2.3).

As can be seen from Table 4, the contamination of sewage sludge by AHTN and HHCB always was at the mg kg^{-1} level. The average concentrations of these polycyclic musks usually ranged from 5000 to 28,000 μg kg^{-1} dm. Only the Swiss sewage sludge samples showed lower musk levels (median of AHTN and HHCB: about 1000 and 4000 μg kg^{-1} dm [14]). This may be due to incomplete extraction of the musk compounds from sludge, because in [14] hexane was used as extraction solvent, whereas in the other studies more polar solvents like acetone or dichloromethane were applied for the extraction of sewage sludge samples (see Sect. 2.2).

In general, the contents of musk xylene and musk ketone in sewage sludge were comparatively low (often below the detection limit, occasionally values up to 60 μg kg^{-1} dm, higher levels only in sewer slime).

3.3.2
Other Nitro Musks and Polycyclic Musks in Sewage Sludge

There are only few data about further musk compounds in sewage sludge which are given in Table 2. ADBI was found in two sludge samples with a mean level of 200 μg kg^{-1} dm [12]. In Swiss sewage sludge [14], the three nitro musk fragrances musk ambrette, musk tibetene, and musk moskene could not be detected (detection limit: 1 μg kg^{-1} dm); DPMI, ADBI, and AHDI were determined with median concentrations of about 70, 110, and 200 μg kg^{-1} dm, respectively.

3.3.3
Musk Levels in Sewage Sludge from Different Waste Water Catchment Areas

The musk contamination of sewage sludge depends on the type of waste water being led to the STP (domestic or industrial origin). As can be seen from several investigations, the concentration of musk fragrances is distinctly higher in sewage sludge from domestic sewage than in samples from industrial waste water.

As already mentioned above, sewer slime was analyzed for micropollutants accumulating from waste water [12]. Here, 17 slime samples were taken from the sewerage in an industrial area and 2 samples from sewers in residential areas. The mean musk contents in the sewer slime samples are given separately for both areas in Table 4. In the residential area, the nitro musk (polycyclic musk) levels were 2- to 3-fold (7- to 11-fold) higher than in the industrial area.

Swiss sewage sludge samples from different catchment areas were investigated in [14]. The selected sludge samples contained no (type A), low (type B), or higher (type C) amounts of industrial waste water (the number of the corresponding samples were 7, 3, and 2, respectively). The mean content of AHTN and HHCB in sludge of type C was approximately 50% lower than in sludge of type A.

In [2, 3] synthetic musk concentrations in sewage sludge from six Hessian industrial STPs are reported. In contrast to the sewer systems in [12] and [14], no private households were connected to the industrial STPs. The mean AHTN or HHCB concentrations in industrial sewage sludge were 10–180 μg kg^{-1} dm (mean: 55 μg kg^{-1} dm) and thus 2–3 orders of magnitude lower than in sewage sludge samples from seven Hessian municipal STPs (see Table 4).

3.3.4
Degradation of Nitro Musks in Sewage Treatment Plants

Besides digested sewage sludge from seven Hessian STPs, samples from two STPs which are not equipped with a digestion step were also analyzed for musk compounds [2, 3]. Musk xylene and musk ketone were not detectable in the digested sewage sludge samples (see Table 4). In six non-digested sludge samples musk xylene was not found either, but musk ketone showed significantly different results, i.e., 170–630 μg kg^{-1} dm (mean: 300 μg kg^{-1} dm). The polycyclic musk levels were comparable in both sludge types.

In six STPs, sludge samples were taken before and after the digestion step [2]. Again a distinct influence of the digestion on the musk ketone content was observed: before digestion the sewage sludge samples contained 80–470 μg kg^{-1} dm musk ketone, after digestion musk ketone was no longer found. These results clearly show that metabolism of musk ketone occurs during the digestion process.

The anaerobic degradation is confirmed by the results of several studies in which samples from STPs were investigated and nitro musk metabolites in the reduced form were detected. Amino metabolites of musk xylene and musk ketone were identified in effluents of municipal STPs, whereas they were not detectable in the influent samples [24, 25]. In the effluent of an STP, the nitro musk concentrations were one to two orders of magnitude lower than in the influent and 4- to 40-fold smaller as compared to the corresponding amino derivatives [24]. The simultaneous decrease of the nitro musk levels and the increasing transformation product contents show the importance of metabolism processes in STPs. In the river Elbe and in some sediment samples, the musk xylene and musk ketone metabolites were measured too [25].

Amino metabolites of musk xylene, musk ketone, and musk moskene were also detected in Swiss digested sewage sludge [14]. Their concentrations were partly higher than those of the parent compounds. Especially musk xylene und musk moskene were found mainly as their amino derivatives in the analyzed sludge samples. For further details about nitro musk metabolites see the chapter about transformation pathways (pp 189).

3.4
Partition of Synthetic Musks Between Water and Solid Matter Phase

3.4.1
Determination of the Partition Coefficients

Partition coefficients of musk compounds have been determined in different solid matter matrices. The literature data are summarized in Table 5.

In several field experiments, the partition of musk fragrances between SPM obtained by flow-through centrifuges and water was studied. The samples were

Table 5 Partition coefficients between water and solid matter phase

Musk ketone	AHTN	HHCB	n[a]	River	Reference
Suspended particulate matter					
Partition coefficient, K_p [L kg^{-1}]					
1500	–	–	12	Rhine	[26]
1700	9550	4490	31	Elbe	[4]
–	8250	5640	17	Ruhr	[18]
1300	3100	2400	3	Three brooks	[8]
–	6310	7240		Calculated values	[27]
Adsorbed portion					
3%	–	–	12	Rhine	[26]
5%	17%	8%	31	Elbe	[4]
3%	8%	7%	3	Three brooks	[8]
–	9%	10%		Calculated values	[27]
Partition coefficient based on organic carbon, K_{oc} [L kg^{-1}]					
10,000	27,900	21,300	3	Three brooks	[8]
Sediment					
Partition coefficient, K_p [L kg^{-1}]					
--	144	124	3	Three brooks	[8]
Partition coefficient based on organic carbon, K_{oc} [L kg^{-1}]					
--	10,400	7200	3	Three brooks	[8]
Activated sludge					
Partition coefficient, K_p [L kg^{-1}]					
–	11,590	13,600		Laboratory experiments	[27]
Partition coefficient based on organic carbon, K_{oc} [L kg^{-1}]					
–	63,660	72,470		Laboratory experiments	[27]

[a] Number of samples.
–: no data.
--: determination not possible.

taken from the rivers Rhine [26], Elbe [4], Ruhr [18], and from three brooks in Baden-Württemberg [8]. The fragrances were quantified in the SPM samples and the corresponding water samples after centrifugation (or filtration [8]). In another study, partition coefficients were calculated for musks in SPM/water [27].

Besides investigations with SPM, there are some results for further solid matter matrices. Some field experiments regarding the partition between sediment and water are reported in [8]. Laboratory experiments with activated sludge were carried out in [27]; in a sorption isotherm experiment, ^{14}C labeled test substances were used.

The partition coefficient K_p is calculated by the equation: $K_p=C_{solid}/C_{water}$ [L kg^{-1}] (C_{solid}: concentration in solid matter phase; C_{water}: concentration in water phase).

When the partition coefficient refers to the TOC content of the solid matter sample, K_{oc} values are obtained: $K_{oc}=K_p/TOC$ [L kg^{-1}].

3.4.2
Partition Coefficients in Different Solid Matter Matrices

The partition coefficients resulting from the experiments described above are compiled in Table 5. Usually the results represent mean values from measurements of several samples.

3.4.2.1
K_p Values in SPM

For musk ketone K_p values of 1300–1700 L kg^{-1} are reported in three studies [4, 8, 26]. The K_p values given in four studies were 3100–9550 L kg^{-1} for AHTN and 2400–7240 L kg^{-1} for HHCB [4, 8, 18, 27]. The results from the different investigations agree very well for musk ketone and are in the same order of magnitude for the polycyclic musks.

From the K_p data the adsorbed portions of the musk compounds were calculated with the following results: 3–5% for musk ketone, 8–17% for AHTN, and 7–10% for HHCB. Thus, the musk fragrances are adsorbed to SPM only to a small extent; the main portion is dissolved in water.

There is a remarkable discrepancy regarding the partition behavior of musk ketone on the one hand and of AHTN or HHCB on the other [1]. The octanol/water partition coefficients differ by two orders of magnitude [4], whereas the corresponding K_p values are in the same range. An explanation of this phenomenon cannot be given so far.

For two further synthetic musks literature data for partition coefficients are available. For ADBI a K_p value of 6630 L kg^{-1} was determined in [4], and for musk xylene a K_p value of 1860 L kg^{-1} was found [26]. In field studies on the partition equilibrium of other musk fragrances, a calculation of K_p values was not possible, because in one or two phases the musk compounds were not detectable [8].

3.4.2.2
K_p Values in Sediment

Partition coefficients between sediment and water were determined for AHTN and HHCB in [8]. The much lower K_p values for sediment (about 134 l kg^{-1}) than for SPM are coincident with similar differences between the musk concentrations in sediment and SPM samples as given in Tables 3 and 1 (see Sect. 3.2).

3.4.2.3
K_{oc} Values in SPM and Sediment

K_{oc} values were calculated for SPM and sediment in [8]. For AHTN and HHCB K_{oc} values of 27,900 L kg^{-1} and 21,300 L kg^{-1} are reported for SPM and for sediment 10,400 L kg^{-1} and 7200 L kg^{-1}, respectively. Thus, after normalization to the TOC the great difference between the K_p values for SPM and sediment (factor 20) is reduced significantly, leading to K_{oc} values which are comparable for both matrices differing only by factor 3.

3.4.2.4
K_p and K_{oc} Values in Sludge

The partition coefficients obtained in the experiments with activated sludge were K_p=11,590 L kg^{-1} and K_{oc}=63,660 L kg^{-1} for AHTN and K_p=13,600 L kg^{-1} and K_{oc}=72,470 L kg^{-1} for HHCB [27]. The K_p as well as the K_{oc} values for sludge are higher than the corresponding partition coefficients for SPM; this is not surprising as sewage sludge and SPM represent different types of solid matter.

3.4.3
AHTN:HHCB Concentration Ratios in Surface Water and Solid Matter Samples

In several studies on musk contents in surface water, a mean AHTN:HHCB concentration ratio of approximately 1:2 was obtained. For instance, this ratio was observed in samples from the river Ruhr [28], Elbe [4, 5], or Berlin's surface waters [29]. It is in agreement with the 2.5-fold higher use volume of HHCB in Europe as compared to AHTN [27]. On the contrary, the AHTN:HHCB ratio found in solid matter samples generally was 1:1. With the exception of one study [10], the AHTN and HHCB concentrations showed nearly identical values in SPM and sediment samples as given in Tables 1 and 3 (of course the corrected HHCB values have to be considered, see Sect. 2.3).

From the shift of the AHTN:HHCB ratio of 1:2 in water to 1:1 in solid matter samples can be concluded that AHTN is stronger adsorbed to particles than HHCB. This is confirmed by the partition coefficients as determined in [4] which are twofold higher for AHTN than for HHCB (K_p of AHTN=9550 L kg^{-1}, K_p of HHCB=4490 L kg^{-1}). Similar K_p ratios were also obtained in laboratory partition experiments which were performed with spiked SPM samples and water (distilled water as well as humic acid solution): The measured K_p values of AHTN were two- to threefold higher than those of HHCB [4].

The partition coefficients for SPM from the other three studies listed in Table 5 gave AHTN:HHCB ratios of 1.5:1 [18] and 1:1 [8, 27]. The deviating ratio in [27] is probably due to the fact that the K_p values were calculated and not measured. The partition coefficients in [8] were determined adequately, but the polycyclic musk pattern in the selected brooks was quite untypical and the K_p determination was based on only three samples. However, K_p values from field experiments showed relatively high variations, e.g., by a factor of 5–8 [4] or 6–7 [8]. As a consequence, the K_p values as given in [4] reflecting the stronger adsorption for AHTN are probably realistic.

4
Conclusions

As shown in the preceding sections, suspended particulate matter, sediment, and sewage sludge are contaminated by synthetic musk compounds. The concentrations of the polycyclic musks in solid matter samples are much higher than those of the nitro musks. The investigated solid matter matrices show a clear order of musk contamination. Obviously the musk concentrations in sewage sludge are the highest, followed by the musk levels in SPM and finally by those in sediment.

There still remain several questions regarding musk fragrances in solid matters. Some of them are given here:

- Are the indications of decreasing synthetic musk concentrations in solid matter samples confirmed by other studies?
- Which factors are responsible for the much lower musk levels in SPM obtained by sedimentation chambers as compared to the centrifuge-type SPM?
- Does AHTN adsorb more strongly to SPM than HHCB?
- How are musks eliminated in STPs (adsorption and/or degradation)?

As a consequence, further studies on the occurrence and fate of musk compounds in the aquatic environment are needed.

5 References

1. Rimkus GG, Gatermann R, Hühnerfuss H (1999) Toxicol Lett 111:5
2. Fooken C, Gihr R, Häckl M, Seel P (1997) Orientierende Messungen gefährlicher Stoffe. Landesweite Untersuchungen auf organische Spurenverunreinigungen in hessischen Fließgewässern, Abwässern und Klärschlämmen, 1991–1996. Schriftenreihe der Hessischen Landesanstalt für Umwelt; Umweltplanung, Arbeits- und Umweltschutz, Heft Nr. 233, Wiesbaden, Germany
3. Fooken C, Gihr R, Seel P (2000) Orientierende Messungen gefährlicher Stoffe. Landesweite Untersuchungen auf organische Spurenverunreinigungen in hessischen Fließgewässern, Abwässern und Klärschlämmen, 1991–1998. Ergänzender Bericht zu 1997–1998 (Bericht der Hessischen Landesanstalt für Umwelt für die hessischen Wasserbehörden), Wiesbaden, Germany
4. Winkler M, Kopf G, Hauptvogel C, Neu T (1998) Chemosphere 37:1139
5. Wiegel S, Harms H, Stachel B, Reincke H (2000) Synthetische Moschus-Duftstoffe in der Elbe. Arbeitsgemeinschaft für die Reinhaltung der Elbe, Hamburg, Germany
6. Lach G, Steffen D (1997) Orientierende Untersuchungen von Gewässersedimenten auf Nitro-/Polymoschusverbindungen und die Flammschutzmittel TCEP und TCPP. Oberirdische Gewässer 3/97. Niedersächsisches Landesamt für Ökologie, Hildesheim, Germany
7. Steffen D (2000) Niedersächsisches Landesamt für Ökologie, Hildesheim, Germany. Personal communication
8. Landesanstalt für Umweltschutz Baden-Württemberg (2001) Untersuchungen zum Vorkommen von Xenobiotika mit toxikologischer oder endokriner Wirkung in Schwebstoffen und Sedimenten. Abschlußbericht Juni 2000. Karlsruhe, Germany
9. Fromme H, Otto T, Pils K (2001) Water Res 35:121
10. Breukel RMA, Balk F (1996) Musken in Rijn en Maas. RIZA Werkdocument 96.197x. National Institute for Inland Water Management and Waste Water Treatment, Lelystad, The Netherlands
11. Rijs GBJ (2000) RIZA, Lelystad, The Netherlands. Personal communication
12. Sauer J, Antusch E, Ripp C (1997) Vom Wasser 88:49
13. Blok J (1998) Measurement of polycyclic and nitromusks in sludges of sewage treatment plants in The Netherlands. Final report to the Research Institute for Fragrance Materials (RIFM). BKH Consulting Engineers, Delft, The Netherlands
14. Herren D, Berset JD (2000) Chemosphere 40:565
15. Landesanstalt für Umweltschutz Baden-Württemberg (2000) Beschaffenheit der Fließgewässer, Jahresdatenkatalog 1998 (CD-ROM). Karlsruhe, Germany
16. Stachel B, Elsholz O, Reincke H (1995) Fresenius J Anal Chem 353:21

17. Arbeitsgemeinschaft für die Reinhaltung der Elbe (1999) Wassergütedaten der Elbe, Zahlentafel 1998. Hamburg, Germany, pp 58–60
18. Eschke HD (2000) Ruhrverband, Essen, Germany. Personal communication
19. Winkler M (2000) Umweltforschungszentrum Leipzig-Halle, Magdeburg, Germany. Personal communication
20. Ruhrverband (1995) Ruhrwassergüte 1994, Essen, Germany, p 73
21. Eschke HD (1996) Ruhrverband, Essen, Germany. Personal communication. In: van de Plassche EJ, Balk F (1997) Environmental risk assessment of the polycyclic musks AHTN and HHCB according to the EU-TGD. RIVM Report 601503 008. National Institute of Public Health and the Environment, Bilthoven, The Netherlands, p 39
22. Gatermann R (2000) University of Hamburg, Germany. Personal communication
23. Balk F, Ford RA (1999) Toxicol Lett 111:57
24. Gatermann R, Hühnerfuss H, Rimkus G, Attar A, Kettrup A (1998) Chemosphere 36:2535
25. Rimkus GG (1999) Toxicol Lett 111:37
26. Breitung V, Woszidlo S, Bergmann H (1996) Vorkommen und Verteilungsverhalten von Nitromoschusverbindungen in Oberflächengewässern, Poster der Bundesanstalt für Gewässerkunde auf dem IKSR Symposium "Leben mit dem Rhein", Koblenz, Germany
27. van de Plassche EJ, Balk F (1997) Environmental risk assessment of the polycyclic musks AHTN and HHCB according to the EU-TGD. RIVM Report 601503 008. National Institute of Public Health and the Environment, Bilthoven, The Netherlands
28. Eschke HD, Traud J, Dibowski HJ (1994) UWSF - Z Umweltchem Ökotox 6:183
29. Heberer T, Gramer S, Stan HJ (1999) Acta Hydrochim Hydrobiol 27:150

The Handbook of Environmental Chemistry Vol. 3, Part X (2004): 49–84
DOI 10.1007/b14174

Synthetic Musks in Fish and Other Aquatic Organisms

Pim E. G. Leonards · Jacob de Boer

Netherlands Institute for Fisheries Research (RIVO), P.O. Box 68, 1970 AB IJmuiden, The Netherlands
E-mail: pim.leonards@wur.nl

Abstract Musk compounds are widely spread environmental pollutants. Musk compounds were found in aquatic organisms from the North Sea, in rivers, lakes and estuaries in Canada, Czech Republic, Germany, Italy, Luxembourg, Japan, Norway, Switzerland, Sweden, and The Netherlands. Two nitro musks, musk xylene (MX) and musk ketone (MK), and two polycyclic musks (HHCB and AHTN) were the major musk compounds determined and observed in freshwater as well as in marine organisms. The main source of nitro and polycyclic musk residues in aquatic organisms are effluents from sewage treatment plants (STPs). The presence of synthetic musk compounds in biota can, therefore, be used as an indicator of the exposure of biota to STP effluents. Synthetic musk compounds have mainly been determined in fish, but some data are also available for mussels and shrimps. In addition, MX was found in eggs of coastal bird species, and HHCB and AHTN were identified in otters. The concentrations of HHCB and AHTN in freshwater organisms from Europe are one to two orders of magnitude higher than MX and MK, and comparable to levels of PCBs in fish. Indications were found that several fish species such as eel (*Anguila anguila*) could metabolise HHCB and AHTN, and that food chain transfer of these musk compounds from prey fish (e.g. roach) to carnivore fish (pike-perch) occurs. Time trend data for MX and MK in eel from the river Elbe (Germany) showed that for some locations a decline in the concentrations from 1994 to 1999 occurred, probably due to the restriction of the use of MX in Germany since 1993. Similar results were observed for MX in eel from the river Rhine.

Nitro musks and polycyclic musks were found in fish samples (e.g. trout, herring, mussels, tuna and mackerel) collected at food markets. In some samples (trout and shrimp) the concentrations of MX and MK were similar to the concentrations of PCBs, while in other samples (e.g. halibut and mussels) the concentrations of MX and MK are one to two orders of magnitude lower than those of PCBs. In general, the highest concentrations of MX were found in trout, and in some tuna samples rather high concentrations of MK were found.

Keywords Fish · Aquatic organisms · Nitro musk · Polycyclic musk · Bioaccumulation

1 Introduction

The presence of the nitro musks, musk xylene (MX) and musk ketone (MK) was shown for the first time in 1981 in the aquatic ecosystem [1]. The nitro musks were found in fish from the Japanese river Tama, and later also in mussels, river water and waste water from the river Tama and Tokyo Bay [2]. Almost a decade later, Rimkus and Wolf [3] initiated new research on the occurrence of nitro musks in the aquatic environment. This was the start of more extensive surveys and monitoring programs of musk compounds in the German environment. The research was extended to water systems in other countries; however, most data on musk compounds have been reported for German rivers, e.g. Elbe and Rhine.

In this chapter an overview of data on nitro musks and polycyclic musks in aquatic organisms is presented and discussed. The analytical methods used for the determination of synthetic musk compounds in organisms will be briefly discussed. Information on concentrations, patterns, trends and species differences of synthetic musks in organisms from natural water systems, systems influenced by water of sewage treatment plants (STPs), and farmed fish will be discussed. Two German monitoring studies on the occurrence of nitro musks and polycyclic musks in the river Elbe and its tributaries, and MX in the river Rhine will be discussed in more detail, because monitoring studies have been performed for several years and various fish species in these rivers. In addition, information on musk compounds in higher trophic organisms (birds and mammals) is given.

2 Analysis of Synthetic Musks in Biota

The analytical procedures used for the analysis of musk compounds in biota are very similar to methods used for the residue analysis of, e.g. PCBs and organochlorine pesticides (OCPs). However, some steps are more critical and cross-contamination is a greater problem than with PCBs or OCPs.

In general, biota samples are extracted using a Soxhlet apparatus, Ultra Turrax extraction or accelerated solvent extraction (ASE), with a mixture of organic solvents, e.g. hexane:acetone, cyclohexane:ethylacetate or water:acetone:petroleum ether [4–6]. These extracts not only contain the contaminants but also a relatively high amount of lipids, which have to be removed before the final analysis by gas chromatography (GC) can be performed. Rimkus et al. [7] critically discussed the importance of the removal of lipids for a reliable analysis, and the necessity of a thorough clean-up of fat-containing samples is emphasised. Without a proper clean-up, matrix effects on capillary GC columns and unstable detector responses can be expected [7]. A "new fast and simple method" for the determination of musk xylene and musk ketone in fish samples without clean-up, using GC with mass spectrometry (MS), published in 1996 [8, 9], therefore, is not recommended. Various clean-up methods for the elimination of lipids and other matrix compounds such as silica gel, florisil, aluminium oxide or gel permeation chromatography (GPC) are used [2, 4, 10, 11]. For the analysis of polycyclic musk compounds absorption column chromatography with silica gel or florisil has resulted in a high variability and losses of these compounds [27]. Consequently, the elution conditions should be optimised for each compound, as these compounds are more retained on silica gel then PCBs and OCPs. Sequential mixtures of organic solvents for the elution of the nitro and the polycyclic musks from silica are used [6]. A first fraction, eluted with hexane, contains PCBs only, and a second fraction eluted with hexane/toluene contains MX. MK, HHCB and AHTN are eluted from silica with toluene and toluene/acetone only [6]. An alternative method is the use of a silica gel solid-phase extraction (SPE) cartridge, and elution with dichloromethane [12]. The latter method also elutes nitro musk amino metabolites. In particular, GPC is a very useful and efficient technique to remove large amounts of lipids with, e.g. cyclohexane:ethyl acetate as a mobile phase [6]. GPC has been combined successfully with HPLC for the routine analysis of fatty samples. This combination delivers very clean extracts for GC/MS analysis [13]. The use of concentrated sulfuric acid to remove lipids is not recommended, as recoveries for MK and several polycyclic musks are too low due to partial destruction [27].

A critical step in the analysis of musks is the evaporation of the bulk amount of organic solvents used in several steps to prevent losses of the volatile synthetic musks. Careful evaporation is necessary. This is generally performed at a water bath temperature of 40 °C to a volume of 1 mL, and followed by further concentration by a gentle stream of nitrogen, which will result in high recoveries. It has been suggested to use the polycyclic musk DPMI as a monitor for recovery losses by evaporation [12]. Another critical step in the analysis of polycyclic musks is the prevention of contamination of the samples from the laboratory environment [14]. As musk compounds are frequently used in soaps and perfumes, contamination easily occurs. The contamination by HHCB during the clean-up steps was discussed by Rimkus and Wolf [14]. Regular analysis of procedural blanks of the analytical method is recommended.

The final step of the analysis of the synthetic musks is the separation of these compounds by gas chromatography (GC). For the separation of nitro and polycyclic musks, 100% methylpolysiloxane, 95% methyl 5% phenyl polysiloxane or 85% methyl, 7% phenyl, 7% cyanopropyl, 1% vinyl polysiloxane have been used as suitable stationary phases. Recently, the diastereometric forms of HHCB in fish and human adipose tissue have been separated on a methylpolysiloxane phase with 12–15% phenyl groups or a polyethylene glycol phase [5, 16]. The enantiomeric composition of HHCB and AHTN in various aquatic species has also recently been determined using a modified cyclodextrin as chiral stationary phase coupled to a high resolution (HR) mass spectrometer (MS) [16] (see chapter of Hühnerfuss et al. in this monograph).

For the detection of nitro musks electron capture detection (ECD) is often used because of the high sensitivity. Polycyclic musks are detected with low resolution (LR) and HRMS in the electron impact mode (EI), or in the MS/MS mode that enhanced the sensitivity remarkably due to the reduction in background noise [17, 18]. The use of the MS in the negative chemical ionisation mode (NCI) is especially recommended for the detection of the nitro musks as these compounds forms easily negative ions [4]. This is in contrast to the polycyclic musks which can be detected with higher sensitivity with EI-MS than with NCI-MS [14].

3 Occurrence, Patterns and Trends of Synthetic Musks in Fish

3.1 Data Collection

This paragraph provides an overview of the levels of five nitro musks -MX, MK, musk ambrette (MA), musk moskene (MM), musk tibetene (MT)-, and five polycyclic musks – HHCB, AHTN, ADBI, ATII, and AHDI- in aquatic organisms published in scientific literature and in reports. Additional information on the origin of the samples, time of sampling, number of individual specimen per sample, and lipid content was collected and used for the evaluation of the data. All data were divided in organisms from i) natural fresh and marine water systems, ii) effluents of domestic sewage treatment plants (STPs), iii) fish farms and iv) samples from the food market. If original data were published, the median and mean values were calculated. Concentrations are given in mg kg^{-1} lipid weight (lw) whenever possible. Some studies reported on wet weight (ww) basis only [2, 19–22] and, therefore, these data will be presented on wet weight basis.

3.2 Synthetic Musks in Fish and Shellfish

3.2.1 *Biota from Natural Fresh and Marine Waters*

A summary of levels of MX, MK, HHCB and AHTN in fish and shellfish from fresh and marine water systems is given in Table 1. Concentrations of MA, MM, MT, ADBI, ATII and AHDI are generally below the limit of detection and, therefore, not reported in Table 1. Most studies focused on the occurrence of MX and MK in freshwater fish species, and the major part of the data results from samples from Germany. Additionally, data from fish samples from Canada, Czech Republic, Italy, Luxembourg, Japan, Norway, Switzerland, Sweden and The Netherlands are given. Synthetic musks have also been determined in a number of fish, mussel, shrimp and crab samples from the marine environment. The global distribution of the nitro musks and the polycyclic musks with regard to concentration and patterns found in aquatic organisms is discussed. Further, information on species specific differences in levels and patterns found in various aquatic species of MX for the Rhine and MX, MK, HHCB, AHTN for the river Elbe and its tributaries is provided. The chapter will end with information on temporal trends for MX and MK in the river Elbe, the Berlin area and the river Rhine.

3.2.1.1 Concentrations and Patterns

MX, MK, HHCB and AHTN were the most frequently found compounds in all samples. MX was present in freshwater species (e.g. eel, pike-perch, bream) at concentrations from below the limit of detection (<0.001) to a concentration of 1.3 mg kg^{-1} lw in pike-perch from the Elbe (Germany), whereas MK was found in concentrations ranging from <0.01 to 1.4 mg kg^{-1} lw in pike-perch from the Elbe. The relatively high concentrations in some fish samples resemble the concentrations observed in fish from sewage ponds (see section below, Table 2), but in general are lower. In effluents of municipal STPs high levels of nitro musks and polycyclic musks have been detected [10]. Samples of brown trout from the river Stör (Germany), a tributary of the Elbe, showed relatively high levels of MX and MK of 0.20–0.24 mg kg^{-1} lw and 1.0–1.2 mg kg^{-1} lw, respectively, which can be related to the sampling site, which was only 3 km downstream from an STP [3, 23].

In 1998 MX, MK, HHCB and AHTN were detected in bream from the Elbe and other waters at the Czech Republic [24]. Relatively high concentrations for MX and MK were found (mean 0.21 mg kg^{-1} lw and 0.19 mg kg^{-1} lw, respectively), yet the concentrations of HHCB and AHTN were even an order of magnitude higher (3.2 mg kg^{-1} lw and 1.3 mg kg^{-1} lw, respectively). These samples were collected downstream of a detergent production plant, which could explain the relatively high levels. More details are described in the chapter of Hajslova et al. in this monograph.

In general, the concentrations of polycyclic musks in freshwater fish from Europe are equal to or exceed the concentrations of nitro musks. The pattern in aquatic organisms is predominantly based on the sampling distance from STPs. The closer samples have been taken to the discharge point of STPs, the higher the ratio of HHCB and AHTN compared to MX and MK. Typical patterns of synthetic musks in eel from the Elbe and Stör are shown in Fig. 1. In Canada a different pattern in aquatic organisms is observed (Fig. 1). The concentrations of MK in various fish and shellfish samples from Nova Scotia, New Brunswick and Lake Ontaria are one to two orders of magnitude higher than the concentrations of MX and polycyclic musks [11], which can be attributed to the different uses of fragrances in both parts of the world. It seemed that in Canada MK is used predominantly, while in Western Europe the polycyclic musks are the major musk compounds used due to the phase-out of MX. Another important factor for the explanation of the synthetic musk pattern found in fish is the metabolism capacity of fish. For MX and MK it is know that fish are able to metabolise these compounds. Several monoamino metabolites have been found in various fish species, with 4-NH_2-MX as the dominant metabolite, with often concentrations higher than the parent compound [25]. More information on the metabolism of synthetic musk compounds can be found in the chapter of Biselli et al. in this monograph.

Table 1 Concentrations of nitro musks (MX and MK) and polycyclic musks (HHCB and AHTN) in mg kg^{-1} lipid weight in aquatic organisms from various freshwater and seawater locations. Location, species, year of sampling, lipid content, mean, median, and ranges of concentrations are given. Concentrations in brackets are in mg kg^{-1} wet weight basis instead of lipid weight basis

Location	Species	Year of sampling	Lipid content (%)	Musk xylene indiv. or min-max	Musk xylene mean/median
Canada					
Lake Ontaria	Lake trout, whole body		4	0.002–0.004	0.003/0.003
New Brunswick (Cap Pele)	Lobster hepatopancreas		3	nd-0.002	
New Brunswick (Miramichi)	Eel liver			0.006	
New Brunswick (Miramichi)	Eel muscle			nd	
New Brunswick (Miramichi)	Flounder liver			d	
New Brunswick (Miramichi)	Flounder muscle			0.049	
Nova Scotia (Halifax harbour)	Blue mussels			nd	
Nova Scotia (Halifax harbour)	Lobster hepatopancreas			0.003	
Nova Scotia (Halifax harbour)	Pollock liver			nd	
Nova Scotia (Halifax harbour)	Soft-shelled clam			0.110	
–	Striped bass liver			d	
–	Striped bass liver			nd	
Czech republic					
Elbe (Hrensko)	Bream			0.38	
Elbe (Pradubice downstream)	Bream			0.21	
Elbe (Pradubice upstream)	Bream			0.19	
Elbe (Strekov)	Bream			0.08	
Germany					
Eider	Bream	1991–1993		0.01	
Eider	Eel	1991–1993		0.03	
Elbe	Bream	1991–1993		0.08	
Elbe	Bream	1991–1993		0.08	
Elbe	Eel	1991–1993		0.06	
Elbe	Eel	1991–1993		0.07	

Musk ketone indiv. or min-max	Musk ketone mean/median	HHCB value, indiv. or min-max	HHCB mean/median	AHTN value, indiv. or min-max	AHTN mean/ median	Refer- ence
0.076–0.730	0.344/0.285	0.029–0.030	0.030/0.030	0.025–0.045	0.033/0.029	11
0.110–0.190	0.143/0.130	0.010–0.018	0.015/0.016	0.007–0.012	0.009/0.009	11
0.190		d		d		11
0.340		0.034		0.017		11
0.460		0.040		0.056		11
2.700		d		d		11
2.200		1.650		nd		11
0.130		0.120		nd		11
nd		nd		nd		11
17.700		3.000		1.100		11
d		nd		nd		11
1.400		0.100		0.070		11
0.15		3.194		1.289		24
0.22		1.0		1.6		24
0.18		0.9		0.6		24
0.21		0.5		0.6		24
						3, 4
						3, 4
0.08						3, 4
0.09						3, 4
0.04						3, 4
0.05						3, 4

Table 1 (continued)

Location	Species	Year of sampling	Lipid content (%)	Musk xylene indiv. or min-max	Musk xylene mean/median
Elbe (various locations)	Eel	1994		<0.010–0.47	
Elbe (various locations)	Eel	1995		<0.020–0.31	
Elbe (various locations)	Eel	1997		<0.020–0.27	
Elbe (various locations)	Eel	1998		<0.020–0.35	
Elbe (various locations)	Eel	1999		<0.005–0.329	
Elbe	Eel		21.1–29.1	0.010–0.070	
Elbe	Pike-perch		0.3–0.7	<0.010–0.090	
Elbe	Pike-perch			<0.01–0.990	
Elbe	Pike-perch			<nd–1.292	
Elbe (Brunsbüttel)	Eel		23.7	0.01	
Elbe (Haseldorf)	Eel		21.1	0.07	
Elbe (Haseldorf)	Eel		23.4	0.01	
Elbe (Haseldorf)	Pike-perch	1994	0.5	0.069	
Elbe (Haseldorf)	Pike-perch	1994	0.5	0.099	
Elbe (Haseldorf)	Pike-perch		0.7	0.08	
Elbe (Haseldorf)	Pike-perch		0.7	0.09	
Elbe (Lauenberg)	Eel		26.2	0.02	
Elbe (Lauenberg)	Eel		29.1	0.01	
Elbe (Lauenberg)	Pike-perch		0.3	<0.01	
Elbe (Lauenberg)	Pike-perch		0.5	0.02	
Elbe (Lauenburg)	Pike-perch	1994	0.3	0.030	
Elbe (Lauenburg)	Pike-perch	1994	0.4	0.023	
Elbe (Wedel)	Pike	1993	0.7	0.046	
Rhine (Grißheim)	Eel	2000		0.005–0.010	0.007/0.007
Rhine (Grißheim)	Roach	2000		0.047	
Rhine (Kembs)	Roach	2000		0.236–0.407	0.308/0.282
Rhine (Mannheim)	Eel	2000		0.005–0.032	0.019/0.021
Rhine (Mannheim)	Roach	2000		0.022–0.068	0.049/0.057
Rhine (Neuburgweier)	Eel	2000		0.010–0.030	0.021/0.021
Rhine (Neuburgweier)	Roach	2000		0.033–0.122	0.063/0.034

Musk ketone indiv. or min-max	Musk ketone mean/median	HHCB value, indiv. or min-max	HHCB mean/median	AHTN value, indiv. or min-max	AHTN mean/ median	Reference
<0.010–0.39						48
<0.020–0.15						36
<0.050–0.050						36
<0.020–0.40						36
<0.055–0.102						36
0.010–0.030		0.030–0.090		0.040–0.120		27
0.010–0.070		0.600–3.840		0.320–0.990		27
<0.01–1.400						48
<nd–0.257						36
0.01		0.03		0.04		35
0.03		0.09		0.12		35
0.01		0.03		0.04		35
0.064						25
0.027						25
0.07		3.84		0.99		35
0.03		2.82		0.56		35
0.01		0.04		0.05		35
0.02		0.05		0.07		35
0.01		0.60		0.32		35
0.03		1.97		0.45		35
0.054						25
0.025						25
0.176						25
						38
						38
						38
						38
						38
						38
						38

Table 1 (continued)

Location	Species	Year of sampling	Lipid content (%)	Musk xylene indiv. or min-max	Musk xylene mean/median
Rhine (Schwörstadt)	Eel	2000		0.008–0.032	0.013/0.012
Rhine (Taubergießen)	Eel	2000		0.005–0.028	0.017/0.015
Rhine (Taubergießen)	White bream	2000		0.024–0.041	0.033/0.035
Rhine (Konstanz)	Eel	2000		nd	nd
Rhine (Konstanz)	Roach	2000		0.013–0.018	0.015/0.015
Rhine (Grenzach)	Eel	2000		0.009–0.026	0.015/0.014
Lake Constance and inlets	fish	1992–1993		<0.005–1.37	0.46
Ruhr (Witten)	Bream		0.45–1.56	0.90/0.68	
Ruhr (Witten)	Chub		0.6		
Ruhr (Witten)	Eel		20.9		
Ruhr (Witten)	Eel		23.7		
Ruhr (Witten)	Perch		0.36		
Ruhr (Witten)	Perch		0.72		
Ruhr (Witten)	Roach		1.13		
Schlei	Bream	1991–1993		0.02	
Schlei	Eel	1991–1993		0.01	
Schlei	Flounder	1991–1993		0.02	
Stör	Bream	1991–1993		0.06–0.35	0.16/0.07
Stör	Eel	1991–1993		0.02–0.07	0.4/0.4
Stör	Ide	1991–1993		0.08–0.13	0.10/0.10
Stör	Pike	1991–1993		0.23	
Stör	Pike-perch	1991–1993		0.13	
Stör (downstream)	Bream	1993	1.7	0.010	
Stör (upstream)	Bream	1993	1.2	0.084	
Trave	Bream	1991–1993		0.05	
Trave	Eel	1991–1993		0.07	
Trave	Flounder	1991–1993		0.07	
Trave	Perch	1991–1993		0.17	
Trave	Pike	1991–1993		0.17	
Italy					
Piave	Trout		4.9		
Piave	Trout		3.0		
Piave	Trout		1.9		

Musk ketone indiv. or min-max	Musk ketone mean/median	HHCB value, indiv. or min-max	HHCB mean/median	AHTN value, indiv. or min-max	AHTN mean/ median	Reference
						38
						38
						38
						38
						38
						38
						49
		2.8–3.8	3.2/2.9	2.2–7.1	4.6/4.4	26
		1.7		3.2		26
		0.40		0.5		26
		0.6		0.7		26
		2.5		3.5		26
		3.3		5.0		26
		1.4		2.6		26
						3, 4
						3, 4
						3, 4
0.07–0.21	0.14/0.14					3, 4
0.01–0.04	0.03/0.03					3, 4
0.06–0.19	0.13/0.13					3, 4
0.38						3, 4
0.07						3, 4
0.085						25
0.217						25
						3, 4
						3, 4
						3, 4
						3, 4
		0.592		0.408		5
		0.300		0.167		5
		<0.2		nd		5

Table 1 (continued)

Location	Species	Year of sampling	Lipid content (%)	Musk xylene indiv. or min-max	Musk xylene mean/median
Piave	Trout		2.8		
Piave	Trout		8.7		
Piave	Trout		2.9		
Piave	Trout		8.2		
Piave	Trout		3.3		
Po	Crucian carp		0.5		
Po	Crucian carp		0.4		
Po	Crucian carp		0.4		
Po	Sheatfish		4.4		
Livenza	Trout		2.7		
Livenza	Trout		3.1		
Livenza	Trout		2.8		
Livenza	Trout		6.4		
Livenza	Trout		2.9		
Ticino	Trout		4.1		
Ticino	Trout		2.2		
Garigilano	Chub		1.7		
Garigilano	Crucian carp		0.8		
Garigilano	Crucian carp		0.1		
Adige	Italian nose		0.7		
Adige	Italian nose		1.3		
Torrente	Trout		1.1		
Torrente	Trout		1.2		
Sompunt	Trout		2.3		
Sompunt	Trout		1.3		
Japan					
Tama river	*Carassius auratus*	1980		(0.2)	
Tama and Tokyo Bay	Carp			(0.0015–0.041)	(0.015)
Tama and Tokyo Bay	Shellfish			(0.0017–0.053)	(0.002)
Luxembourg					
Sûre	Perch	1998/1999	0.24	0.158	
Sûre	Eel	1998/1999	21.5	0.057	
Moselle	Perch	1998/1999	0.7	0.130	

Musk ketone indiv. or min-max	Musk ketone mean/median	HHCB value, indiv. or min-max	HHCB mean/median	AHTN value, indiv. or min-max	AHTN mean/ median	Refer-ence
		<0.1		nd		5
		0.391		0.276		5
		0.379		0.207		5
		0.573		0.329		5
		0.485		0.242		5
		1.000		1.000		5
		1.250		1.000		5
		1.250		1.250		5
		0.773		2.386		5
		0.222		nd		5
		0.161		nd		5
		0.393		nd		5
		0.125		nd		5
		0.207		nd		5
		0.146		0.098		5
		0.182		0.182		5
		0.294		0.235		5
		1.125		0.500		5
		5.000		4.000		5
		<0.6		nd		5
		<0.3		nd		5
		<0.4		nd		5
		<0.3		nd		5
		<0.2		nd		5
		<0.3		nd		5
(0.05)						1
(nd–0.027)	(0.002)					2
(0.0009–0.026)	(0.002)					2
						32
						32
						32

Table 1 (continued)

Location	Species	Year of sampling	Lipid content (%)	Musk xylene indiv. or min-max	Musk xylene mean/median
Moselle	Eel	1998/1999	22.3	0.091	
The Netherlands					
Haringvliet-oost	Eel	1995		0.013	
Haringvliet-oost	Eel	1996	7.9	0.063	
Haringvliet-west	Eel	1995		0.07	
Hollands Diep	Eel	1995		0.13	
Hollands Diep	Eel	1996	18	0.039	
River IJssel (Deventer)	Eel	1996	18	0.094	
IJssel Lake	Pike-perch	1996	1.2	<0.042	
IJssel lake (Medemblik)	Eel	1995		<0.02	
IJssel Lake (Medemblik)	Eel	1996	27	<0.007	
IJssel Lake (Urk)	Eel	1995		<0.02	
Ketelmeer	Eel	1995		0.11	
Lek (Culemborg)	Eel	1995		0.27	
Meuse (Eijsden)	Eel	1996	7.2	0.625	
Meuse (Keizersveer)	Eel	1995		0.18	
Nieuwe Merwede	Eel	1995		0.18	
Nieuwe Merwede	Eel	1996	19	0.074	
Rhine (Lobith)	Eel	1995		0.43	
Rhine (Lobith)	Eel	1996	19	0.084	
Roer (Vlodrop)	Eel	1995		0.31	
Roer (Vlodrop)	Eel	1996	15	0.113	
Twentekanaal (Enschede)	Eel	1996	9.3	<0.032	
Waal (Tiel)	Eel	1995		0.16	
Waal (Tiel)	Pike-perch	1996	0.9	<0.056	
Norway					
Trondheim, inner harbour	Thornback ray, filet	1999	0.86	0.008	
Trondheim, inner harbour	Thornback ray, liver	1999	40	0.001	
Trondheim, inner harbour	Haddock, filet	1999	0.3–1.3	nd–0.078	
Trondheim, inner harbour	Haddock, liver	1999	64–67	nd	

Musk ketone indiv. or min-max	Musk ketone mean/median	HHCB value, indiv. or min-max	HHCB mean/median	AHTN value, indiv. or min-max	AHTN mean/ median	Reference
						32
0.06						33
0.051						34
<0.02						33
0.07						33
0.039						34
0.117						34
<0.042						34
0.06						33
<0.019						34
0.07						33
0.05						33
0.19						33
0.153						34
0.05						33
0.10						33
0.058						34
0.29						33
0.084						34
0.24						33
0.220						34
<0.054						34
0.08						33
<0.056						34
0.0013		0.073		0.089		30
0.001		0.0021		0.003		30
nd		0.112–0.574		0.0135–0.373		30
nd		0.073–0.374	0.0243	0.024–0.034	0.024	30

Table 1 (continued)

Location	Species	Year of sampling	Lipid content (%)	Musk xylene indiv. or min-max	Musk xylene mean/median
Tromso, inner harbour	Atlantic cod, filet	1999	0.4–1.3	nd–0.004	
Tromso, inner harbour	Atlantic cod, liver	1999	15–29	nd–0.004	
Tromso, inner harbour	Saith, filet	1999	2.3	nd	
Tromso, inner harbour	Saith, liver	1999	37	nd	
Oslofjord	Atlantic cod, liver	1997/98	29–42	0.035–0.070	0.051
Larvik	Atlantic cod, liver	1998	12–45	0.0003–0.004	0.003
Sweden					
South Sweden	Pike		0.38		
South Sweden	Pike		0.19		
Switzerland					
Lake Ceresio/ Lake Verbano	Fish Fish			(–0.002) (<0.001–0.090)	
North Sea					
Southern part	Cod liver	1996	45	<0.022	
Southern part	Cod	1995		<0.02	
Southern part	Fint	1996	1.9	<0.053	
Southern part	Mackerel	1996	7.6	<0.026	
Southern part	Whiting	1996	1.0	<0.050	
Southern part	Sole	1996	1.1	<0.045	
Wadden Sea					
Wadden Sea (Spiekeroog)	Various species	1992/1993		<0.005–0.120	
Wadden Sea (Dutch part)	Blue mussels	1996	1.6	<0.031	
Wadden Sea (Dutch part)	Shrimps	1996	1.5	<0.033	
Wadden Sea (Nordstrand)	Shrimps		1.0	0.01	
Hörnum	Blue mussels	1991–1992		0.02	
List	Blue mussels	1991–1992		<0.01	
Wadden Sea (Lower Saxony)	Blue mussels			<0.01–0.02	
Schleswig-Holstein	Shrimps	1993–1994		0.01	0.01
North Sea coast	Blue mussels	1991–1993		0.01–0.04	0.02

Musk ketone indiv. or min-max	Musk ketone mean/median	HHCB value, indiv. or min-max	HHCB mean/median	AHTN value, indiv. or min-max	AHTN mean/ median	Refer- ence
nd–0.003		0.025–0.056	0.041	0.013–0.047	0.026	30
nd–0.001		0.029–0.043	0.035	0.006–0.010	0.008	30
nd		0.225		0.093		30
nd		0.007		0.001		30
0.014–0.042	0.026	0.132–1.510	0.561	0.081–0.380	0.199	30
0.0001–0.002	0.001	0.009–0.031	0.020	nd–0.0018	0.016	30
		nd		nd		26
		nd		nd		26
(0.001–0.020)						23
						20
<0.044						34
<0.02						33
<0.105						34
<0.053						34
<0.050						34
<0.045						34
						48
<0.063						34
<0.067						34
0.02		0.16		0.06		11, 35
0.03		0.11		0.06		35
0.01		<0.03		<0.03		35
						22
0.03–0.05	0.04					4
0.01–0.04	0.03					4

Table 1 (continued)

Location	Species	Year of sampling	Lipid content (%)	Musk xylene indiv. or min-max	Musk xylene mean/median
North Sea	Shrimps		1.0	0.01	
North Sea	Shrimps		1.0	<0.01	
North Sea (Heligoland)	Blue mussels	1991–1992		0.01	
North Sea (Heligoland-Sylt)	Shrimps		0.9	<0.01	
Estuaria					
Westerschelde Estuary	Blue mussels	1996	1.4	<0.036	
Oosterschelde Estuary	Blue mussels	1996	1.7	<0.029	
Weser Estuary	Soft-shelled clam			0.01–0.08	

indiv. = concentration of individual sample.
nd = not detected.
d = detected but not quantified.

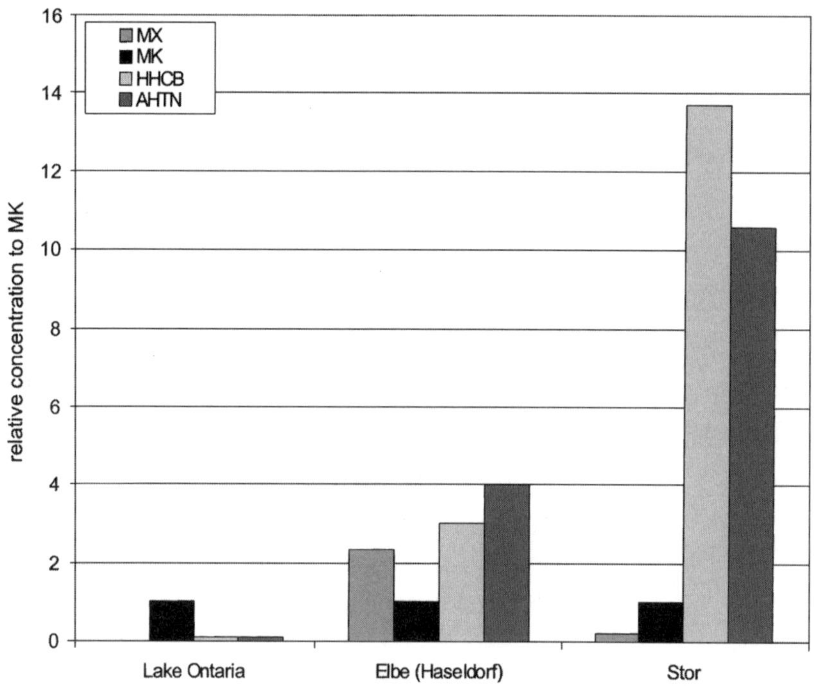

Fig. 1 Relative concentration of MX, HHCB and AHTN compared to MK in Canadian lake trout (*Salvelinus namaycush*) from Lake Ontario [11] and eel from the European rivers Elbe and Stör [11, 35]

Musk ketone indiv. or min-max	Musk ketone mean/median	HHCB value, indiv. or min-max	HHCB mean/median	AHTN value, indiv. or min-max	AHTN mean/ median	Reference
0.02		0.17		0.06		11, 35
0.43		0.37		0.05		11, 35
0.01		0.07		0.04		35
0.03		<0.04		<0.04		11, 35
<0.071						34
<0.059						34
						51

HHCB and AHTN have been determined in various fish species (bream, chub, eel, perch and roach, see Fig. 2) from the river Ruhr (Germany) [26]. The concentrations are similar to the concentrations found in the Czech Republic, and are strongly correlated with the relatively high levels of these compounds detected in the water of the river Ruhr [26]. This study is useful to compare the concentrations of polycyclic musks between various fish species. Concentrations have been normalized to a lipid weight basis (Figs. 1 and 2) to correct for the different lipid contents of the various species. Interesting are the relatively low concentrations of HHCB and AHTN in both eel samples compared to those in the other fish species. The same pattern was observed, even more pronounced, for eel and pike-perch samples from the Elbe [35]. It has been reported that the levels of organochlorine compounds were also low in the eels from the Elbe, which could indicate that these eels have not lived long in this relatively polluted river [27]. However, the relatively low concentrations of HHCB and AHTN in eel have been observed for several locations in the Elbe (Haseldorf and Lauenberg), and also in the Ruhr. This observation is a first indication that eels have the capacity to metabolise both polycyclic musk compounds. Signs for metabolism of HHCB and AHTN by bluegill sunfish under laboratory conditions have been reported before [28, 29]. In this study unidentified polar metabolites of HHCB and AHTN were observed. In the environment HHCB-lactone is the major transformation product of HHCB, which has been found in various fish species (e.g. haddock, cod and saith) [30]. The higher metabolic capacity of eels compared to other fish have been demonstrated by the fact that eels can metabolise PCB 77, whereas other fish species cannot metabolise PCB 77 [31]. It would be interesting to investigate if eel can metabolise HHCB and AHTN, and which metabolites are formed. The chapter of Biselli et al. in this monograph will provide more information of the metabolism of synthetic musks by aquatic organisms. Another interesting fact from Fig. 2 is the relatively high levels of HHCB and AHTN in pike-perch compared to eel from the Elbe for both locations (Haseldorf and Lauenberg). The ratio of the concentration of these compounds between pike-perch and eel is much higher than observed between eel and other fish from the Ruhr (Fig. 2). As pike-perch is a carnivore fish, this could indicate that, apart from the uptake of HHCB and AHTN from water by gills, also the uptake of residues of these compounds by food (biomagnification) is an important factor. This phenomenon is further supported by the levels of HHCB and AHTN in eel and pike-perch found at other locations in the Elbe [11, 36].

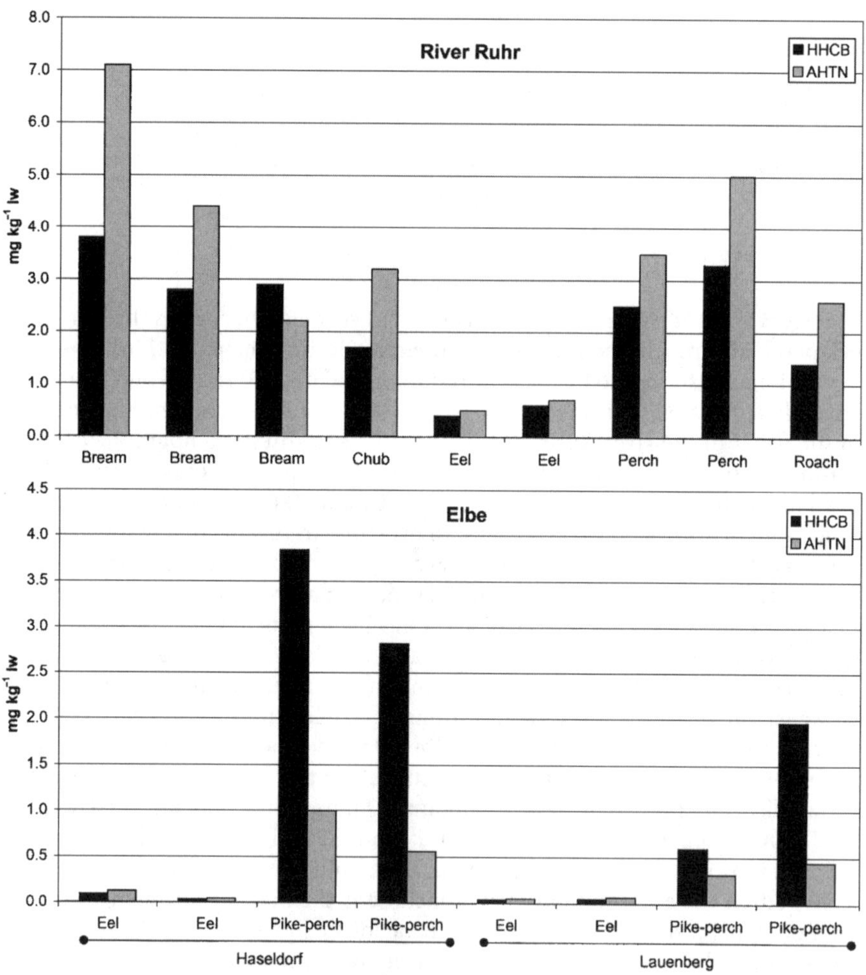

Fig. 2 Concentrations (mg kg^{-1} lw) of HHCB and AHTN in various freshwater fish species from the river Ruhr (Witten) [26] and in eel and pike-perch from the river Elbe (Haseldorf and Lauenberg) [35]

In the Netherlands MX and MK have been determined in yellow eel and pike-perch from various rivers and lakes collected in 1994 [33] and 1996 [34]. Important to note is that none of the sample sites was influenced by local STPs. In the study of 1994 also MA, MM and MT have been determined. However, all concentrations were below the limit of detection. Median concentrations in eel in rivers and lakes for MX and MK were 0.13 mg kg^{-1} lw and 0.07 mg kg^{-1} lw, respectively. In the IJssel Lake both nitro musks could not be detected in eel and pike-perch. The highest concentration of MX (0.625 mg kg^{-1} lw) was observed in eel from the Meuse at location Eijsden, which is at the border of Belgium. This high concentration, also compared to other countries, is possibly the result of untreated sewage water from the Meuse from Belgium, as in this country not all waste water is treated by STPs. A relative high concentration of MX was also observed in eel from the Rhine at location Lobith (border with Germany). For MK the highest concentrations were found in eel from the rivers Lek, Rhine, Meuse and Roer. In general, the concentrations of MX and MK in eel from 1994 were higher than in 1996, which could indicate a decrease in contamination of MX and MK.

In Italy freshwater fish has been collected in the northern part of the country from rivers (Piave, Po, Livenz, Ticino, Adige, Torrent Gadera) and a lake (Sompunt), and in the southern part from river Garigliano [5]. Mainly trout (*Salmo trutta fario* L. and *Salmo trutta lacustris* L) has been collected and analysed for the polycyclic musks HHCB, AHTN, ADBI, AHDI and ATII. Similar to other countries, the dominant musks were HHCB and AHTN, whereas AHDI was found in 3 of the 28 samples, all of which are trout from river Piave. HHCB was found in concentrations ranging from <0.1 to 5 mg kg^{-1} lw, and AHTN in concentrations below the detection limit to 4 mg kg^{-1} lw. The highest concentration for HHCB and AHTN was found in crucian carp from river Garigilano of 5 mg kg^{-1} lw and 4 mg kg^{-1} lw, respectively. Concentrations of AHDI and ATII were below the limit of detection. In fish from the rivers Adige, Torrent and lake Somput none of the samples contained detectable residue levels of polycyclic musks.

Recently MX has been determined in fish species frequently consumed in the Grand Duchy of Luxembourg, to investigate the human intake and distribution of contaminants [32]. Samples were collected in the Sûre reservoir and the river Moselle. The mean concentration of MX in perch from the Sûre reservoir (0.158 mg kg^{-1} lw) was similar to the concentration found in perch from the river Moselle (0.130 mg kg^{-1} lw). For both locations the concentration in eel were lower, 0.057 and 0.091 mg kg^{-1} lw, respectively. The concentrations of MX were in the same order of magnitude as PCB 153 in perch and eel from the Sûre reservoir. In the Sûre reservoir the concentration of MX in fish was much lower than in fish from the river Moselle. The river Moselle is highly contaminated with PCBs due to industry (mining) upstream in France, even after the factories have been redeveloped [32].

In the marine environment nitro musks and polycyclic have been determined in a number of fish and shellfish samples from the North Sea, Wadden Sea and five Norwegian marine sites. In cod, mackerel, twaite shad, sole and whiting from the North Sea, MX and MK were detected in concentrations from 0.022 to 0.053 mg kg^{-1} lw and 0.02 to 0.105 mg kg^{-1} lw [33, 34] respectively, which in general are lower than observed in freshwater fish. In blue mussels and shrimps collected at the Dutch and the German Wadden Sea, MX and MK have been detected in concentrations similar to North Sea fish [33, 34]. In addition, polycyclic musks have also been detected in some of these samples, showing concentrations higher than MX and MK, which is consistent with the data of various fish species for the Norwegian sites [30] and herring from Denmark [35]. However, the relative concentrations of HHCB and AHTN in fish from the Norwegian sites are higher (0.003–1.5 mg kg^{-1} lw) than in fish from the North Sea (<0.03–0.37 mg kg^{-1} lw). The difference is mainly due to the fact that the Norwegian sites are related to densely populated areas. The highest concentration of HHCB was found in Atlantic cod liver from the Oslo area, a densely populated area. In addition, the Norwegian study showed the presence of HHCB-lactone and ATII in many samples. The concentrations of HHCB-lactone in Atlantic cod from the Oslo area were in the same order of magnitude as HHCB. ATII concentrations were approximately ten times lower than AHTN. Interesting to note are the relative high concentrations of MX to the total burden of musks (sum of polycyclic and nitro musks) in some Norwegian fish. This indicates that nitro musks are probably used in larger portions in fragrances in products in Norway than other European countries [30].

The pattern for marine organisms is similar to the pattern found in freshwater fish, except that the difference in concentration between the polycyclic and nitro musks in general is larger in freshwater species, due to a much larger influence of effluent water of STPs at freshwater locations. In general, the pattern in marine fish is MX, MK<HHCB, AHTN.

3.2.1.2
Temporal Trends

In Germany (1993) and in Switzerland (1994) the fragrance industry recommended their members not to use MX anymore in washing and cleaning agents. To obtain information on temporal trends for MX and MK the German ARGE ELBE (Arbeitsgemeinschaft für die Reinhaltung der Elbe) monitoring program for fish in the river Elbe can be used [36]. In total 431 muscle tissue samples of eel, bream, and pike-perch were collected during the period 1994 to 1999 at various locations in the Elbe, from the border with the Czech Republic (Hafen Prossen) to the estuary of the German Bight (Brunsbüttel). Sediment and water samples have also been collected and analysed for MX, MK, HHCB, AHTN, ADBI, AHDI and ATII. As illustration of temporal trends for nitro musks in fish, some information of the extensive dataset of the ARGE ELBE study will be presented for locations Hafen Prossen and Gorleben only (Fig. 3 and Table 1). More information of all sampling locations of the ARGE ELBE program can be obtained from the report [36]. For MX a sharp decrease in the concentration in eel was observed at the location Gorleben, with mean concentrations of 0.127 mg kg^{-1} lw in 1994 to below the limit of detection (<0.005 mg kg^{-1} lw) in all samples (n=15) in 1999. This decrease is also well demonstrated by decreasing number of samples in which MX could be detected (98%, 97%, 60% and 0% of the samples in 1994, 1995, 1997 and 1999, respectively). The sharp decline of MX in a period of five years showed that MX could be eliminated relatively fast from the aquatic ecosystem, and is not as persistent as, e.g. PCBs. For location Hafen Prossen, however, a small increase in MX between 1994 and 1999 was observed. Prossen is situated directly at the Czech borderline and, therefore, MX is probably still used and/or produced in the Czech Republic. A decline in the mean concentration of MK in eel for location Gorleben from 1994 to 1999 has also been observed; 0.093 mg kg^{-1} lw and <0.005 mg kg^{-1} lw, respectively.

The Internationale Kommission zum Schutze des Rheins (IKSR) and the International Commission to Protect the Rhine (CIPR) investigated the levels of organochlorine compounds and nitro musks (MX) in fish species from the Rhine [37, 38]. Data of MX in eel and other freshwater fish were collected in 1995 in the Rhine over a distance of 590 km [37]. In 2000 a greater number of eel samples per location (n=15) than in 1995 (n=2–5) has collected for the analysis of MX [38] (see Table 1), but only a restricted number of exact overlapping geographical locations exist between 1995 and 2000. For the overlapping locations the mean concentrations of MX in eel for 1995 and 2000 have been calculated and presented in Fig. 4. at location Grenzach (km 160) the mean concentration of MX in

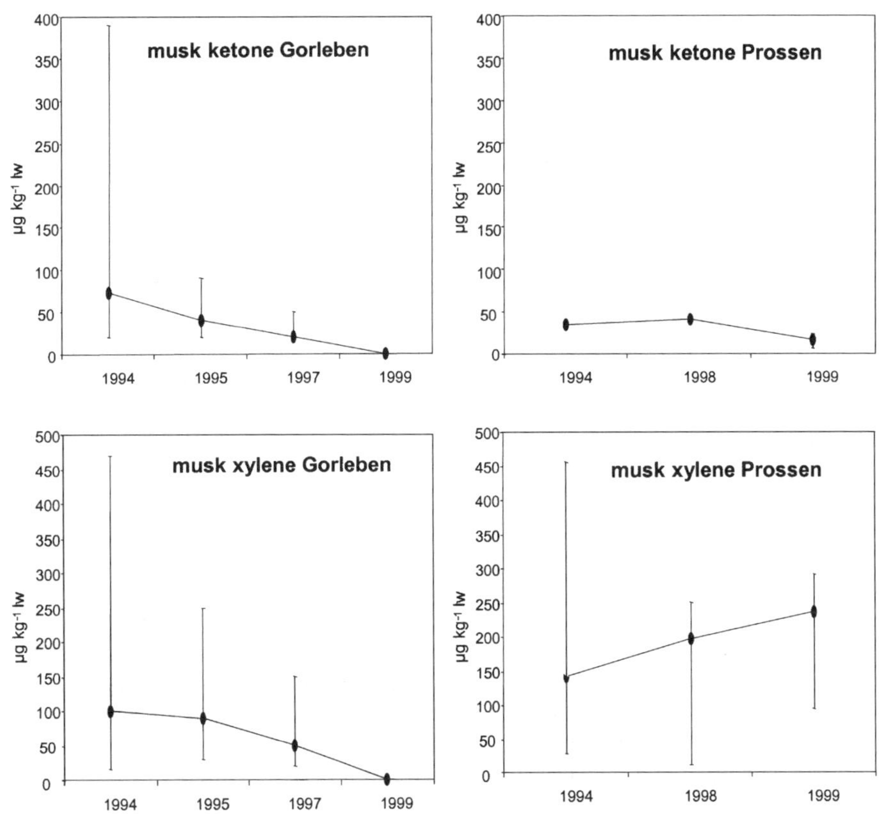

Fig. 3 Musk ketone and musk xylene in eel from the river Elbe at the locations Gorleben and Prossen from 1994 to 1999 (data from [35])

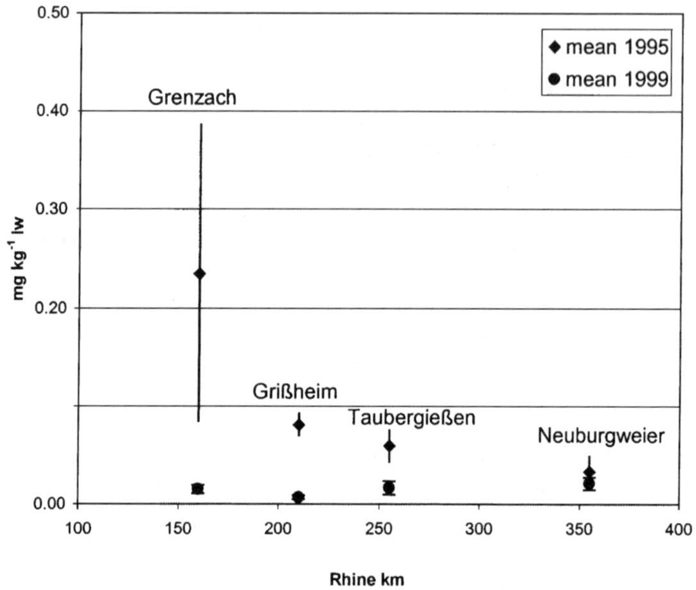

Fig. 4 Concentrations of musk xylene in eel from the river Rhine collected at four locations in 1995 and 2000 [37, 38]

1995 decreased from 0.235 mg kg^{-1} lw to 0.015 mg kg^{-1} lw in 2000 with a factor of 16. For other locations, Grißheim (km 210), Taubergießen (km 255) and Neuburgweier (km 355), the concentrations decreased with factors of 11, 4 and 2, which indicates that the decrease is less in downstream areas of the Rhine than in upstream areas. In general, the decreasing concentrations of MX in fish from the Rhine are comparable with the results of the Elbe.

In several rivers and lakes in the area of Berlin eel has been collected in 1995 and 1996 for the determination of MX and MK [6]. This study showed a decline in the concentration of MX but not for MK. For more details about these investigations see the chapter of Heberer et al. in this monograph.

3.2.2
Biota from STPs Effluents

A number of studies have focussed on the importance of effluents of STPs as source for the accumulation of synthetic musks by biota [4, 6, 14, 39] (Table 2). Nitro musks and polycyclic musks have been determined in water, sediment and eel from three areas of Berlin [6, 39]. One area (rivers Spree and Dahme, 'A') has a low input of municipal effluent relative to the total water content. The second area ('B') has a moderate input of STP effluent of around 15% of the total water content. The third area receives a very high input of STP effluents ('C'). In 1996 and 1997 samples have been collected in these areas and MX, MK, HHCB, AHTN, ADBI, ATII and AHDI were determined. All musk compounds shows strong positive correlations between the relative input of STP effluents and the concentration in water, sediment and eel. This shows that musk compounds, especially HHCB and AHTN, can be used as an indicator of the exposure of biota to STP effluents. The concentrations of HHCB and AHTN in eel exceeded the concentration of MX and MK with a factor of 10–30. Concentrations of HHCB and AHTN in eel from area C (high STP input) were 6 and 3 mg kg^{-1} lw, respectively [6, 39]. The polycyclic musks ADBI, ATII and AHDI were also found in eel samples from area C at concentrations of 0.021, 0.53 and 0.48 mg kg^{-1}, respectively. For more details about the Berlin area investigations see the chapter of Heberer et al. in this monograph.

A number of studies has been devoted to the determination of synthetic musks in biota collected from ponds directly connected to effluent of STPs in the Federal States of North-Rhine Westfalia and Schleswig-Holstein (Germany) [18, 26, 52] (Table 2). Musks have been studied in various fish species (crucian carp, eel, chub, rudd, tench) and zebra mussels from these ponds. Concentrations of HHCB and AHTN in fish from the pond in Schleswig-Holstein are in general higher than in fish from the pond in North-Rhine Westfalia, except eel. MK concentrations in fish from the pond in Schleswig-Holstein are higher than in the Berlin high contaminated area (C), while the concentration of MX was similar in both areas. Very high concentrations of HHCB and AHTN were found in all fish and mussel samples; highest concentration 160 and 58 mg kg^{-1}, respectively. Concentrations of these compounds are much higher than found in biota from natural water systems. This contrasts to MX and MK, for which the concentrations in biota from the sewage ponds are similar to the highest concentrations, found in biota from natural freshwater systems. Rudd and eel seems to be able to metabolise HHCB and AHTN, because the mean concentrations in these species are 15–20 times lower than in the other analysed fish species from the same pond. Further details are discussed in the chapters of Biselli et al. and Hühnerfuss et al. in this monograph.

In conclusion, high concentrations of synthetic musks (especially HHCB and AHTN) are found in biota collected close to the effluents of STPs or in sewage settlement ponds.

3.2.3
Biota from Fish Farms

Synthetic musk compounds have been determined in farmed fish (rainbow trout and carp) [3, 4, 21, 22, 27, 35, 40] (Table 3). In the German studies trout samples were collected directly at fish farms, or from the German food market (e.g. German Food Monitoring System). The origin (country) of these samples is sometimes known (Germany, Denmark, Austria or Spain), but not the exact fish farms. Nitro musks and polycyclic musk residues in farmed fish were found at concentrations comparable to fish from natural water systems. Interesting is the fact that the concentrations of nitro musks were in the same order of magnitude as the polycyclic musks, while in natural systems concentrations of polycyclic musks are generally much higher. Concentrations of HHCB (0.11–0.65 mg kg^{-1} lw) and AHTN (0.20–0.59 mg kg^{-1} lw) are comparable with the concentrations found in less polluted natural freshwater systems.

Table 2 Concentrations of nitro musks and polycyclic musks in mg kg^{-1} lipid weight in fish and mussels collected from ponds directly connected to a STP (North Rhine Westfalia and Schleswig-Holstein), or collected in water that received a high proportion of effluent from STPs (Berlin area and Stör). All samples have been taken in Germany. Concentrations in brackets in mg kg^{-1} wet weight

Origin	Species	Year	Lipid (%)	Musk xylene indiv. or min-max	Musk xylene mean
Sewage pond					
North Rhine Westfalia	Crucian carp		0.87		
	Crucian carp		0.98		
	Crucian carp		1.13		
	Crucian carp		1.74		
	Eel		16.5		
	Eel		25.3		
	Eel		30.2		
	Eel		26.9		
	Chub		1.9		
	Chub		2.3		
	Chub		2.8		
Schleswig-Holstein	Rudd	1997	0.6	0.028	
	Rudd	1997	1.1	0.050	
	Rudd	1997	0.8	0.018	
	Tench	1997	0.8	0.230	
	Tench	1997	1.3	0.360	
	Tench liver	1997	4.0	0.100	
	Tench	1997	0.8	0.220	
	Tench	1997	0.6	0.190	
	Crucian carp	1997	1.1	0.220	
	Crucian carp	1997	1.7	0.280	
	Crucian carp	1997	2.0	0.370	
	Crucian carp	1997	2.6	0.320	
	Crucian carp	1997	2.5	0.360	
	Crucian carp	1997	1.7	0.310	
	Crucian carp	1997	3.5	0.440	
	Crucian carp liver	1997	10.4	0.292	
	Eel	1997	15.7	0.260	
	Eel	1997	18.1	0.220	

Musk ketone indiv.	Musk ketone mean	HHCB indiv.	HHCB mean	AHTN indiv.	AHTN mean	Reference
		6.6		13.0		26
		14.7		34.1		26
		6.0		12.2		26
		19.8		37.2		26
		7.6		14.5		26
		6.1		10.1		26
		63.6		57.9		26
		49.1		49.1		26
		8.9		15.3		26
		1.0		3.0		26
		6.9		16.4		26
0.320		6.2		5.0		52
0.430		7.1		5.7		52
0.300		7.5		6.1		52
1.400		150		30		52
1.200		160		32		52
0.588						25
1.300		160		35		52
1.300		150		42		52
0.920		39		26		52
1.400		59		31		52
1.400		91		31		52
1.200		71		30		52
0.990		50		34		52
1.100		66		40		52
1.800		84		34		52
1.090						25
0.420		4.8		2.6		52
0.360		4.6		2.7		52

Table 2 (continued)

Origin	Species	Year	Lipid (%)	Musk xylene indiv. or min-max	Musk xylene mean
	Mussels	1997	1.4	0.130	
Berlin area					
Area A (low input of STP effluents in river)	Eel	1996–1997			
Area B (moderate input of STP effluents in river)	Eel	1996–1997			
Area C (high input of STP effluents in river)	Eel	1996–1997			
Area A, B, C	Eel	1995	23	(0.001–0.170)	0.104
Area A, B, C	Eel	1996	21	(0.001–0.079)	0.057
Area A (low input of STP effluents in river)	Eel	1995/1996	21		0.011
Area C (high input of STP effluents in river)	Eel	1995/1996	23		0.171
Stör					
3 km from outlet STP	Brown trout		3.5	0.20	
3 km from outlet STP	Brown trout		1.3	0.24	

indiv. = concentration of individual sample.

Musk ketone indiv.	Musk ketone mean	HHCB indiv.	HHCB mean	AHTN indiv.	AHTN mean	Reference
1.400		120		45		52
			0.445		nd	39
			0.658		0.287	39
			6.470		3.060	39
	0.178					6
	0.185		2.819		1.276	6
	0.018		0.433		0.167	6
	0.428		5.930		2.734	6
1.00		20.30		10.60		35
1.15		13.70		8.74		35

Table 3 Concentrations of nitro musks and polycyclic musks in mg kg^{-1} lipid weight in rainbow trout (*Oncorhynchus mykiss*) and carp (*Cyprinus caprio*) from fish farms. Concentrations in brackets in mg kg^{-1} wet weight

Country	Year	Lipid (%)	Musk xylene indiv. or min-max
Rainbow trout			
Austria[a]	1997	3.9	0.100
Denmark[a]	1995	2.5	0.207
Denmark[a]	1997	2.4	0.011
Denmark[a]	1997	3.5	0.018
Denmark[a]	1997	4.2	0.056
Denmark[a]	1997	2.8	0.036
Denmark[a]			
Denmark[a]		2.6	0.21
Denmark[a]		2.9	0.18
Denmark[a]		3.3	0.12
Denmark[a]		3.2	0.10
Denmark/Spain[a]	1990–1992		0.01–1.06
Germany (South Bavaria)[b]			(<0.0002–0.031)
Germany (Schleswig-Holstein)[b]	1990–1992		<0.01–0.10
Germany (Tübingen)[b]			0.3–0.78
Germany (Tübingen)[b]			0.34–1.8
Germany[c]	1991/1992		(0.0001–0.0952)
Germany[c]	1992		(0.0001–1.0220)
Germany[c]	1995		(nr–0.0140)
Germany[c]	1996		(nr–0.0110)
Import[a]			<0.005
Spain[a]	1997	2.3	0.340
Rainbow trout and other species			
Germany (Tübingen)[b]	1991–1992		0.11–1.80
Carp			
Germany (Schleswig-Holstein)[b]	1990–1992		<0.01–0.05

[a] Samples collected at the German food market.
[b] Samples directly taken from fish farms.
[c] Samples from the German Food Monitoring System (a mixture of samples from German food market and fish farms).
indiv. = concentration of individual sample.
nr = not reported.

Musk xylene mean	Musk ketone indiv. or min-max	Musk ketone mean	HHCB	AHTN	Reference
	0.022				25
	0.111				25
	0.050				25
	0.049				25
	0.049				25
	0.029				25
(0.013)					21
	0.11		0.25	0.32	35
	0.10		0.43	0.32	35
	0.12		0.65	0.59	35
	0.03		0.11	0.20	35
0.33	0.02–0.33	0.14			3, 4
(0.003)					22
0.03	0.01–0.11	0.04			4
0.39					40
0.68					40
					41
					41
(0.0016/0.008)	(nr–0.0220)				42
(0.0011/0.0002)	(nr–0.0290)	(0.0015/0.0002)			43
					40
	0.111				25
					40
0.01	0.01–0.07	0.02			4

In some samples relative high concentrations of MX (above 1 mg kg^{-1} lw) were observed, which is probably a consequence of the use of STP effluent water by some fish farms. In general, trout from fish farms of the Northern part of Germany (Schleswig-Holstein) show lower concentrations of MX and MK than from fish farms from other parts of Germany (Tübingen, South Bavaria), and from the food market samples with origin Spain and Denmark. In some trout samples the concentrations of MX and MK are similar to the concentrations of PCB 153 (data not shown) [42, 43].

3.2.4
Samples from Food Market

The German Food Monitoring System reports yearly the levels of various organic and inorganic contaminants in food products including fishery products which are collected from producers, shops, supermarkets, wholesaler, fish markets etc. [41–46]. For every species pooled samples are analysed and the data of contaminants are presented as mean, median, percentiles and maximum concentrations. Determination of MX and MK were included since 1992, in the scope of the PCB/OCP analysis. Data on levels are presented in Table 4, excluding. The trout samples collected at the German market have already been discussed in the former paragraph.

Concentrations on wet weight basis for MX varied between 0.00003 mg kg^{-1} ww in mussels to 0.72 mg kg^{-1} ww in carp. For MK the conserved tuna samples are interesting because of the high mean and maximum concentrations (mean: 0.029 mg kg^{-1} lw, and a maximum of 2.0 mg kg^{-1} lw). Mean concentrations of MK in other market samples are much lower and range between 0.0009 mg kg^{-1} ww in mussels to 0.0024 mg kg^{-1} ww in herring. In most samples the mean concentrations of MX and MK are one to two orders of magnitude lower than the mean concentration of PCB 153, except the pollack samples from 1996 which had a higher mean concentration of MK than PCB 153. Herring collected at the German food market had concentrations for MX and MK between <0.01–0.1 mg kg^{-1} lw.

In conclusion, the concentrations of synthetic musks in food market samples are lower than observed in fish species from natural freshwater water systems, which is mainly due to the fact that most of the samples came from the marine environment which is less contaminated then the freshwater environment. In some conserved tuna samples high levels of MK have been found.

3.3
Synthetic Musks in Other Aquatic Organisms

A long-term program to monitor contaminants in birds breeding in the Wadden Sea has been established since the 1980s for oystercatcher and common tern [47, 50]. In 1993 eggs of eight species of birds (shelduck, eider, oystercatcher, avocet, redshank, black-headed gull, herring gull, common tern) were collected at the island of Spiekeroog (German Wadden Sea), and dunlin eggs from Northern Norway as a reference location. Various contaminants were determined including MX. This is the first study that showed that residue levels of a synthetic musk compound could be found in organisms higher in the aquatic food chain. MX was

Table 4 Concentrations of nitro musks and polycyclic musks in mg kg^{-1} lipid weight and in brackets in mg kg^{-1} wet weight basis in fishery products from food markets in Germany. Indiv. = concentration of individual sample; nr = not reported

Food market	year	lipid (%)	Musk xylene indiv or min-max	Musk xylene mean/median	Musk ketone indiv. or min-max	Musk ketone mean/median	HHCB value, indiv. or min-max	AHTN value, indiv. or min-max	PCB 153 mean/median	Reference
Germany										
Herring	1995		nr-0.050		nr-0.090					42
Herring	1996		nr-0.1100	0.088	nr-0.0360	0.00240				43
Herring	1996		(nr-0.0020)	(0.00018)	(nr-0.0060)	(0.00130)				43
Pollack	1995		(nr-0.057)		(nr-0.0100)					42
Pollack	1996		(nr-0.0025)	(0.00007)	(nr-0.0280)	(0.00056)			(0.00039/0.0002)	43
Crab	1995		(nr-0.0028)		(nr-0.0083)					42
Carp	1997		(nr-0.7200)	(0.00490)	(nr-0.0190)	(0.00048)			(0.00290/0.0002)	44
Carp	1998		(nr-0.0130)	(0.00030)	(nr-0.0043)	(0.00018)			(0.00310/0.0003)	45
Eel	1997		nr-0.0820	0.0024	nr-0.0160	0.00087				44
Halibut	1998		(nr-0.0020)	(0.00006)	(nr-0.0031)	(0.00030)			(0.0020/0.0004)	45
Halibut	1998		(nr-0.0022)	(0.00016)	(nr-0.0047)	(0.00060)			(0.00390/0.0021)	45
Mussel	1998		(nr-0.0002)	(0.00003)	(nr-0.0049)	(0.00009)			(0.0046/0.0037)	45
Tuna conserved	1999		nr-0.0580	0.00210	nr-2.000	0.0290				46
Mackerel	1999		nr-0.1200	0.00190	nr-0.0950	0.00230				46
Denmark										
Herring		7.0	0.01		0.02		0.12	0.07		35
Herring		12.4	<0.01		<0.01		<0.01	<0.01		35
Ireland										
Herring		7.7	<0.01		0.01		<0.01	<0.01		11
Baltic Sea										
Herring		8.6	0.01		0.05		0.75	0.53		35
Arabian Sea										
Shrimps		0.5	0.02		0.02		<0.06	<0.06		35
Shrimps		0.7	0.05		0.10		0.16	0.07		35

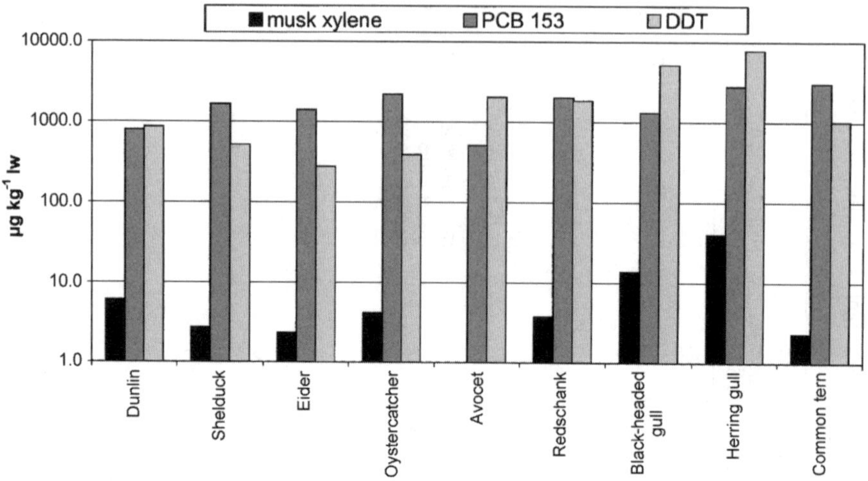

Fig. 5 Concentrations of musk xylene in eggs from coastal bird species (waders and waterfowls) breeding in Northern Norway (Dunlin) and German Wadden Sea in 1993 [47, 50]

found in 29% of the egg samples. Highest concentrations of PCBs and OCPs were found in common tern and herring gull eggs, while the highest concentrations of MX were found in herring and black-headed gull eggs (Fig. 5). Mean concentrations of MX ranged from 0.002 mg kg^{-1} lw to 0.040 mg kg^{-1} lw in herring gull eggs. Mean concentration MX in eggs of Dunlins from Northern Norway (0.006 mg kg^{-1}lw)was similar to the concentrations found in bird eggs from the Wadden Sea. The ratio between PCB 153 and MX varied between the species, in gulls the concentration of MX was two orders of magnitude lower than for PCB 153 and for common tern three orders of magnitude.

The occurrence of polycyclic musks has been reported in the liver of a Danish otter. Otters have a diet of mainly fish. In this otter sample, high concentrations

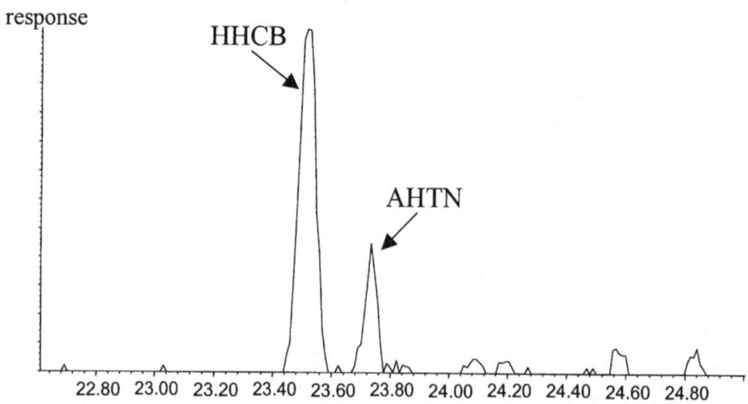

Fig. 6 GC-MS selective ion chromatogram (m/z 243) of HHCB and AHTN, analysed in a liver sample from a Danish otter (*Lutra lutra*)

of PCBs (150 mg kg^{-1} lw, sum of seven congeners) has been found, but also high concentrations of HHCB and AHTN of 140 and 95 mg kg^{-1} lw, respectively (Fig. 6).

In conclusion, almost no data on the occurrence of synthetic musk compounds in higher organisms are available. The preliminary data showed that synthetic musks can be found at sometimes high concentrations.

4 General Conclusions

In summary, it can be concluded that synthetic musk compounds are widely spread environmental pollutants in the aquatic ecosystem. The nitro musks MX and MK and the polycyclic musks HHCB and AHTN are the major compounds in aquatic organisms and found in freshwater as well as in marine organisms. The main source of musk compounds in aquatic organisms are effluents from STPs. Concentrations of HHCB and AHTN in freshwater organisms from Europe are one to two orders of magnitude higher than MX and MK, and often comparable to levels of PCBs in fish. The presence of synthetic musks in aquatic top predators needs further investigation, as data are very limited. Temporal trend data showed that for some locations concentrations for MX and MK in fish are declining.

Nitro musks and polycyclic musk were also found in fish samples (e.g. trout, herring, mussels, tuna and mackerel) collected at food markets. In some samples (trout and shrimp) the concentrations of MX and MK were similar to the concentrations of PCBs, while in other samples (e.g. halibut and mussels) the concentrations of MX and MK are one to two orders of magnitude lower than for PCBs. In general, the highest concentrations of MX were found in several trout samples, and in some tuna samples rather high concentrations of MK were found.

5 References

1. Yamagishi T, Miyazaki T, Horri S, Kanenko S (1981) Bull Environ Contam Toxiol 26:656
2. Yamagishi T, Miyazaki T, Horri S, Akiyama K (1983) Arch Environ Contam Toxicol 12:83
3. Rimkus G, Wolf M (1993) Dtsch Lebensm-Rundsch 89:171
4. Rimkus G, Wolf M (1995) Chemosphere 30:641
5. Draisci R, Marchiafava C, Ferretti E, Palleschi L, Catellani G, Anastasio A (1998) J Chromatogr A 814:187
6. Fromme H, Otto T, Pilz K, Neugebauer F (1999) Chemosphere 39:1723
7. Rimkus G, Butte W, Geyer HJ (1997) Chemosphere 35:1497
8. Fernandez C, Carballo M, Tarazona JV (1996) Chemosphere 32:1805
9. Boleas S, Fernandez C, Tarazona JV (1996) Bull Environ Contam Toxicol 57:217
10. Eschke H-D, Traud J, Dibowski H-J (1994) UWSF-Z Umweltchem Ökotox 6:183
11. Gatermann R, Hellou J, Hühnerfuss H, Rimkus G, Zitko V (1999) Chemosphere 38:3431
12. Herren D, Berset JD (2000) Chemosphere 40:565
13. Rimkus G, Rummler M, Nausch I (1996) J Chromatogr A 737:9
14. Rimkus G, Wolf M (1996) Chemophere 33:641
15. Müller S, Schmidt P, Schlatter C (1996) Chemosphere 33:17
16. Franke S, Meyer C, Heinzel N, Gatermann R, Hühnerfuss H, Rimkus G, König WA, Francke W (1999) Chirality 11:795
17. Eschke H-D, Dibowski H-J, Traud J (1995) Dtsch Lebensm-Rundsch 91:375
18. Eschke H-D, Dibowski H-J, Traud J (1995) UWSF-Z Umweltchem Ökotox 7:131

19. Ceschi M, De Rossa M, Jäggli M (1996) Trav Chim Aliment Hyg 87:189
20. Kokot-Helbig K, Schmidt P, Schlatter C (1995) Mitt Gebiete Lebensm Hyg 86:1
21. Green M (1994) Lebensmittelchemie 48:8
22. Geyer H-J, Rimkus G, Wolf M, Attar A, Steinberg C, Kettrup A (1994) UWSF-Z Umweltchem Ökotox 6:9
23. Rimkus G, Brunn H (1996) Ernährungs-Umschau 43:442
24. Hajslova J, Gregor P, Chladkova V, Alterova K (1998) Organohalogen Compd 39:253
25. Rimkus GG, Gatermann R, Hühnerfuss H (1999) Toxicol Lett 111:5
26. Eschke H-D, Traud J, Dibowski H-J (1994) Vom Wasser 83:373
27. Rimkus GG (1999) Toxicol Lett 111:37
28. Van Dijk A (1996) Report to RIFM, RCC Umweltchemie AG, Project 364825, cited in van de Plassch and Balk (1997), RIVM report no. 601503008 Netherlands
29. Van Dijk A (1996) Report to RIFM, RCC Umweltchemie AG, Project 381418, cited in van de Plassch and Balk, RIVM report no. 601503008, Netherlands
30. Kallenborn R, Gatermann R, Nygard T, Knutzen J, Schlabach M (2001) Fres Environ Bull 10:832
31. De Boer J, Stronck CJN, Traag WA, Meer van der J (1993) Chemosphere 26:1823
32. Dauberschmidt C, Hoffmann L (2001) Bull Environ Contam Toxicol 66:222
33. Wiertz J (1995) Nitro musks in fish samples. Technical report GmbH, Hamburg, Germany
34. De Boer J, Wester P (1996) RIVO report, IJmuiden, The Netherlands
35. Rimkus G (1997) Meeting ICES, March 1997, Oostende, Belgium
36. Reincke H, Wiegel S, Harms H, Stachel B (2000) Report Arbeitsgemeinschaft für die Reinhaltung der Elbe, Synthetische Moschus-Duftstoffe in der Elbe, http://www.arge-elbe.de/wge/Download/Berichte/00Moschus.pdf
37. Rhein Aktuell (1995) IKSR and CIPR report no. 81, Schadstoffgehalte in Rheinfischen 1995
38. IKSR (2000) Kontamination von Rheinfischen 2000. Bericht Nr. 124-d
39. Fromme H, Otto T, Pilz K (2001) Water Res 35:121
40. Hahn J (1993) Dtsch Lebensm-Rundsch 89:175
41. German Food Monitoring System (1992) Bericht über die Monitoringergebnisse des Jahres 1992. Bundesinstitut für gesundheitlichen Verbraucherschutz und Veterinärmedizin, Zentrale Erfassungs- und Bewertungsstelle für Umweltchemikalien, Berlin
42. German Food Monitoring System (1995) Bericht über die Monitoringergebnisse des Jahres 1995. Bundesinstitut für gesundheitlichen Verbraucherschutz und Veterinärmedizin, Zentrale Erfassungs- und Bewertungsstelle für Umweltchemikalien, Berlin
43. German Food Monitoring System (1996) Bericht über die Monitoringergebnisse des Jahres 1996. Bundesinstitut für gesundheitlichen Verbraucherschutz und Veterinärmedizin, Zentrale Erfassungs- und Bewertungsstelle für Umweltchemikalien, Berlin
44. German Food Monitoring System (1997) Bericht über die Monitoringergebnisse des Jahres 1997. Bundesinstitut für gesundheitlichen Verbraucherschutz und Veterinärmedizin, Zentrale Erfassungs- und Bewertungsstelle für Umweltchemikalien, Berlin
45. German Food Monitoring System (1998) Bericht über die Monitoringergebnisse des Jahres 1998. Bundesinstitut für gesundheitlichen Verbraucherschutz und Veterinärmedizin, Zentrale Erfassungs- und Bewertungsstelle für Umweltchemikalien, Berlin
46. German Food Monitoring System (1999) Bericht über die Monitoringergebnisse des Jahres 1999. Bundesinstitut für gesundheitlichen Verbraucherschutz und Veterinärmedizin, Zentrale Erfassungs- und Bewertungsstelle für Umweltchemikalien, Berlin
47. Mattig FR, Roesner HU, Giessing K, Becker PH (2000) J Ornithol 141:361
48. Gaumert T (1996) Arbeitsgemeinschaft für die Reinhaltung der Elbe (Hrsg), Hamburg, Juli 1996
49. Gluck B, Hahn J (1995) Fleischwirtschaft 75:92
50. Mattig FR, Ballin U, Bietz H, Giessing K, Kruse R, Becker PH (1997) Arch Fish Mar Res 45:113
51. Kruse R (1992) Arbeitstagung des Arbeitsgebietes Lebensmittelhygiene der Deutschen Veterinärmedizinischen Gesellschaft eV, 29.9–2.10.1992, Garmisch-Patenkirchen, p 115
52. Gatermann R, Biselli S, Hühnerfuss H, Rimkus GG, Hecker M, Karbe L (2002) Arch Environ Contam Toxicol 42:437

The Handbook of Environmental Chemistry Vol. 3, Part X (2004): 85–104
DOI 10.1007/b14128

Synthetic Musks in Ambient and Indoor Air

Roland Kallenborn[1] · Robert Gatermann[2]

[1] Norwegian Institute for Air Research (NILU), The Polar Environmental Centre, NO-9296 Tromsø, Norway
E-mail: roland.kallenborn@nilu.no

[2] Dr. Wiertz – Dipl. Chem. Eggert – Dr. Jörissen GmbH Analytical Laboratory (WEJ), Stenzelring 14b, 21107 Hamburg, Germany

Abstract For the first time, the presence of synthetic musks is confirmed and investigated in air samples. New analytical methods are described developed for the purpose of ultra trace analysis of synthetic musks in low contaminated samples including ambient air. In order to achieve the lowest possible detection limits and avoid sample and laboratory contamination, a rigid quality control/quality assurance program must be followed for the analytical methods of synthetic musks. Standard analytical methods for the determination of persistent organic pollutants (POP) in ambient air were adapted to the need of quantifying synthetic musks. In addition, a new method was developed and applied in a first pilot study for the determination of synthetic musk in Norwegian indoor air. High resolution gas chromatography coupled to low resolution mass spectrometry (HRGC/LRMS) was used for the quantification of the Norwegian outdoor and indoor air samples. Considerable levels of synthetic musks found in outdoor air from Kjeller (taken close to the Norwegian capital Oslo) confirm the importance of densely populated human settlements as primary source for synthetic musks. As already found for other matrices, polycyclic musks were dominating the air samples whereas the nitro musks musk xylene and musk ketone were lower concentrated contaminants. Surprisingly high concentrations of synthetic musks were found in Norwegian indoor air samples with the highest concentration values for HHCB and AHTN in the atmosphere of a hairdresser shop (HHCB: 44 ng m^{-3}). Clear differences in the distribution patterns were found for nitro musks in outdoor and indoor air samples. In outdoor air samples, MX was found to contribute with 7–11% to the total synthetic musks burden whereas 0.5–3% only could be related to MX in indoor air. Based on the data presented here, a first assessment of uptake into the human body from the surrounding air was undertaken. For HHCB a 10–40 times higher average uptake was calculated compared to MX.

Keywords Indoor air · Ambient air · Quality control · Synthetic musks · Human exposure

1 Introduction

Musk compounds belong to the oldest used natural fragrances known in human societies. Natural musk is derived from the rump glands of the Asian musk deer (*Moschus moschiferus* L.), responsible for secreting the sexual pheromone musk into the environment. During the mating season these glands often reach chicken egg size. The strong smelling secretion of the glands is important for territorial marking and attraction of the female individuals during the mating season. Thus, the natural musk is a typical mammal sexual pheromone and must be considered as an important chemical communication signal aimed for the attraction of females and warning for male individuals crossing the territorial borders at the same time. According to standard textbooks [1] pheromones are chemical substances produced as messengers influencing the behavior of other individuals of the same (intraspecific) or other species (interspecific). The role of pheromones and their effects on human behavior has been underestimated for a long time. However, recent investigations show that the olfactory system has certain influences on human behavior as well especially with respect to the long discussed physiological role of the vomeronasal organ (VMO) or Jakobson's organ which is assumed to be responsible for certain non-smelling olfactory induced behavioral effects also in humans [2–4].

One basic physico-chemical property common for all pheromone-like chemicals is their relatively good solubility in their respective transport medium (water for aquatic and air for terrestrial organisms). Thus, natural or artificial pheromones of terrestrial organisms must be sufficient volatile in air in order to be functioning sufficiently as messenger chemicals between individuals.

Synthetic musks are designed as artificial pheromones to transfer intraspecific chemical information between human individuals. This type of chemical is not naturally occurring in the environment and must therefore be considered as xenobiotic. Thus, certain pheromone-like properties can be attributed to this type of artificial fragrance. These specific pheromone-like properties furnish synthetic musks with unique qualities compared to other anthropogenic environmental contaminants especially valuable for monitoring of environmental pollution. Artificial fragrances like synthetic musks are used as odorous ingredients of perfumes and hygienic products. Therefore, they are supposed to create a favorable odorous image of a human individual and thus, as a consequence, influence the behavior of human individuals nearby. Since the definition as artificial pheromones applies for synthetic musks, two additional consequences in addition to those of conventional environmental contaminants are obvious and should be included in future ecotoxicological assessments of this type of compounds released into the environment:

1. Investigations about unwanted pheromone-like effects on higher organisms (e.g., behavioral abnormalities, reduced reproduction success etc.)
2. Considerations of pheromone triggered endocrine reactions in higher organisms (leading to, e.g., embryological anomalies, dysfunction of reproductive organs), as already known for aquatic and terrestrial organisms

Today synthetic musks belong to the most used artificial fragrances in hygienic and toiletry products, as well as cosmetics and washing powders and can be divided roughly into three subtypes: 1 – nitro musks with musk xylene (MX) and musk ketone (MK) as main contributors; 2 – polycyclic musks with HHCB (e.g., Galaxolide) and AHTN (e.g., Tonalide) as most abundant compounds; and 3 – the complex mixtures of macrocyclic musks. In terms of production, application, and use, nitro musk and polycyclic musks are still dominating (see Chap. 1). Therefore, this chapter will focus on these two types of synthetic musk compounds. For a more detailed insight into the importance of synthetic musks as artificial fragrance the reader is referred to the introduction and the comprehensive overview in Chap. 1.

Pheromone-like properties in connection with the social meanings of artificial fragrances like synthetic musks and their direct connection to human activities and population density makes synthetic musk valuable as modern monitoring compounds for the evaluation and assessment of anthropogenic influences in the environment. A comprehensive discussion about this issue is given in a recent viewpoint publication [5].

Airborne anthropogenic messenger chemicals are designed to transfer odorous signals between human individuals. Thus, physical properties like solubility and vapor pressure are important parameters in order to describe their efficiency as fragrances. A comprehensive list of physico-chemical properties for selected synthetic musks is given in Table 1.

In addition to their pheromone like behavior, the majority of the synthetic musks are lipophilic expressed by a log K_{OW}≥4 (see Table 1).

As already outlined, for synthetic musks as artificial fragrances, ambient air is important as transfer medium for the transmission of the intended chemical sig-

Table 1 Physical properties of the four most abundant synthetic musks compounds musk xylene (MX), musk ketone (MK), HHCB (e.g., Galaxolide), and AHTN (e.g., Tonalide) according to [6]

Compounds	CAS-No.	MW [m/z]	Log K_{OW}	Water solubility [mg/L]	Vapor pressure [Pa]
MX	81-15-2	297.3	4.9	0.49	0.00003
MK	81-14-1	294.3	4.3	1.9	0.00004
HHCB	1222-05-5	258.4	5.9	1.75	0.0727
AHTN	1506-02-1	258.4	5.7	1.25	0.0608

Abbreviation: MW=Molecular weight.

nal to another individual. In order to produce the intended odorous effect, the chemical must enter the organism via a respiratory organ and interact with the odorous receptor system. This interaction, in turn, produces the nervous signals leading to the musky odorous impression in humans. Therefore, the introduction of synthetic musks via respiration must be considered as an important entrance mode and, thus, should not be neglected compared to the evaluation of uptake mechanisms and accumulation processes via food and direct skin contact, only because of an estimated low intake rate.

Due to the direct application of synthetic musks in hygiene and cosmetic products, an elevated contamination risk must be assumed for sample storage and preparation procedures caused by human presence and activities in the laboratory prior to analysis and quantification. Therefore, analytical methods must be adapted to the special needs of synthetic musk trace analysis. In particular, the development of analytical method for the determination of synthetic musks in air samples must include effective sample treatment procedures (e.g., effective storage facilities, reducing amount of solvents used, exclusive usage of fresh solvents, extensive cleaning procedures, etc.) in order to reduce contamination risk to a minimum, since ultra trace concentrations are expected in outdoor air.

2
Analytical Methods and Quality Control

Although special care has to be taken when analyzing environmental samples for synthetic musks, the gas chromatographic analysis and mass spectrometric detection of this compound group is not unusually difficult and does not require particularly sophisticated analytical instrumentation in the laboratory. Usually a standard trace analytical laboratory will be able to determine synthetic musk levels by multiresidue methods if an effective quality assurance program is followed which includes eliminating laboratory contamination, defining background levels, and analyzing representative amounts of field, transport, and laboratory blank samples. This includes also atmospheric samples like indoor and outdoor samples.

2.1
Sample Preparation and Quantification

2.1.1
Outdoor Air Samples

A first endeavor to trace synthetic musks in ambient air samples was undertaken by the Norwegian Institute of Air Research in 1998 [7]. To the best knowledge of the authors, this study today still represents the only published effort to identify synthetic musks in ambient air samples. A new trace analytical method designed for the simultaneous quantification of polycyclic and nitro musk compounds in ambient air was developed based on a well-established standard sampling and quantification method for the determination of persistent organic pollutants (POPs) in ambient air. The general design of the sampling and quantification method is described in Fig. 1 (sampling device) and Fig. 2 (method description).

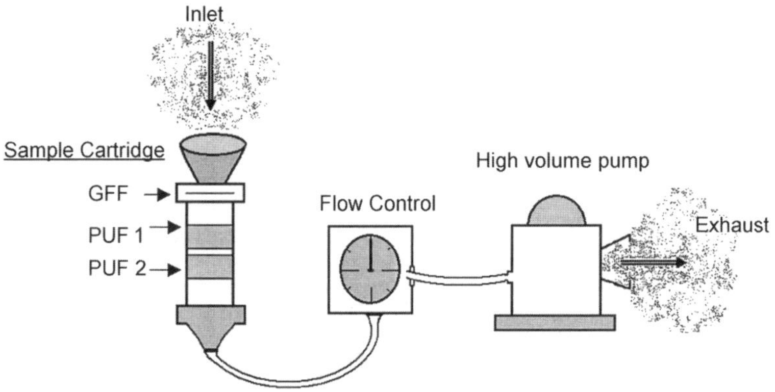

Fig. 1 Schematic presentation of the sampling device for indoor and outdoor air samples. Polyurethane foam plugs and glass fiber filter were subject for further clean-up prior to analysis and quantification. GFF: glass fiber filter, PUF: polyurethane foam plug

As a common sampling design for both outdoor air and indoor air sampling, one glass fiber filter and two polyurethane foam plugs were placed in a glass cartridge. The sampling cartridge is connected to a flow meter and a high volume pump. An exactly recorded air volume is pumped through the glass fiber filter/polyurethane foam (GFF/PUF) plug sampling set. The GFF collected the particulate phase whereas the PUF plugs adsorb contaminants from the gaseous phase (Fig. 1). This sampling method is well-established for tracing of various persistent organic pollutants including chlorinated dibenzo-*p*-dioxins in ambient air collected in pristine remote regions [8, 9] as well as contaminated city air [10]. The exposed GFF/PUF filter are then subjected to further clean-up and analysis (Fig. 2). The main differences between the sampling devices for outdoor and indoor air sampling devices are the dimensions of the sampling equipment and the air volumes collected.

During the above-mentioned pilot study, [7] high-volume air samples were collected in South Norway (Lista fyr) and at Kjeller (near Oslo) in 1997/1998 using standard sampling procedures (Fig. 1). A more detailed discussion about the results will be given in Sect. 3 (Concentrations and Levels). For outdoor air samples a high volume sampling device was employed. Glass fiber filters (Gelman type AE, No. 61635) and PUF-plugs (100 mm id, 50 mm thickness, and density of 25 kg m^{-3}) were exposed to 500–1000 m^3 air during a 24–48 h sampling period. Silica fractionation with a mixture of *n*-hexane and ethyl acetate was used for sample clean-up prior to separation with high resolution gas chromatography and quantification with mass selective detection as described elsewhere [7]. As part of the comprehensive quality control program for the assessment of laboratory contamination one indoor air sample was collected and analyzed for synthetic musk. This sample proved to be highly contaminated and directed the authors' interest towards indoor air musk contamination in general (see Sect. 3).

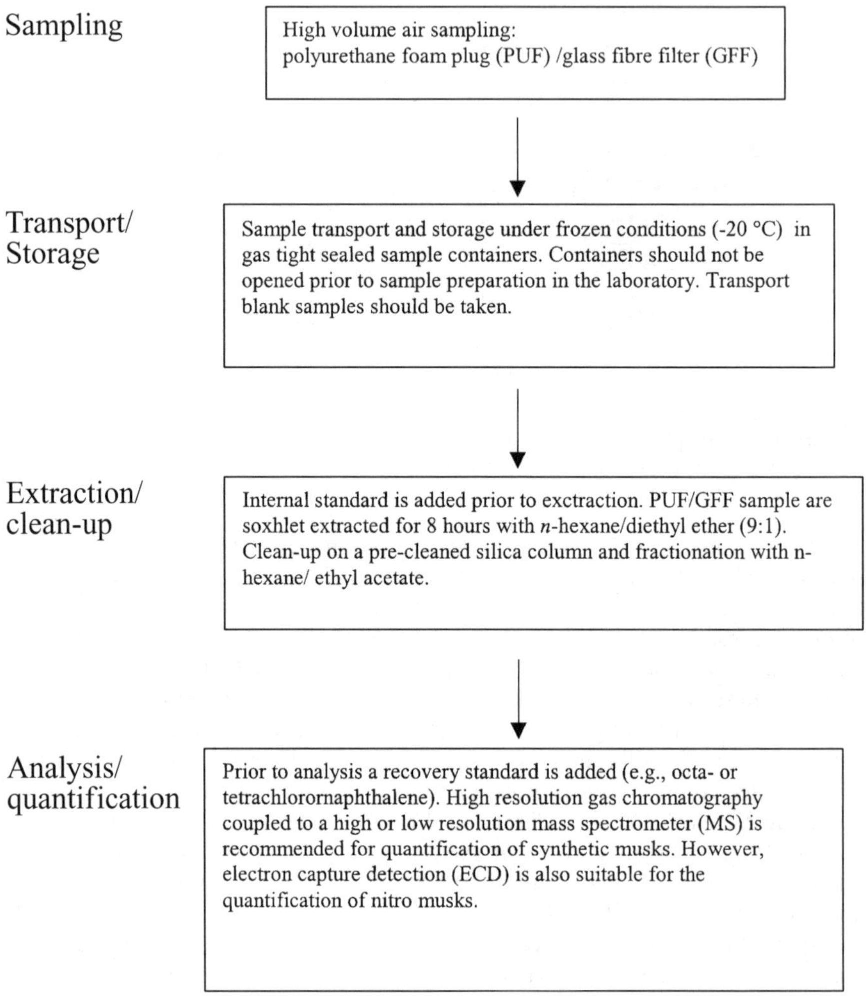

Fig. 2 Standardized sample preparation and analysis method for the quantification of synthetic musk traces in air samples based on accredited methods for the analysis of persistent organic pollutants in ambient air

2.1.2
Indoor Air Samples

Due to the clear indications for elevated synthetic musk levels in indoor air given in the above-mentioned first Norwegian study on ambient air levels, a first indoor air measurement campaign on synthetic musk contamination was initiated in 1999 designated to gather first hand indications about the role of indoor air as synthetic musk exposure source for men in Norway. For this indoor air sampling campaign, a well established low-volume sampling device was used based on requirements for the PCB (polychlorinated biphenyl) and pesticide analysis in in-

Table 2 Sampling sites, sample volumes [m^3] and concentration levels [ng m^{-3}] determined during the first indoor air synthetic musk measuring campaign in Norway. The indoor air levels determined in the 1997–98 sampling campaign (Table 6) are included for comparison

Sample	Sampling location	Site status	Volume	ATII	AHTN	HHCB	MX	MK
Laboratory 1998 [7]	Norwegian Institute for Air Research	Restricted access	108	0.4	0.6	2.5	0.5	0.1
Laboratory 1999	Norwegian Institute for Air Research	Restricted access	70.47	0.3	1.9	5.6	0.3	0.1
Rest facilities	Norwegian Institute for Air Research	Accessible for all employees	76.12	0.8	5.8	19.0	0.6	0.2
Hair dresser	"Downtown" Kjeller	Public accessible	36.27	5.2	13.4	44.3	1.0	0.3
Toilet	Norwegian Institute for Air Research	Accessible for all employees	77.44	0.8	6.2	18.9	0.4	0.1
Cafeteria	Norwegian Institute for Air Research	Accessible for all employees	81.89	4.8	11.6	35.3	n.d.	n.d.
Method blank	–	–	Calc. for 100	0.001	0.1	0.3	0.005	0.001

Abbreviation: n.d.=not detected.

door air [11] according to the general principles for air sample treatment and quantification (Figs. 1 and 2). Sampling and analysis was adapted to the needs for synthetic musk analysis.

Low volume indoor air samples were taken on different locations within the Norwegian Institute for Air Research (NILU) and at the workplace of a local hairdresser situated in the nearby town of Kjeller (Akershus county, Norway). All samples were taken in July 1999. The sampling sites and concentration values are listed in Table 2.

A volume of 36–108 m^3 indoor air (Table 2) was pumped through the GFF/PUF sampling set during a period of 24–48 h (Fig. 1). After exposure and storage at –25 °C prior to analysis, the PUF/GFF sampling sets were Soxhlet extracted with a *n*-hexane/diethyl ether mixture (9:1, v/v) for 8 h according to the method used in [7]. As internal standard, 10 µL of a 1 ng µL^{-1} solution of $^3H_9$-musk xylene and ATTN was added prior to the Soxhlet extraction. The resulting extract was concentrated to 500 µL. The same clean-up method was used as described for the first ambient air study (1997/1998). Column chromatographic silica clean-up with *n*-hexane and ethyl acetate was used. Four fractions were collected: 0–30 mL, *n*-hexane (100%), 30–60 mL, *n*-hexane/ethyl acetate (99:1, v/v), 60–90 mL, *n*-hexane/ethyl acetate (9:1, v/v), 90–120 mL, ethyl acetate 100%. All synthetic

Table 3 Important physical parameters for the low resolution mass selective detection of synthetic musks according to [7]

Compounds	Target ion (m/z)	Reference ion (m/z)	Ionization mode	Dwell time [ms]	Ion source pressure [Pa]	Ion Source temperature [0 °C]
MX	267	268	NCI	50[a]/100[c]	500[a]/0.03[c]	200[a]/180[c]
MK	264	265	NCI	50[a]/100[c]	500[a]/0.03[c]	200[a]/180[c]
HHCB	243	213	EI[b]	100	0.006[b]	160
AHTN	243	258	EI	100	0.006[b]	160
ATII	215	258	EI	100	0.006[b]	160

[a] NCI mode with a HP 5989 low resolution mass spectrometer.
[b] EI mode with a Fisons MD 800 low resolution mass spectrometer.
[c] NCI mode with a Fisons MD 800 low resolution mass spectrometer.

musks of concern were found in fraction 3. Fraction 4 was collected as "safety fraction" to avoid unexpected losses due to matrix related retention delay during the silica clean up. For more details see [7].

After the silica clean up, fraction 3 was concentrated to 500 μL with a Turbovap 500 (Zymark, Palo Alto, CA, USA). Octachloronaphthalene (recovery standard OCN, 10 μL added from a 1 ng μL^{-1} solution) was added prior to analysis. Gas chromatographic separation was performed with a Fisons CE high resolution gas chromatograph Mega II 8560 (Milan, Italy) using helium as carrier gas (He 5.0 quality, purchased from Hydro, Porsgrunn, Norway) at a flow rate of 1 mL min^{-1}. A volume of 2 μL of the sample was injected on-column by a Fisons AS800 autoinjection system into a DB 5MS column (J&W, Folsom, CA, USA, column parameter: i.d. 0.25 mm, length 30 m, film thickness 0.2 μm). The following temperature program was used for the gas chromatographic separation: 70 °C (2 min) – 6 °C min^{-1} – 155 °C (5 min) – 1 °C min^{-1} – 195 °C – 10 °C min^{-1} – 230 °C (10 min). The gas chromatograph was connected to an MD 800 low resolution quadrupole mass spectrometer at a interface temperature of 250 °C and a ion source temperature of 180 °C for electron impact mode (EI) and 160 °C for negative ion chemical ionization (NCI). The determination of polycyclic musks was performed in EI and for nitro musks in NCI. For NCI, methane (5.0 quality, purchased from Hydrogas, Porsgrunn, Norway) was used as reactant gas. All quantifications were performed in selected ion monitoring (SIM). For each compound two masses (m/z) were monitored (target ion and reference ion). The quantification was performed using target ion traces whereas isotope ratios (ratio between target and reference traces) were used for identification. Parameters for the mass selective detection modes used for the determination of synthetic musks in EI and NCI mode are given in Table 3.

Typical SIM chromatograms for the determination of synthetic musk are presented in Fig. 3 for both a standard solution and a typical indoor air sample.

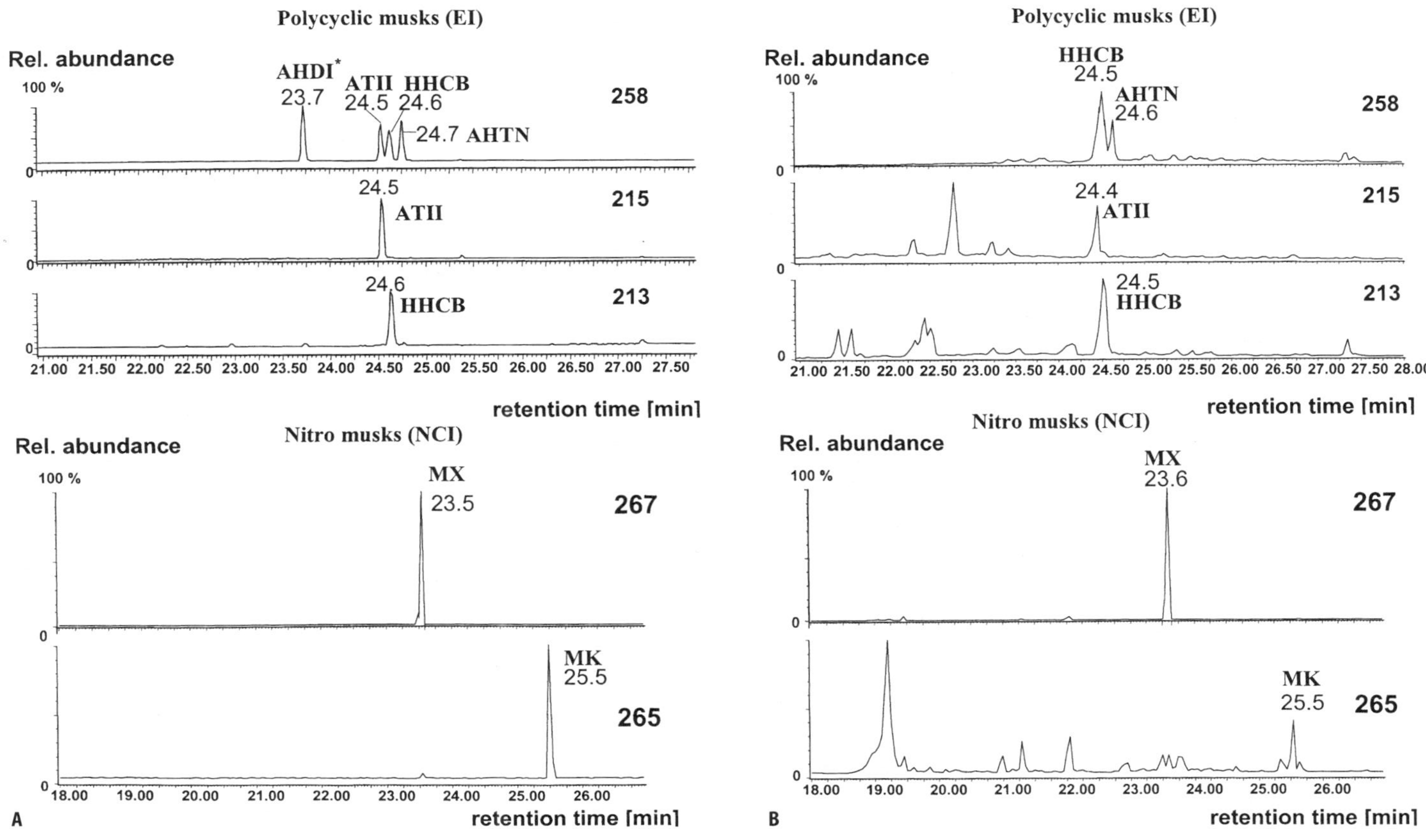

Fig. 3A, B Typical SIM fragmentograms of the synthetic musk determination in indoor air samples: **A** standard solution; **B** typical indoor air sample.
* AHDI, e.g., Phantolide, only in standard solution

2.2 Quality Assurance

As an online performance check of the trace analytical method applied the application of an internal standard is to be preferred compared to often used external standard quantification of synthetic musks referred to in the available scientific literature, for obvious reasons. For external standard quantification, information about possible losses and recovery rates can only be obtained when the final volume of sample and standard solution is exactly the same. In addition, small changes in the sensitivity of the detection system between two chromatographic runs (HRGC or HPLC) may have strong influences on the accuracy of the quantification results obtained.

According to the standard textbooks of environmental analytical chemistry [12, 13], internal standards should be chosen as similar to the compounds to be analyzed as possible but, in addition, the standard compound must not be found in environmental samples. Thus, an ideal internal standard must:

1. Follow the sample preparation procedure in the same chromatographic time and separation windows as the analytes
2. Be adapted to the quantification methods used in terms of high sensitivity and selectivity

A list of recommended standard compounds for the quantification of synthetic musks is presented in Table 4.

When choosing high resolution gas chromatography and mass spectrometric detection methods, isotope labeled internal standards should be preferred if available. For nitro musks the synthesis of a ^{3}H labeled musk xylene is described [14]. The use of a homemade deuterated $^{2}H_9$ musk xylene was used as internal standard for the determination of nitro musks in Norwegian air samples [7]. In addition, $^{2}H_3$ isotope labeled AHTN (e.g., Tonalide) and $^{2}H_{15}$ isotope labeled MX are commercially available on the market from Dr. Ehrensdorfer (Augsburg, Germany). Usually, isotope labeled standards used to be relatively expensive to purchase and thus are often not achievable without considerable increase of costs for the analytical and quantification procedure chosen. In addition, for the de-

Table 4 List of recommended internal and recovery standards for the quantification of synthetic musks (nitro and polycyclic compounds)

Standard type	Nitro musks		Polycyclic musks	
	Isotope labeled	Not labeled	Isotope labeled	Not labeled
Internal standard	$^{2}H_9$-musk xylene; $^{2}H_{15}$-musk xylene	–	$^{2}H_3$-AHTN	ATTN
Recovery standard	No information available or published yet	Octachloronaphthalene (OCN) Tetrachloronaphthalene (TCN)	No information available or published yet	Octachloronaphthalene (OCN) Tetrachloronaphthalene (TCN)

termination of nitro musk compounds with high resolution gas chromatography coupled to an electron capture detector (ECD), isotope labeled internal standards cannot be used. This type of very selective detector is generally not able to discriminate between isotope labeled and non-labeled compounds. Therefore, if an ECD detection method is chosen for quantification, non-labeled chemicals, with structure similarities to those chemicals to be analyzed but at the same time not present in the environment, must be chosen as internal standards. For the GC/MS analysis of polycyclic musks, often a related polycyclic musk compound with similar chemical structures is chosen as internal standard (e.g., ATTN). However, as mentioned, isotope labeled polycyclic musk standards are available commercially for GC/MS analysis. In our recent investigations [7, 15] and the indoor air sampling campaign discussed herein (see Sect. 3 Concentrations and Levels) ATTN (e.g., Versalide) was used and is also recommended as reliable internal standard based on our scientific experiences. An additional recovery standard, added to the sample extract prior to quantification, is recommended for determination of recovery and robustness of the method (see Table 4). This type of standard is supposed to be recovered 100% and, thus, can be used to calculate the recovery of the internal standard for both GC/ECD and GC/MS separation and quantification methods [15].

As already published in several previous scientific papers, trace analysis of synthetic musks demands a thoroughly organized and controlled clean-up and analytical procedure in order to avoid contamination during sample storage and handling in the laboratory [7, 17–20]. As one of the major contamination risks within the laboratory, hygienic and cleaning products as well as soaps used to clean laboratory facilities were identified. In addition, bottles containing organic solvents (e.g., *n*-hexane, acetone, dichloromethane, diethyl ether, etc.) used over a longer period for extraction and clean-up in the laboratory are often opened and closed and, thus, the solvents tend to accumulate synthetic musks from the surrounding laboratory air and therefore contribute directly to the elevated laboratory contamination during the various sample preparation steps. A comprehensive and detailed discussion about contamination risks and limitations of standard analytical methods are given in Chap. 2. Due to the very low concentration levels in the pg m^{-3} range usually expected in outdoor air, laboratory contamination must be controlled efficiently and reduced to a minimum during sample preparation and analysis of ambient air samples for the quantification of synthetic musks.

In Table 5, typical contamination risks for air sampling, storage, and laboratory treatment are listed. Although special emphasis is laid upon the risks related to air sampling, the different topics are also valid for other types of samples and sample treatment. As a matter of a basic good laboratory practice for the ultra trace analysis of every type of environmental contaminant sufficient number of blank samples (field blanks, transport blanks, laboratory blanks, and method performance tests) should be prepared using the same methods as for the sample set. A detailed quality control program including sample protocols especially adapted to the analytical program used for the analysis of synthetic musks must be developed and documented in order to control and register the reliability of the final analytical data set obtained [16, 21].

Table 5 Basic measures to reduce background levels of synthetic musks during sample storage and laboratory treatment

Synthetic musk contamination source for ambient air analysis	Proposed counter measure
Contamination during sampling (direct skin contact, cross contamination etc.)	Use fragrance free examination gloves for sampling if needed. Avoid direct skin contact with sampling equipment. Minimize cosmetic use of laboratory personal
Contamination of the sampling device	Extensively cleaning with contamination free solvents. Check solvent blanks. Use fragrance free tissue
Contamination during the transport of the samples	Seal every sample in gas tight, pre-cleaned containers under transport (e.g., aluminum boxes, glass container)
Repeated thawing and freezing of sample materials (cross contamination)	Store all samples in separate, gas tight containers at a minimum of –20 °C and avoid opening sample container prior to preparation for analysis
Cleaning and hygiene products permanently used in the laboratory	Record the back ground level at the working place with sufficient laboratory and method blanks. Reduce exposure of the samples to the laboratory air to an absolute minimum
Toiletry products on the skin surface of laboratory personnel	Avoid skin contact with the ambient air samples and equipment used for sample preparation. Use always fragrance free examination gloves for sample handling
Solvent contamination due to repeated surface exposure (e.g., open bottles, vials)	Reduce amount of solvent used to an absolute minimum. Check and document always the solvent blank before using solvents for extraction and clean-up

Based on comprehensive measurements including field and laboratory blanks limit of detection (LOD) values were determined for all synthetic musk compounds analyzed in air samples. The LOD was calculated as three times of the signal to noise ratio in both laboratory and field blanks. The maximum values were used for LOD calculation (worst case scenario). For polycyclic musks the best results were achieved with GC/EI-MS quantification methods (LOD: HHCB= 45 pg m^{-3}, AHTN=12 pg m^{-3}, and ATII=5 pg m^{-3}). For nitro musks GC/NCI-MS methods were used for LOD determination (LOD: MX=12 pg m^{-3} and MK=4 pg m^{-3}). More details on the determination and evaluation of LOD values for synthetic musks in ambient air can be found in an earlier publication [7].

3 Concentrations and Levels

3.1 Outdoor Air

Although air as transfer medium is of essential importance for artificial fragrances like synthetic musks, concentration levels and pattern distribution in air were not published until 1999 [7], 18 years after the first report of synthetic musks in the environment [22,23]. Until today, this investigation about synthetic musk patterns in Norwegian outdoor air still remains the only published work about this research issue so far. The authors intended to present a trace analytical method for the simultaneous quantification of nitro and polycyclic musk fragrances. The presence of nitro and polycyclic musks in Norwegian outdoor air could be clearly confirmed although elevated blank levels reduced the outcome of the study. Therefore, considerable efforts are still necessary to improve the current analytical method described for the analysis of synthetic musks in outdoor air. Slightly elevated levels from Kjeller close to Norwegian capital Oslo compared to the South Norwegian background station Lista fyr were determined (Table 6). These results indicate an increase of synthetic musk levels in ambient air towards the densely populated region around Oslo (Norway). However a significant correlation could not be established based on these 14 air samples investigated. Therefore, a more comprehensive monitoring program is necessary to determine significant correlations.

3.2 Indoor Air

The surprisingly high synthetic musk levels found in the single indoor air sample analyzed during the first air sampling campaign in 1997–98 directed the interest of the same group of researchers towards possible and still underestimated exposure routes via elevated synthetic musks values in indoor air. Thus, a preliminary measuring campaign was performed by the Norwegian Institute for Air Research using institute facilities as indoor air sampling sites including the laboratory sampled in the 1997–98 study as well as a hairdresser shop at the nearby city of Kjeller as external sampling site (Table 2).

The laboratory sample collected in the 1999 study (Table 2) expressed a comparable level distribution as the sample taken in 1998 (Table 6). However HHCB and AHTN were about 50% higher in the 1999 laboratory air sample as the levels found in laboratory one year before. Even so, the concentration levels are still in the same order of magnitude. The same pattern distribution for the five synthetic musk compounds investigated which was found for the first indoor air sample in 1998 was confirmed for all indoor air samples collected in 1999. This finding supports the assumption, that one major synthetic musk source contributes mainly to the synthetic musk levels found in indoor air at the Norwegian Institute for Air Research. The most abundant compound was HHCB (e.g., Galaxolide) in all samples analyzed, followed by AHTN (e.g., Tonalide), ATII (e.g., Traseolide), musk

Table 6 Results of a first screening for synthetic musks in 14 Norwegian ambient air samples and one indoor air sample, according to [7]

Sample no.	1	2	3	4	I	A[a]
Sampling site	Kjeller					
Sample type	Ambient air	Ambient air	Ambient air	Indoor air	Blank I	Ambient air
Sample volume [m^3]	1110	1020	1190.5	108	Calc. for 1000	907
Sampling date	Week 47 (1998)	Week 49 (1998)	Week 50 (1998)	Week 51 (1998)	Week 51 (1998)	Week 49 (1998)
MX conc. [pg m^{-3}]	17	17	23	530	4	35
MK conc. [pg m^{-3}]	6	5	6	120	1	24
AHTN conc. [pg m^{-3}]	44	41	46	600	4	76
HHCB conc. [pg m^{-3}]	130	110	116	2470	15	223
ATII conc. [pg m^{-3}]	10[b]	19[b]	6[b]	430	1[b]	15[b]

Sample no.	5	6	7	8	9	10	11	12	13	14
Sampling site	Lista fyr									
Sample volume [m^3]	570	570	571	571	566	570	578	575	571	571
Sampling date	Week 19 (1997)	Week 20 (1997)	Week 21 (1997)	Week 22 (1997)	Week 23 (1997)	Week 24 (1997)	Week 25 (1997)	Week 26 (1997)	Week 27 (1997)	Week 32 (1997)
MX conc. [pg m^{-3}]	8	15	15	15	15	13	10	19	19	13

Blanks	II	III	IV	V	Average (n=4)	SD
MX conc. [pg m^{-3}]	2	2	3	2	2	0.4

[a] 1 GFF and 2 PUF plugs were analyzed separately, the three resulting single concentration values were then summarized to one concentration value for the entire sample.

[b] Isotope ratio deviates more than 20% from standard measurements.

Table 7 Comparison of concentration levels for synthetic musks and halogenated contaminants in laboratory air [ng m^{-3}] collected at the same sampling site. PCBs, hexachlorocyclohexane derivatives (HCH) and hexachlorobenzene (HCB) are published in [11]. Synthetic musk values were presented in [7]; see also Table 6

Sampling date	Week 13, 1997	Week 51, 1998
Sum PCB	0.2	–
α-HCH	0.1	–
γ-HCH	3.0	–
HCB	0.1	–
ATII	–	0.4
AHTN	–	0.6
HHCB	–	2.5
MX	–	0.5
MK	–	0.1

xylene, and musk ketone. This pattern was also repeatedly found for other environmental samples reported in literature (e.g., biota, sediment, human milk). For more information about synthetic musk patterns in biota and other environmental samples please consult the previous chapters of this monograph. In the presented indoor air study, the highest synthetic musk values were found in the hairdresser facility with 44 ng m^{-3} HHCB. However, the air collected in the institute cafeteria also contained high synthetic musk burden with 35 ng m^{-3} HHCB and 12 ng m^{-3} AHTN, respectively (Table 2).

The concentrations found are considerably higher than the PCB levels found in highly contaminated households in Norway during a study performed by NILU in 1997 [11]. A comparison of the musk concentration values with selected chlorinated contaminants is given in Table 7.

The HHCB level found in the 1998 laboratory air sample is in the same concentration range as determined for the most abundant compound in the 1997 laboratory air sample, namely γ-hexachlorocyclohexane (HCH) and about ten times higher than the levels found for PCB (Table 7). Nitro musks (MX and MK) were found in the same concentration range as determined for α-HCH and hexachlorobenzene (HCB: 0.1 ng m^{-3}).

These relatively high concentration values found for synthetic musks, in particular for polycyclic compounds, in indoor air demonstrates again the need for a comprehensive long-term monitoring of this type of compounds in indoor air. In general, based on these preliminary results, it can be assumed that a correlation between public accessibility of the facilities and the concentration of synthetic musks in indoor air exist. However, in order to draw general conclusion, a more comprehensive and statistically supported study must be performed.

In addition, due to the striking similarities of the synthetic musk patterns found in all indoor air samples regardless of different concentration levels, one main source for synthetic musks might be assumed with several other minor sources contributing to the general distribution as already outlined above (Table 2). Thus, our main working hypothesis today is that a few cleaning products containing similar patterns and levels of synthetic musks are used extensively in the institute facilities and deliver continuously high background levels into the indoor air. Depending on the accessibility of the different sampling facilities individual sources like hygienic products and perfumes contribute more or less significantly to this already elevated synthetic musk levels determined in indoor air.

Another possible explanation should also be mentioned. It cannot be ruled out that all products used in the facilities (hygienic, cleaning, perfumes, etc.) are characterized by the same distribution pattern of synthetic musks. Therefore, all sources might contribute more or less equally to the pattern; only the concentrations found are different depending on the accessibility of the sampling facility. However, due to the large differences known in composition and content of fragrances in the various products already investigated in other European countries, it is unlikely, that this would be the case for the presented indoor air study [24].

The results of the study described above still represent a very early stage for the investigation of synthetic musks in air samples. Considerable efforts are still needed to answer all the remaining questions and more comprehensive monitoring is needed on international and national levels. In the near future, a new study will be performed by the Norwegian Institute for Air Research in order to elucidate possible sources and their contribution to the general distribution pattern and the high concentration levels of synthetic musks found in Norwegian indoor air.

4
Discussion and Perspectives

In scientific research the answer of one important question usually raises a large number of additional new scientific questions to be answered during subsequent research efforts. This is also true for synthetic musk investigations in the environment. The first analyses in indoor and outdoor air confirmed that synthetic musks are present in both compartments. In particular the indoor environment is contaminated with relatively high concentrations of synthetic musks. Polycyclic musk compounds are dominating in both indoor and outdoor air. Although the concentration levels are varying strongly, the percentage distribution pattern shows interesting similarities between outdoor air and indoor air samples (Fig. 4) indicating similar sources for synthetic musks. HHCB and AHTN dominate in both investigated sample types whereas the nitro musks MX and MK are minor contributors. However, certain differences can be seen when comparing distribution patterns of synthetic musks in indoor air and outdoor air samples in detail as presented in Fig. 4.

Whereas HHCB contributes in indoor air with about 68–72% of the total synthetic musk content determined, the HHCB contribution is in outdoor air considerably lower ranging between 58 and 62% (Fig. 4). In addition, the MK contribution in indoor air is very low, ranging from 0.1 to 0.9%. In outdoor air the MK contribution is surprisingly high, varying between 2.7 and 3.1%. Also MX contributes considerably more to the overall musk contamination in outdoor air samples (between 7 and 11%) compared to indoor air samples (between 0.5 and

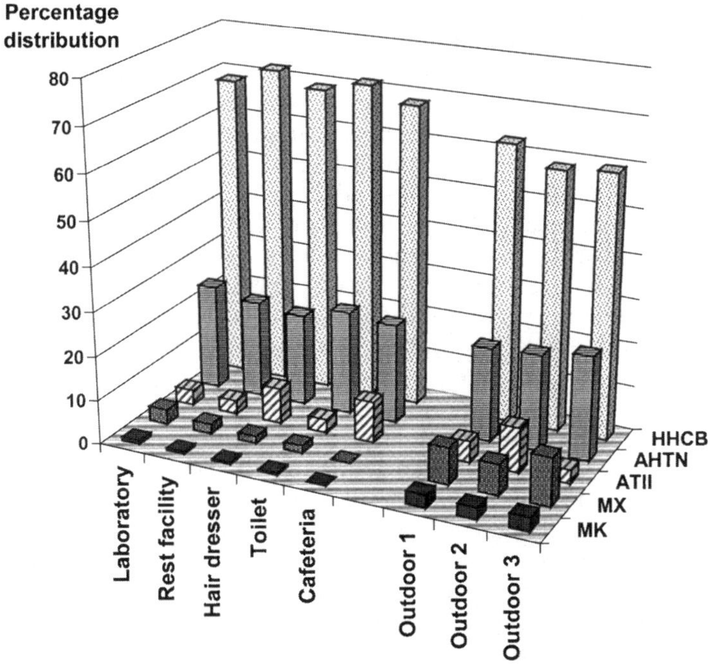

Fig. 4 Comparison of the percentage distribution of synthetic musks in indoor and outdoor air (concentrations, see Tables 2 and 6, respectively). The sum of all five analyzed synthetic musks is set 100%. Outdoor 1–3=Kjeller, ambient air samples 1–3 (see Table 6)

3%). The percentage distribution of ATII in indoor and outdoor air samples is characterized by a relatively high variation (between 0.3 and 5%); therefore, no clear tendencies can be determined for this compound based on the sample set investigated. AHTN has a similar contribution both in indoor and outdoor air, between 20–23% of the total synthetic musk contamination (Fig. 4). These differences in synthetic musk distribution between indoor and outdoor air indicate different mechanisms contributing to the contamination pattern. In indoor air samples, due to the high concentration levels, a permanent supply of synthetic musks into the air from few major sources can be assumed, resulting in a continuously high contamination. Therefore, the synthetic musk patterns found in indoor air always represent a relatively fresh synthetic musk contamination with minimal influence from biochemical or photochemical transformation processes.

The low concentration levels found for synthetic musks in outdoor air compared to indoor air are a direct result of the contribution from various sources (e.g., emissions from municipal and private facilities, direct contribution from densely populated regions, etc.). In addition, photochemical and microbial degradation processes as well as re-emission from municipal sewage treatment plants may have important influences on the final synthetic musk contamination in outdoor air.

At present, no degradation products of synthetic musks, neither for nitro nor polycyclic musks, were analyzed in ambient air samples. Due to the ultra trace concentrations expected, considerable effort is still needed to improve the performance of the present methods and reduce the detection limit for a reliable analysis of transformation products for synthetic musks in air. In addition, transformation products of synthetic musks found in environmental samples are more polar than their parent compounds. Thus, the analytical methods described here and applied to synthetic musk analysis in air need substantial methodological adaptations and optimization in order to include polar transformation products as additional target compounds.

Within the indoor air sample set taken at different sample sites a surprising similarity concerning pattern distribution of the synthetic musk compounds analyzed was found (Fig. 4). However, in the air sample taken in the institute cafeteria, no nitro musks were analyzed; therefore no concentration values given in Table 2 and for Fig. 4 no contribution of nitro musk could be calculated for this sample. Based on the presented results, it can be assumed that further investigations are urgently needed to elucidate sources and contamination patterns of synthetic musks in indoor air in more detail.

Based on the results presented here a first estimation of the role of air as medium for the uptake of synthetic musks by the human organism via respiration only was undertaken. For an average human individual of 70 kg weight a first estimation was made for the uptake of MX and HHCB (e.g., Galaxolide). These preliminary results are summarized in Table 8.

As a first rough estimation, an average daily respiration volume of 15 m^3 and a 10% resorption capability of respiratory pulmonary system is assumed. For the following calculations average as well as maximum values were calculated to describe both average and worst case scenarios for respiration as an uptake route of synthetic musks into the human organism.

Table 8 Estimated contribution of air as contamination pathway for the uptake of MX and HHCB (e.g., Galaxolide) into the human organism

	Indoor air uptake			
	Average HHCB	Average MX	Maximum HHCB	Maximum MX
Indoor air concentration [ng m^{-3}]	20.90	0.56	44.30	1.00
Estimated daily uptake [ng day^{-1}]	31.40	0.84	66.54	1.50
	Outdoor air uptake			
	Average HHCB	Average MX	Maximum HHCB	Maximum MX
Outdoor air concentration [ng m^{-3}]	0.14	0.02	0.22	0.04
Estimated daily uptake [ng day^{-1}]	0.21	0.03	0.33	0.06

This first comparison (Table 8) indicates that for MX, indoor and outdoor air is a minor input way compared to HHCB. For HHCB, a 10–40 times higher average uptake is calculated compared to MX. These preliminary results confirm that the respiration of surrounding air is probably of minor importance as uptake route for synthetic musks compared to other contamination pathways (e.g., food, skin) into the human body. However, for the comprehensive assessment of the total uptake of synthetic musks the contribution of polluted indoor must also be considered. As outlined earlier, beside hazardous properties generally monitored for conventional environmental pollutants, synthetic musks add new important aspects to the discussion on ecotoxicologically relevant properties due to their specific design as artificial pheromones. As shown for natural pheromones, a behavioral response of an organism can be seen at very low concentrations of the respective pheromone. In addition, the presence of pheromones can trigger hormonal reaction in the respective organisms at much lower levels as usually shown for anthropogenic contaminants. Although no experimental results are yet available today to support or falsify this hypothesis, it is important to include these aspects in future research on the ecotoxicological potential of synthetic musks and other fragrances.

5 Conclusions

The sampling and quantification methods presented for synthetic musk analysis in ambient air still need considerable improvement. In particular, the relatively high background levels mostly caused by elevated laboratory blanks and storage contamination must be reduced for reliable low-concentration determination of synthetic musk in outdoor air. One major source for elevated blank values is the extensive use of nonpolar solvents during extraction and clean-up. The reduction of solvent use during sample preparation will immediately lead to a decrease of contamination. Therefore the Norwegian Institute for Air Research is currently performing a method test on solvent-free sampling and quantification methods for the determination of synthetic musks in ambient air. First tests with passive sampling collectors and thermodesorption as solvent free injection method into the GC coupled to a mass spectrometric detector seem to be a good substitution for the PUF/GFF air sampling currently used for synthetic musks air sampling [25].

Acknowledgements The realization of the investigations presented here would not have been possible without the help and the generous support of many colleagues.

The authors appreciate the support and help of the editor Dr. Gerhard G. Rimkus during the final work with the manuscript. We thank Prof. Dr H. Hühnerfuss, Dr. Martin Schlabach, Dr. Ole-Anders Braathen, Norbert Schmidbauer, and Dr. I.C. Burkow for their generous support, as well as for open minded and continuous discussions about the topic. The help of Sissel Planting, Dr. Dorte Herzke, and Gro Hammerseth during sample preparation and analysis is highly appreciated. Dr. Scarlett Biselli and Dipl Chem. Frank Hoffmann helped during the discussion of the data. Dr. Robert Gatermann is grateful for the financial support of the Norwegian Research council (NFR 125745/720 "Synthetic musks in the Norwegian Environment"). The investigations presented were partly financed by "NILU's Strategic Institute Program for the investigation of persistent organic pollutants in Arctic air".

6
References

1. Van der Meer R, Breed MD, Esplie KE, Winston ML (1998) Pheromone communication in social insects: ants, wasps, bees, and termites. Westview Press, Scranton PA
2. Døving K, Trotier D (1998) J Exp Biol 201:2913
3. Stern K, McClintock MK (1998) Nature 392:177
4. Berliner DL, Monti-Bloch L, Jenning-White C, Diaz-Sanchez V (1996) J Steroid Biochem Mol Biol 58:259
5. Kallenborn R, Gatermann R, Rimkus GG (1999) J Environ Monit 1:70 N
6. Simonich SL, Begley WM, Debaere G, Eckhoff WS (2000) Environ Sci Technol 34:959
7. Kallenborn R, Gatermann R, Planting S, Rimkus GG, Lund M, Schlabach M, Burkow IC (1999) J Chromatogr A 846:295
8. Oehme M, Schlabach M, Kallenborn R, Haugen JE (1996) Sci Total Environ 186:13
9. Oehme M, Haugen JE, Schlabach M (1996) Environ Sci Technol 30:2294
10. Halsall CJ, Lee RGM, Coleman PJ, Burnett V, Harding Jones P, Jones KC (1995) Environ Sci Technol 29:2368
11. Braathen O-A, Schlabach M (1999) PCB in indoor air of a laboratory and office building in Norway. In: Indoor air 99. Proceedings of the 8th International Conference on Indoor Air Quality and Climate, Edinburgh, Scotland, 8–13 August 1999. Vol 2, Construction Research Communications, London
12. Thier H-P, Frehse H (1986) Rückstandsanalytik von Pflanzenschutzmitteln. In: Hulpke H, Hartkamp H, Tölg G (eds) Analytische Chemie für die Praxis, Georg Thieme Verlag, Stuttgart, Germany
13. Fischer R, Siebers J, Blacha-Puller M (1997) Methodenbuch Rückstandsanalytik, Blackwell, Oxford UK
14. Fukuoka M, Nambaru S, Tanaka A (1991) J Labelled Comp Radiopharm 29:1207
15. Kallenborn R, Gatermann R, Nygård T, Knudzen J (2002) Fresenius Environ Bull (in press)
16. Vogelsang J (1991) Fresenius J Anal Chem 340:384
17. Gatermann R, Huehnerfuss H, Rimkus G, Attar A, Kettrup A (1998) Chemosphere 36:2535
18. Franke S, Meyer C, Heinzel N, Gatermann R, Huehnerfuss H, Rimkus G, König WA, Francke W (1999) Chirality 11:795
19. Rimkus GG, Butte W, Geyer HJ (1997) Chemosphere 35:1497
20. Helbling K, Schmid P, Schlatter C (1994) Chemosphere 29:477
21. Anonymous (1994) AMAP Method performance criteria and quality control measures for the determination of persistent organic compounds (POP) in ambient air, precipitation and water, Working group "quality control programme", Workshop on techniques of POP measurements in Northern environments, Waterloo, Ontario, Canada, 16–18 June 1994
22. Yamagishi T, Myazaki T, Horii S, Kaneko S (1981) Bull Environ Contam Toxicol 26:656
23. Yamagishi T, Myazaki T, Horii S, Kaneko S (1983) Arch Environ Contam Toxicol 12:83
24. Sommer C (1993) Dtsch Lebensm-Rundsch 89:108
25. Schmidbauer N (2001) NILU. Personal communication

The Handbook of Environmental Chemistry Vol. 3, Part X (2004): 105–121
DOI 10.1007/b14127

Synthetic Musks in House Dust

Werner Butte

Fachbereich Chemie, Carl von Ossietzky University Oldenburg, PO Box 2503, 26111 Oldenburg, Germany
E-mail: werner.butte@uni-oldenburg.de

Abstract House dust is a sink and repository for semi-volatile organic compounds and particle bound organic matter. Although data are available on the occurrence of pesticides, plasticizers, endocrine disrupters, and other organics that are persistent in house dust, reports on analytical methods and concentrations of synthetic musks in house dust are limited.

The development of methods to analyze synthetic musks in house dust and results obtained are discussed in this chapter. For nitro musks capillary gas chromatography with an electron capture detector proved to be the method resulting in the lowest limit of determination. However this detector has the disadvantage of a rather low selectivity. For polycyclic musks capillary gas chromatography with tandem mass spectrometry was favored. From the analysis of some 30 house dust samples it may be concluded that concentrations of nitro musks are much lower that those of polycyclic musks. Musk xylene and musk ketone were the nitro musks most prominent in house dust with concentrations up to some milligrams per kilogram. Regarding polycyclic musks HHCB (e.g., Galaxolide) and AHTN (e.g., Tonalide) were present in nearly every dust sample with concentrations up to about 0.1 g kg^{-1}. This underlines the fact that synthetic musks are widespread in the indoor environment with higher concentrations of polycyclic musks compared to nitro musks.

Keywords House dust · Semi-volatile organic compounds · Particulate organic matter · Nitro musks · Polycyclic musks

List of Abbreviations

ADBI	4-acetyl-1,1-dimethyl-6-tert.butyldihydroindene (e.g., Celestolide, Crysolide)
AHDI	6-acetyl-1,1,2,3,3,5-hexamethyldihydroindene (e.g., Phantolide)
AHTN	7-acetyl-1,1,3,4,4,6-hexamethyltetrahydronaphthalene (e.g., Tonalide, Fixolide)
CID	collision induced dissociation
DDT	2,2′-bis(4-chlorophenyl)-1,1,1-trichloroethane [dichlorodiphenyltrichloroethane]
ECD	electron capture detector
EI	electron ionization
GC	gas chromatography
GC-MS	gas chromatography – mass spectrometry
GC-MS/MS	gas chromatography – tandem mass spectrometry
HHCB	1,3,4,6,7,8-hexahydro-4,6,6,7,8,8-hexamethylcyclopenta[g]-2-benzopyrane (e.g., Galaxolide, Abbalide, Pearlide)
IS	internal standard (2-methyl-4,6-dinitronanisol)
MA	1-*tert*-butyl-3,5-dinitro-2-methoxy-4-methylbenzene (musk ambrette)
MK	4-acetyl-1-*tert*-butyl-3,5-diemthyl-2,6-dinitrobenzene (musk ketone)
MM	4,6-dinitro-1,1,3,5,5,-pentamethylindane (musk moskene)
MS	mass spectrometry
MSD	mass selective detector
MT	1-*tert*-butyl-2,6-dinitro-3,4,5-trimethylbenzene (musk tibetene)
MX	1-*tert*-butyl-3,5-dimethyl-2,4,6-trinitrobenzene (musk xylene)
m/z	mass to charge ratio
POM	particulate organic matter
SIM	single ion (detection) mode
SVOC	semi-volatile organic compound(s)
VVOC	very volatile (gaseous) organic compound(s)
VOC	volatile organic compound(s)

1 Introduction

On average, people spend around 95% of their time indoors, and most of this time is spent in their homes [1]. Regarding contamination, indoor air and dust are, besides food and workplace, significant sources of exposure for the general population, especially children. Indoor pollution has been ranked by the United

States Environmental Protection Agency Advisory Board and the Center for Disease Control as a high environmental risk [2].

Adverse health effects or impairments observed in housings may have physical, biological, or chemical origin; however public discussion focuses on chemical contamination, mostly organic pollutants. Analyses of organic compounds in house dust are predominantly a measure for an indoor contamination but may also contain valuable information on human indoor exposure.

A classification of organic indoor contaminants, according to their volatility, was given by a WHO working group on organic indoor air pollutants [3]; this group initiated the common practice of dividing organic chemicals according to boiling points (see Table 1).

VVOC and VOC are transitory and predominantly found in air; analyses of these compounds in indoor air are a direct measure for indoor exposure. Organics with a low volatility or high polarity are expected to partition more to dust than to air; they are adsorbed to particles. SVOC are supposed to stay in air as well as in dust, whereas POM are exclusively found in dust. House dust is a sink and repository for semi-volatile organic compounds and particle bound organic matter. Regarding synthetic musks, boiling points are not available, but vapor pressures classify them as semi-volatile, as these are rather similar to the vapor pressures of some well characterized semi-volatile biocides like pentachlorophenol or lindane. The vapor pressure of pentachlorophenol is 7.4 mPa (25 °C) with a boiling point of ~310 °C, that of lindane 7.3 mPa (25 °C) with a boiling point of ~320 °C, respectively [4]. Compared to these biocides vapor pressures of HHCB and AHTN are about ten times higher (HHCB: 60.8 mPa (25 °C), AHTN: 72.7 mPa (25 °C)) [5], those of MK and MX are about a 100 times lower (MX: 0.04 mPa, MK: 0.03 mPa) [4]. Both polycyclic as well as nitro musk compounds are expected to be found predominantly in house dust and to a much lower extent in indoor air.

Table 1 Classification of organic indoor pollutants (after [3])

Abbreviation	Description	Boiling point range [°C]	Examples
VVOC	Very volatile (gaseous) organic compounds	<0 to 50–100	Carbon monoxide, carbon dioxide, formaldehyde
VOC	Volatile organic compounds	50–100 to 240–260	Solvents (aliphatic, aromatic), terpenes
SVOC	Semi-volatile organic compounds	240–260 to 380–400	Pesticides (e.g., chlorpyrifos, lindane, pentachlorophenol) plasticizers (e.g., phthalates), synthetic musk compounds
POM	Particulate organic matter	>380	Pesticides (e.g., pyrethroids), polycyclic aromatic hydrocarbons

2
Definitions, Origin, and Properties of House Dust

"Dust" is the description for the disperse distribution of solid material in the gas phase especially in air [6]. There is no final definition for house dust, but, according to a German norm, house dust is the material deposited indoors (floor dust) [7]. As floor dust is mainly obtained by vacuuming, the contents of vacuum cleaner bags is often denoted "house dust".

House dust is a heterogeneous material consisting of a variety of inorganic and organic particles and fibers of different size. Such particles and fibers can for example be soil, fiber, and abrasion from textiles, combustion products, dead organic material, pollen, and living organisms (bacteria, fungi, pollen). Sources of house dust are diverse. They include material tracked indoors from the outdoor environment, compounds deposited after entry of contaminated outdoor air, as well as indoor sources. Indoor sources are the occupants themselves (hair, skin scales), their pets, and their activities, debris of combustion processes, building materials, and furnishings as well as biological material (pollen, insect parts). Quantity and composition of house dust varies greatly with environmental factors and with factors like age of the house, building material, furniture, quantity

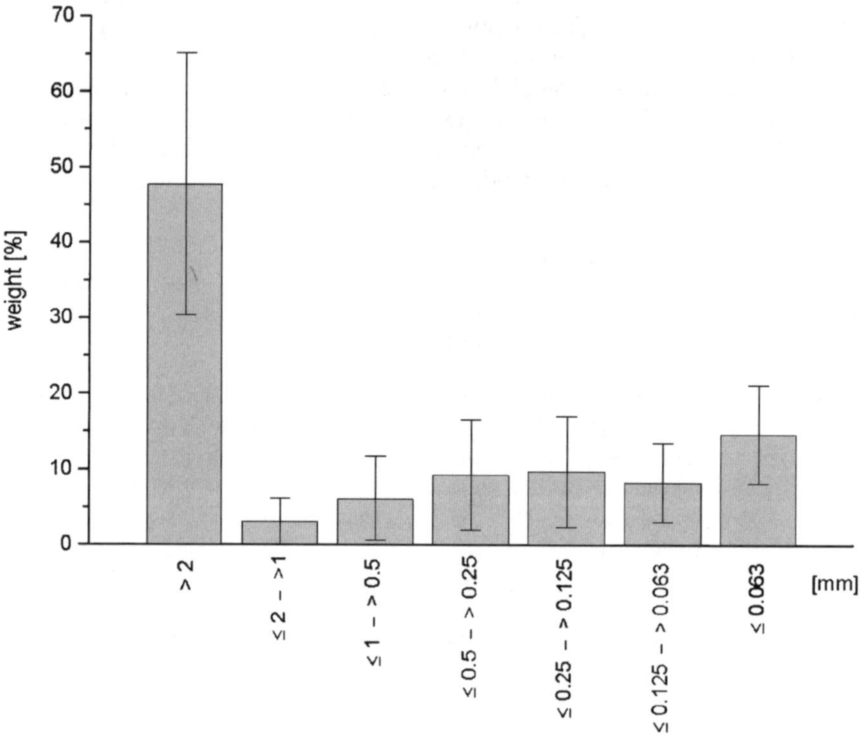

Fig. 1 Particle size distribution of house dust (n=10); the *error bars* represent the standard errors of the classes (after [10])

of carpets, as well as their state of preservation. It further varies for example with ventilation systems and cleaning habits.

House dust may predominantly consist of inorganic or of organic matter. Taking the loss of combustion as a measure for the organic content it may vary between 5% and 95% [8]. Pollutants released from activities and materials in the home or carried in from outside are adsorbed to house dust particles and fibers. Adsorption is governed on the one hand by the properties of the dust (size of the particles, content of organic matter, polarity, etc.) and on the other hand by the physico-chemical properties of the chemical adsorbed (volatility, lipophilicity, polarity).

Besides differences in origin and chemical composition of house dust there is a high variation in particle size, shape, density, and porosity. Examples for the particle size distribution of house dust are given by Que Hee et al. [9] and in Fig. 1 [10].

3 Factors Influencing Analytical Results

3.1 Sampling Methods for House Dust

As already mentioned house dust may be of different origin and of different composition. Furthermore, samples referred to as "house dust" vary significantly. Sampling of house dust may be performed using passive or active methods. Passive sampling is performed in the homes with stationary placed beakers [11, 12] or plates [13] harvesting the dust deposit after a certain time (up to one year [11]).

Active sampling may either be performed by vacuuming or by collecting the dust deposit from window sills and other higher sited horizontal surfaces by wiping [14], brushing [15], or sweeping [7]. In contrast to floor dust passively deposited suspended particulate contains no coarse material; over 99% of the particles are reported to be <50 μm in diameter [13].

Standard methods for sampling floor dust have been issued in the United States [16] and in Germany [7]. For the ASTM (American Society for Testing and Materials) method a "High Volume Small Surface Sampler" (HVS3) is used collecting dust samples of 2–100 g from carpets and bare floors [16]. With this procedure particles >5 μm are obtained, but smaller particles are also collected if stuck to greater matter. With the German standard method several strategies to sample dust for chemical analysis are possible. Methods include passive and active sampling; for active sampling wiping, sweeping, and vacuuming floor dust are feasible. For samples of floor dust a method using a vacuum cleaner with a sampling device built in the vacuum cleaner pipe is favored. With this method exactly seven-days-old dust is collected onto a glass fiber filter.

Other methods of sampling house dust are compiled by Seifert [17]. It is obvious that different sampling methods result in different analytical results for the compounds of interest [18]. In any case house dust analyzed as an indicator for an indoor contamination should be at least 14 days old, because only after that period an equilibrium between house dust and the indoor environment is assumed [19].

3.2
Processing the Samples and Analytical Methods

Neither for processing house dust samples nor for the analyses of the contaminants standard are methods available. Methods published include those for either the complete house dust sample obtained, or a certain subsample in order to make the inhomogeneous dust more suitable for analysis. The latter may consist of either the house dust after discarding larger particles like hair and coarse material or "pieces of fluff" or a certain sieve-fraction of the dust. On sieving coarse material is removed and a separation of particles from fiber is performed. For instance the dust after sieving to ≤2 mm [20], ≤1 mm [21], ≤300 μm [22], ≤150 μm [23], ≤106 μm [24], and ≤63 μm [19] has been used for residue analysis.

Analytical results vary significantly with the material denoted as "house dust". This has been reported for pollutants like polycyclic aromatic hydrocarbons (PAHs) [25] as well as for biocides like permethrin [8, 26] or pentachlorophenol [27]. In general concentrations in house dust increase with decreasing particle size [25–27], with a dramatic increase of PAH for very small particle sizes like 4–25 μm and <4 μm [25]. The analytical precision tends to increase with decreasing particle size of the dust fraction [10].

There are no approved methods to analyze residues or pollutants in house dust. However, techniques applied are often those known to be reliable for sediments or sewage sludge. For analysis of inorganic compounds samples are for example dissolved by nitric acid or aqua regia quantifying the elements in solution (after filtration) by atomic absorption spectrometry (AAS) or atomic emission spectrometry with inductively coupled plasma (ICP-AES). For analysis of organic compounds house dust is mostly extracted with organic solvents like hexane, toluene, etc. The suspension is then either filtered or centrifuged and the dissolved organic components of interest are quantified by chromatographic techniques, either with or without sample clean-up. A review of analytical methods to determine biocides in house dust has been recently compiled by Butte [27].

4
House Dust: a Sink for Semi-Volatile Organic Compounds

4.1
Xenobiotics in House Dust

House dust is an ideal material to screen for an indoor contamination as it is a sink and reservoir for compounds of both natural and man-made origin. Organic compounds of natural origin include fatty acids and their esters, aliphatic hydrocarbons, aldehydes and ketones, and phenols [24], as well as for example steroids produced by fungi [28].

Man-made contaminants investigated so far are biocides (mostly insecticides and fungicides), plasticizers, flame retardants, as well as detergents and cleaning agents.

Biocides may be tracked into the indoor environment after an outdoor application [29], they might be brought directly into the homes as a constituent of impregnated textiles, carpets, or timber, and they may have formed residues after fighting pests. Biocides most often found in house dust are chlorinated hydrocarbons like DDT, lindane, methoxychlor, and chlordane [8, 19, 30–33], pyrethroids like cyfluthrin, cypermethrin, and permethrin [19, 20, 26, 32, 33], organophosphorus compounds like chlorpyrifos, diazinon, dichlorvos, isofenfos, and malathion [14, 19, 32–34], carbamates like propoxur [19, 32, 33], and chlorophenols like pentachlorophenol [8, 19, 20, 32, 35, 36]. The latter is detectable in more than 95% of dusts taken from German households [19].

Only a few data have been published on flame retardants like tris-(2-chloroethyl)-phosphate (TECP) [37] and detergents [21].

Plasticizers like phthalates are man-made compounds that might form residues in house dust of rather high concentrations, i.e., in the order of magnitude of some grams per kilogram [32, 38].

Besides compounds connected to a certain application or outfit of home equipment, residues of persistent and ubiquitously distributed environmental compounds have been of interest in house dust, e.g., polychlorinated biphenyls (PCBs) [32], PAHs [39, 40], and polychlorinated dioxins and furans [41].

Concentrations of biocides in house dust observed after certain applications and concentrations indicating the ubiquitous indoor occurrence, i.e., the widespread distribution of biocides in all media, are given by Butte [27].

4.2 Health Effects Associated with Contaminants in House Dust

On the one hand indoor contamination is one source of human exposure to toxic pollutants and has been classified as a high environmental risk [2, 42]; on the other hand house dust may serve as an indicator for indoor contamination. It is thus not surprising that health effects have been associated with contaminants in house dust. These include childhood leukemia [43], developmental inhibition [2], reduction in motor skills, and coordination and attention disorders [44]. The hypothesis that high residues of biocides in house dust are linked to acute and chronic lymphatic leukemia could neither be rejected nor proven up to now [45].

Compounds adsorbed to house dust may enter the human body by ingestion of particles adhering to food and the skin, non-dietary ingestion (especially infants, toddlers), absorption through the skin, and inhalation of suspended and re-suspended particles. Walking on carpets, cleaning, and vacuuming are typical human activities for re-suspending house dust with particles sizes of 5–25 μm [25].

Particles with an aerodynamic diameter of <10 μm are known to be inhalable, those of <2.5 μm are respirable [25]. Depending on the aerodynamic diameter particles are deposited in different regions of the alveolar tract as shown in Fig. 2.

However, exposure to house dust does not exclusively and may not even predominantly occur via inhalation. For instance ingestion of house dust particles adhering to food objects and the skin or direct absorption through the skin may be primary routes of exposure [36]. These pathways of intake are especially important for small children as they have the tendency to be very tactile and handle and place non-food objects into their mouth [14]. Data from the "National Human Exposure Assessment Survey" of Arizona support the importance of

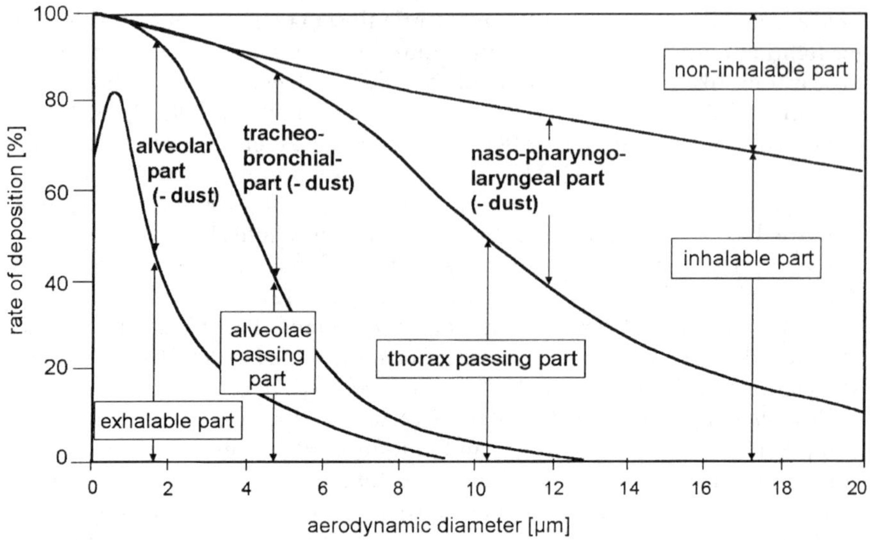

Fig. 2 Deposition rate of dust in the respiratory tract depending on the aerodynamic diameter (after [6])

dermal penetration of semi-volatile compounds as a route of residential human exposure [34]. Especially particles of <100–200 µm most efficiently transfer to and are retained by skin [25].

It may be assumed that indoor air and house dust are not the only sources for contaminants that contribute to an elevated intake for the residents. With other diffuse sources in the indoor environment as well, household dust can be considered as the most important indicator for the state of the indoor contamination by semi-volatile and non-volatile compounds.

5 Synthetic Musks in House Dust

5.1 Analytical Methods for Synthetic Musks in House Dust

5.1.1 *Nitro Musks: Detection by GC-ECD and GC-MS*

Many methods have been published on the analysis of synthetic musk compounds in the outdoor environment [46, 47], but analytical methods for indoors are limited. Up to now only one method for synthetic musks in indoor air [48] and one for nitro musks in house dust [49] are available. The method to quantify nitro musks in house dust is based on gas chromatography with an ECD [49]. This detector proved to be rather selective and very sensitive with a limit of determination of 5 µg L^{-1} in the final extract for each of the nitro musks (calculated by the calibration graph method [50]), i.e., a limit of determination of 0.1 mg kg^{-1} with respect to house dust. For routine analysis 50 mg of the size fraction ≤63 µm were suspended in 1 mL of *n*-hexane; 2-methyl-4,6-dinitronanisol served as internal standard. The sample was extracted in an ultrasonic bath and centrifuged; 1 µL of the supernatant was used for GC. Examples for chromatograms of a calibration solution of 5 µg L^{-1} compared to a chromatogram of a dust extract and a dust extract spiked with 100 µg L^{-1} of each of the nitro musks are given in Fig. 3.

Although the ECD method was sensitive enough to analyze ubiquitous concentrations of nitro musks in house dust, problems arose from the co-extraction and co-elution of other contaminants, especially phthalates. Thus, it seemed advisable to use a more specific detector like an MSD to prove the results obtained by ECD. For the evaluation of the GC-MS method nearly identical chromatographic conditions were applied, but a mass selective detector (quadrupole MSD) with EI in the single ion mode (SIM) instead of the ECD was used. Two to three characteristic fragment ions (m/z) for each of the nitro musks were chosen for quantification, i.e., MA: 253, 254, and 268; MX: 128 and 282; MM: 263, 264, and 278; MT: 251, 255, and 266; MK: 279, 280, and 294; and IS: 165, 182, and 212. Analyses were performed by a Hewlett Packard GC-MSD system (HP 5890 Series II GC; 5971A-MSD and autosampler HP 7673) equipped with a 30-m HP 5

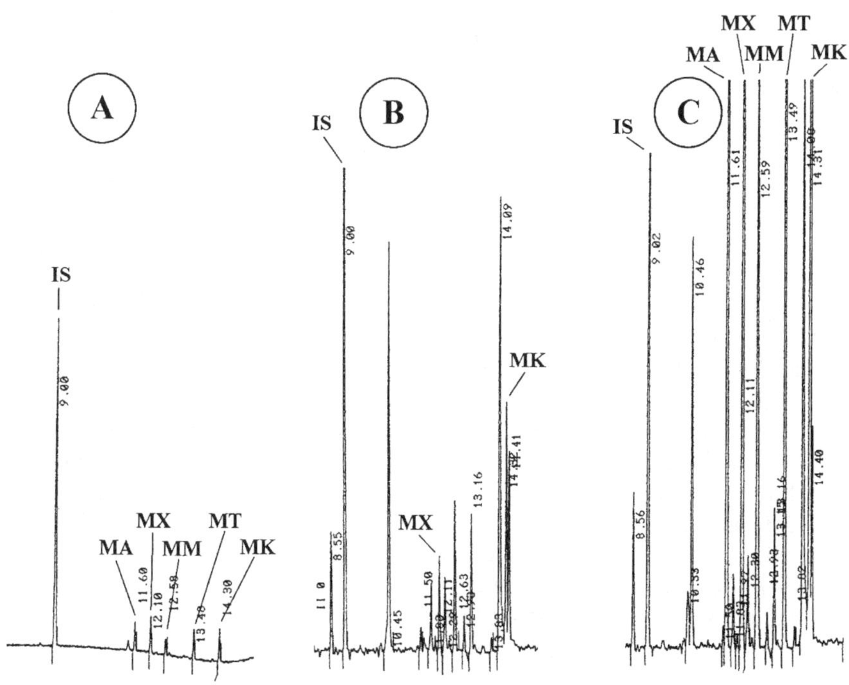

Fig. 3 A–C ECD chromatograms of nitro musks in standard solutions and house dust extracts: A calibration solution of 5 µg L^{-1}; B dust extract; C dust extract spiked with a 100 µg L^{-1} solution of nitro musks (IS: internal standard)

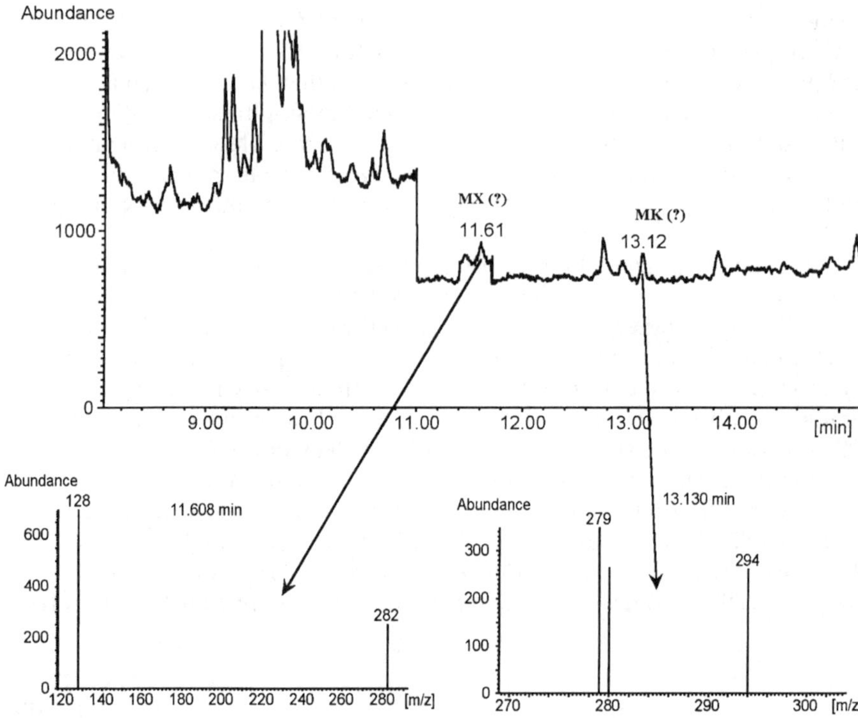

Fig. 4 MSD chromatogram of a dust extract (*upper part*); abundance of the fragment ions for the two peaks (*lower part*)

MS capillary column. The following temperature program for the GC oven was used: start:150 °C (5 min), heat at 7 °C min^{-1} to 290 °C, hold 5 min.

Examples of chromatograms obtained by the GC-MS method are displayed in Figs. 4 and 5.

Comparing the chromatograms it is evident that the compounds supposed to be MK and MX show nearly (but not exactly) the same retention times as the standards, and the abundance of the fragment ions is not identical to that of the standards. Without any further information, for example chromatography with capillary columns of other polarity, it could not be decided unambiguously whether these peaks really have to be identified as MX and MK. Furthermore the signal to noise ratio, a multiple of it taken as limit of detection [51, 52], proved to be inferior for EI-MS compared to ECD.

As the analytical method using a quadrupole MS in the SIM and EI mode neither improved the limit of detection nor the specificity for nitro musks in house dust samples it was not used in routine analysis.

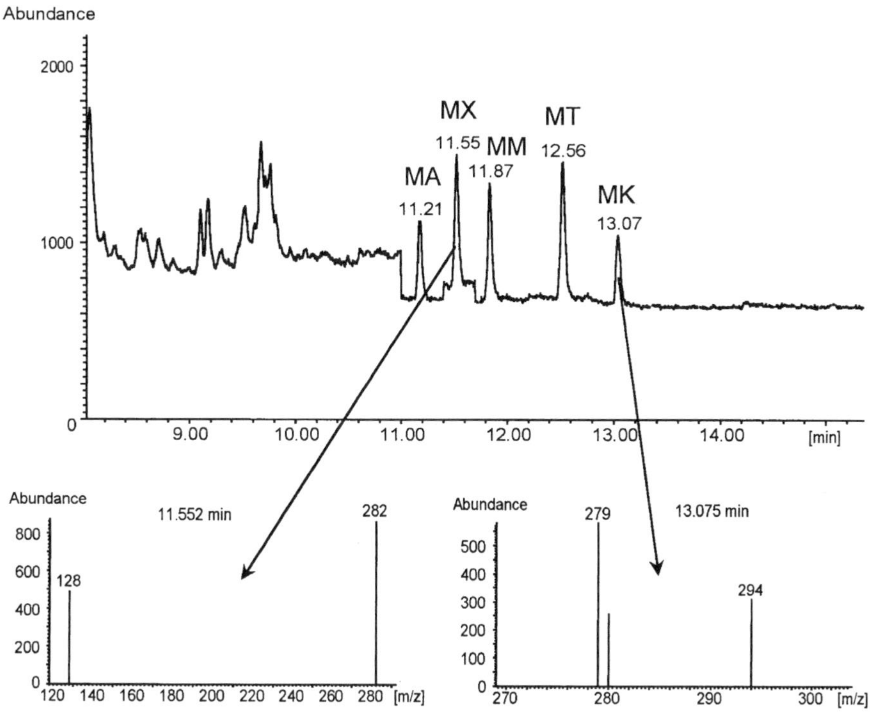

Fig. 5 MSD chromatogram of a dust extract spiked with 100 µg l^{-1} of nitro musks (*upper part*); abundance of the fragment ions for MX and MK (*lower part*)

5.1.2
Nitro Musks and Polycyclic Musks: Detection by GC-MS/MS

GC-MS with a quadrupole MS did not match the sensitivity of the ECD method; thus one could not analyze a background contamination of nitro musks in house dust. On the other hand an ECD is not applicable for the analysis of polycyclic musks in house dust. They do not contain electron capturing groups or elements and, therefore, are "invisible" for an ECD. To overcome these drawbacks it seemed to be advisable to use an MS/MS detector for both groups of musk compounds as already reported for the analysis of polycyclic musks in human fat and milk [53]. MS/MS detection has several advantages such as the high specificity of the ion trap MS detector and the improvement of the signal to noise ratio by a second fragmentation step. Analyses were performed by a Varian Saturn 3 system (ion trap detector with MS/MS option, Varian Series 3400 GC and autosampler Varian 8200) equipped with a 30-m Optima δ-6 capillary column (Macherey & Nagel, Germany). The following temperature program for the GC oven was used: start: 50 °C (1 min), heat at 30 °C min^{-1} to 140 °C and then at 5 °C min^{-1} to 260 °C, hold for 5 min. Injections were performed splitless at 260 °C (transfer-line: 280 °C). Table 2 compiles the conditions used for the MS/MS method.

Table 2 Conditions for the MS/MS detection of synthetic musks in house dust

Substance	Parention mass (m/z)	CID resonant (V)	Recorded mass range (m/z)
ADBI, AHDI	229	1.8	100–250
IS	182	1.6	50–200
HHCB, AHTN	243	1.8	100–300
MA	253	2.5	100–260
MX	282	2.7	210–290
MM	263	1.8	150–270
MT	251	2.5	100–260
MK	279	2.8	100–280

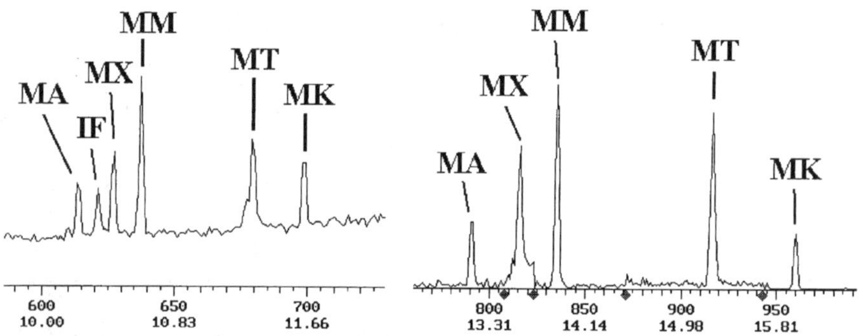

Fig. 6 Total ion chromatograms of nitro musks (50 μg L^{-1} each) using an ion trap system (Varian Saturn 3); *left*: MS mode, *right* MS/MS mode (IF: interfering substance)

The mass window for each parent ion was set to 3 m/z. For the analysis of nitro and polycyclic musks in house dust all the daughter ions obtained from the parent ions were recorded. The improvement of the signal to noise ratio changing from the MS to the MS/MS mode is shown in Fig. 6.

The complete daughter ion mass spectrum that can be obtained even for peaks representing low concentrations leads to a specificity not matched by any other method. Figures 7 and 8 display the daughter ions of ADBI (parent ion: m/z=229) and AHTN (parent ion: m/z=243) comparing a calibration solution (Fig. 7) to a dust extract (Fig. 8). Even though the concentration of ADBI in this house dust sample is just 0.53 mg kg^{-1}, the fragment ion pattern in ADBI of dust is identical to that of the standard.

Validation data of the analytical method including capillary GC separation and MS/MS detection are compiled in Table 3.

MA is not included in this table as all measurements showed an inexplicable high variance. As MA has been forbidden in Germany since 1997 [54] and thus is not supposed to be present in house dust, reasons for the high variance were not further evaluated.

Comparing the three analytical methods described for synthetic musks in house dust, it can be concluded that the MS/MS method is to be preferred to the

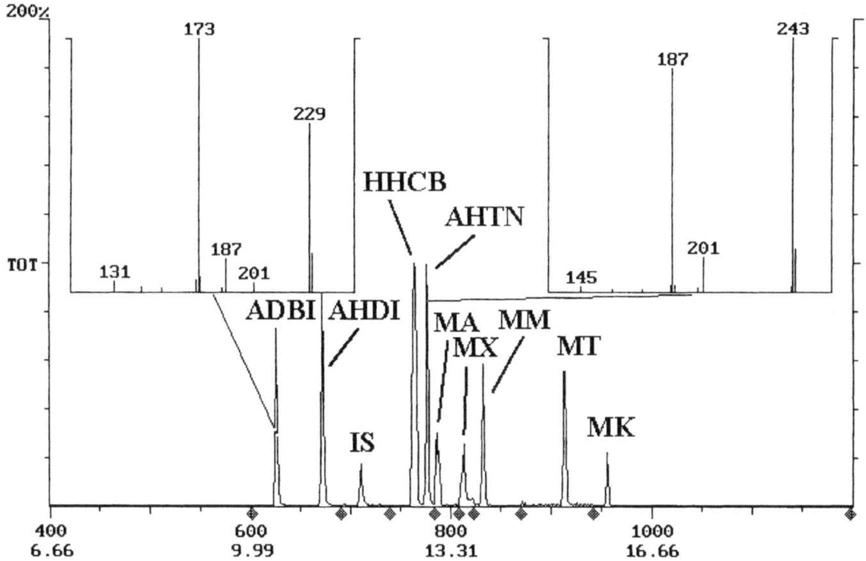

Fig. 7 Total ion chromatograms of a calibration solution of synthetic musks (100 μg L^{-1} each) in MS/MS mode; *enclosed left*: daughter ions of ADBI, *enclosed right*: daughter ions of AHTN

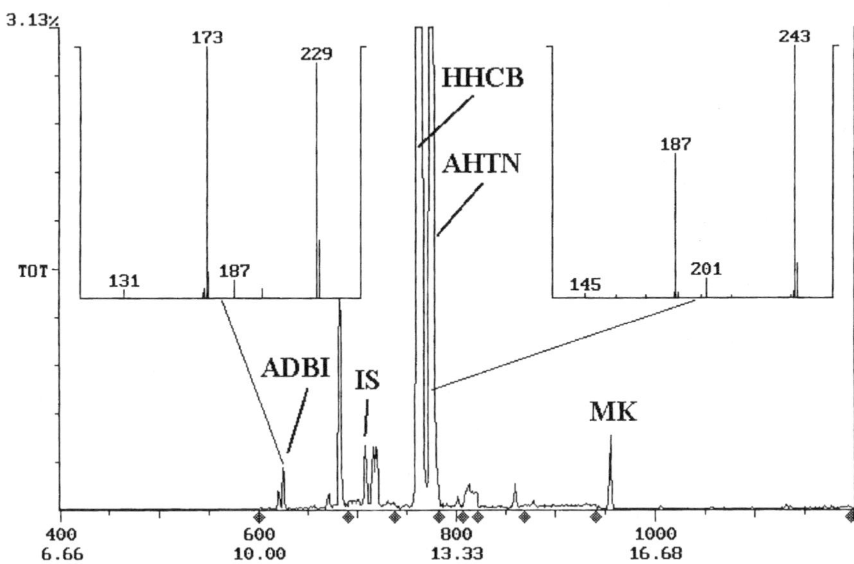

Fig. 8 Total ion chromatogram of a dust extract containing high amounts of HHCB and AHTN and a low amount of ADBI; *enclosed left*: daughter ions of ADBI, *enclosed right*: daughter ions of AHTN

Table 3 Validation of the GC-MS/MS method to detect synthetic musks in house dust

Compound	Limit of detection[a] [mg kg^{-1}]	Between day precision[b] [%]	Recovery rate[c] [%]
ADBI	0.18	8.1	81.5
AHDI	0.16	7.1	89.7
HHCB	0.2	5.7	86.3
AHTN	0.2	5.1	82.9
MX	0.44	9.8	92.4
MM	0.22	4.7	90.1
MT	0.12	6.1	94.4
MK	0.24	5.0	97.1

[a] Calculated by the calibration graph method [50].
[b] Coefficient of variation for a concentration of 2 mg kg^{-1} (n = 10).
[c] For samples fortified to a concentration of 2 mg kg^{-1} (n = 6).

other methods as it is applicable for polycyclic as well as for nitro musks. However, regarding the limits of detection or determination, an ECD analysis is still superior for nitro musks.

5.2 Occurrence of Synthetic Musks in House Dust

With house dust as an indicator for an indoor contamination, an impression of the today's occurrence of synthetic musks in house dust is reported here. At random, 35 samples were chosen from more than 250 house dust samples (vacuum cleaner bags) obtained in connection with the "Norddeutsche Leukämie- und Lymphomstudie" (NLL, North German Leukemia and Lymphoma Study) in 1998 – 1999. Sample processing consisted of sieving the content of the vacuum cleaner bags to <63 μm, discarding coarse matter and fiber. Results listed in Table 4 are thus related to the <63 μm fraction of house dust.

Concerning nitro musks, only MX and MK were present in nearly each sample, but concentrations never exceeded a few milligrams per kilogram, whereas polycyclic musks, especially HHCB and AHTN, showed rather high residues with concentrations up to nearly 0.1 g kg^{-1}. Concentrations of endocrine disrupting substances in samples of the NLL were reported elsewhere [38]; compounds evaluated included pesticides like pentachlorophenol (PCP), DDT, methoxychlor, permethrin, phenols like octylphenol, nonylphenol, and bisphenol A, as well as phthalates like dibutylphthalate, benzylbutylphthalate, and di(2-ethylhexyl) phthalate (DEHP). Regarding the 35 house dust samples for which results for synthetic musks were obtained (Table 4), no correlation was found between either biocides, phenols, or phthalates and polycyclic musks. However, perhaps just by chance, the house dust sample in this subset of samples containing the highest concentration of HHCB (77 mg kg^{-1}) was that with the highest concentration of AHTN (94 mg kg^{-1}), DEHP (2800 mg kg^{-1}), and permethrin (110 mg kg^{-1}). However, on the other hand concentrations of other compounds measured in this sample were not pecu-

Table 4 Concentrations of synthetic musks in house dust [mg kg^{-1}]

Compound	Minimum	Median	Maximum
ADBI[a]	<0.18	<0.18	0.53
AHDI[a]	<0.16	<0.16	0.31
HHCB[a]	<0.2	0.59	77
AHTN[a]	0.34	0.69	94
MA[b]	<0.1	<0.1	0.14
MX[b]	<0.1	0.26	1.4
MM[b]	<0.1	<0.1	<0.1
MT[b]	<0.1	<0.1	0.2
MK[b]	<0.1	0.45	3.8

[a] GC-MS/MS (n = 25).
[b] GC-ECD (n = 10).

liar, i.e., the concentrations of PCP, DDT, methoxychlor, etc. were lower than 1 mg kg^{-1}, and the concentrations of the other phthalates were up to about 100 mg kg^{-1}.

Synthetic musks are widespread in the environment [46, 47, 56], but data describing residues of musks the indoor environment are rare. There are some results for indoor air [48, 57] but regarding house dust no data are available. Kallenborn et al. [48], analyzing one sample of indoor air taken in a laboratory, reported an MX and an MK level of about 0.5 ng m^{-3} and of about 0.1 ng m^{-3}; HHCB and AHTN concentrations were about 4.8 ng m^{-3} and about 1.1 ng m^{-3}, respectively.

Fromme and co-workers [57], who analyzed six polycyclic musks and the nitro musks MX and MK in 38 air samples from kindergartens in Berlin taken in October to December 2000 found much higher concentrations. Whereas only one air sample each contained MX or MK (both about 10 ng m^{-3}), ADBI could be detected in about one-third of the air samples (concentrations up to 21 ng m^{-3}) and AHDI, AHTN, and HHCB were present in almost every air sample. AHDI showed concentrations up to 49 ng m^{-3}, AHTN up to 90 ng m^{-3}, HHCB up to 299 ng m^{-3}, respectively. Unfortunately, measurements for synthetic musks in house dust are not available, and thus a correlation of results for air and dust is not possible.

In general, a change in the utilization of synthetic musks from nitro musks to polycyclic musks is assumed; e.g., Käfferlein and Angerer [55] reported that nitro musks concentrations in human plasma like MX decreased from 0.24 µg L^{-1} in 1992 to <0.1 µg L^{-1} in 1998 (medians). The authors explained this by the discontinued use of MX in Germany since 1993. Synthetic musks are also present in human adipose tissue [58] with highest concentrations of MX (up to 288 µg kg^{-1}) regarding nitro musks and highest concentrations of HHCB (up to 171 µg kg^{-1}) with respect to the polycyclic musks. The authors conclude that food is the main source for these residues, but there is evidence that the percutaneous resorption from cosmetics and from washing powder or fabric softeners adhering to cloths is the major source [46]. Data supporting the amount of chemicals absorbed by dermal contact of contaminated house dust are still lacking, although results obtained for chlorpyrifos and diazinon suggest the importance of dermal penetration [59].

Up to now data on synthetic musks in house dust are limited; a correlation of results for house dust and indoor air is still lacking. However, synthetic musks present indoors may contribute significantly to an intake by the occupants, as the indoor environment is an important source for the exposure of the general population [2]. Studies have to be performed to describe the indoor contamination and to obtain more information on sources for the occurrence of synthetic musks indoors as well as on possible indoor exposure paths.

Acknowledgements The help of Dipl. Chem. Sven Heekmann (University of Basel), Dr. Angelika Mraz (University of Lodz), and Dr. Anke Schmidt (EUKOS GmbH, Plön) in performing the experiments described is gratefully acknowledged. Some of the work on nitro musks is part of the diploma thesis of Sven Heekmann; the evaluation of the MS methods was performed by Angelika Mraz and Dr. Anke Schmidt.

6 References

1. WHO (1999) Environment and health research for Europe. Background Document to the 3rd Ministerial Conference of Environment and Health (London, 16–18 June 1999). WHO Regional Office for Europe, Copenhagen
2. Roberts JW, Dickey P (1995) Rev Environ Contam Toxicol 143:59
3. WHO (1989) Indoor air quality: organic pollutants. Euro Report and Studies No 111. WHO Regional Office for Europe, Copenhagen
4. Rippen G (2000) Handbuch Umweltchemikalien, ecomed, Landsberg
5. van de Plassche EJ, Balk F (1997) Environmental risk assessment of the polycyclic musks AHTN and HHCB according to the EU-TGD. Rep 601503 008, RIVM, Bilthoven, The Netherlands
6. Deutsche Forschungsgemeinschaft (1999) MAK- und BAT-Werte Liste, Wiley-VCH, Weinheim, p 157
7. Verein deutscher Ingenieure (2001) Messen von Innenraumluftverunreinigungen. Probenahme von Hausstaub, VDI 4300, part 8
8. Butte W, Walker G (1994) VDI-Berichte 1122:535
9. Que Hee SS, Peace B, Clarke CS, Boyle JR, Bornschein RL, Hammond PB (1985) Environ Res 38:77
10. Butte W, Heinzow B, Petzold G, Hensen D (2000) Hausstaub: Indikator zum Erkennen von chemischen Kontaminationen in Innenräumen. In: Beyer A, Eis D (eds) Praktische Umweltmedizin. Springer, Berlin Heidelberg New York
11. Aurand K, Drews M, Seifert B (1983) Environ Technol Lett 4:433
12. Seifert B, Becker K, Hoffmann K, Krause C, Schulz C (2000) J Expos Anal Environ Epidemiol 10:103
13. Edwards RD, Yurkow EJ, Lioy PJ (1998) Sci Total Environ 224:69
14. Lioy PJ, Edwards RD, Freeman N, Gurunathan S, Pellizzari E, Adgate JL, Quackenboss J, Sexton K (2000) J Expos Anal Environ Epidemiol 10:327
15. Meißner T, Schweinsberg F (1996) Toxicol Lett 88:237
16. American Society for Testing and Materials (1994) Standard practice for collection of dust from carpeted floors for chemical analysis. ASTM Method D 5439–94
17. Seifert B (1998) Bundesgesundhbl 41:383
18. Reynolds SJ, Etre L, Thorne PS, Whitten P, Selim M, Popendorf WJ (1997) Am Ind Hyg Assoc J 58:439
19. Walker G, Hostrup O, Hoffmann W, Butte W (1999) Gefahrst Reinh Luft 59:33
20. Friedrich C, Helm D, Becker K, Hoffmann K, Krause C, Nöllke P, Schwabe R, Seiwert M (1998) Gesundheitswesen 60:95
21. Vejrup KV, Wolkoff P, Madsen JØ (1999) Proceedings of Indoor Air '99 (4), p 244

22. Allermann L, Poulsen OM (1999) Proceedings of Indoor Air '99 (2), p 719
23. Chuang JC, Callahan PJ, Menton RG, Gordon SM, Lewis RG, Wilson NK (1995) Environ Sci Technol 29:494
24. Papadopoulos A, Karayannis M, Knoeppel H (1999) Proceedings of Indoor Air '99 (4), p 107
25. Lewis RG, Fortune CR, Willis RD, Camann DE, Antley JT (1999) Environ Health Perspect 107:721
26. Stolz P, Meierhenrich U, Krooß J, Weis N (1996) VDI-Berichte 1257:789
27. Butte W (1999) Occurrence of biocides in the indoor environment. In: Salthammer T (ed) Organic indoor air pollutants. Wiley-VCH, Weinheim, p 233
28. Saraf A, Larsson L, Burge H, Milton D (1997) Appl Environ Microbiol 63:2554
29. Lewis RG, Nishioka MG (1999) Proceedings of Indoor Air '99 (2), p 416
30. Roßkamp E, Horn W, Ullrich D, Seifert B (1999) Umweltmed Forsch Prax 4:354
31. Baudisch C, Prösch J (2000) Umweltmed Forsch Prax 5:161
32. Pöhner A, Simrock S, Thumulla J, Weber S, Wirkner T (1998) Z Umweltmed 6:337
33. Roinestad KS, Louis HN, Rosen JD (1993) J Assoc Off Anal Chem Int 76:1121
34. Gordon SM, Callahan PJ, Nishioka MG, Brinkman MC, O'Rourke MK, Lebowitz MD, Moschandreas DJ (1999) J Expos Anal Environ Epidemiol 9:456
35. Liebl B, Mayer R, Kaschube M, Wächter H (1996) Gesundheitswesen 58:332
36. Lewis RG, Fortmann RC, Camann DE (1994) Arch Environ Contam Toxicol 26:37
37. Sagunski H, Ingerowski G, Mattulat A, Scheutwinkel M (1997) Umweltmed Forsch Prax 2:185
38. Butte W, Hoffmann W, Hostrup O, Schmidt A, Walker G (2001) Gefahrst Reinh Luft 61:19
39. Dieckow P, Ullrich D, Seifert B (1999) Vorkommen von polyzyklischen aromatischen Kohlenwasserstoffen (PAK) in Wohnungen mit Parkettböden. WaBoLu Hefte 2/99, Umweltbundesamt, Berlin
40. Simrock S (1998) Z Umweltmed 6:243
41. Wittsiepe J, Ewers U, Mergner H-J, Lahm B, Hansen D, Volland G, Schrey P (1997) Zbl Hyg Umweltmed 199:537
42. Ott WR, Roberts JW (1998) Sci Am 278:72
43. Daniels JL, Olshan AF, Savitz DA (1997) Environ Health Persp 105:1068
44. Guilette LJ, Meza MM, Aquilar MG, Soto AD, Garcia IE (1998) Environ Health
45. Hoffmann W, Hostrup O (1997) Umwelt Gesundh 8:140
46. Rimkus GG (1998) Umwelt Forsch Prax 3:341
47. Rimkus GG (1999) Toxicol Lett 111:37
48. Kallenborn R, Gatermann R, Planting S, Rimkus G, Lund M, Schlabach M, Burkow IC (1999) J Chromatogr A 846:295
49. Heekmann S, Butte W (2000) Analytik von synthetischen Nitromoschusverbindungen im Hausstaub, Poster, Analytica Conference 2000, München
50. DIN 32645 (1994) Nachweis-, Erfassungs- und Bestimmungsgrenze (Detection Limit, Decision Limit, and Determination Limit) Beuth, Berlin
51. Büttner J, Borth R, Boutwell HJ, Broughton PMG, Bowyer RC (1980) J Clin Chem Clin Biochem 18:78
52. Angerer J, Schaller KH (eds) (1985) Analyses of hazardous substances in biological material, vol 1. DFG German Science Foundation. VCH, Weinheim, p 9
53. Eschke H-D, Dibowski H-J, Traud J (1995) Dtsch Lebensm-Rdsch 91:375
54. Bundesinstitut für Arzneimittel und Medizinprodukte (1997) BAnz 243:15,235
55. Käfferlein HU, Angerer J (1999) Umweltmed Forsch Prax 4:204
56. Kallenborn R, Gatermann R, Rimkus GG (1999) J Environ Monit 1:70 N
57. Fromme H, Lahrz T, Piloty M (2001) Personal communication
58. Müller S, Schmid P, Schlatter C (1996) Chemosphere 33:17
59. Gordon SM, Callahan PJ, Nishioka MG, Brinkman MC, O'Rourke MK, Lebowitz MD, Moschandreas MJ (1999) J Exposure Anal Environ Epidemiol 9:456

The Handbook of Environmental Chemistry Vol. 3, Part X (2004): 123–150
DOI 10.1007/b14122

Synthetic Musks in the Aquatic System of Berlin as an Example for Urban Ecosystems

Thomas Heberer[1] · Susanne Jürgensen[2] · Hermann Fromme[3]

[1] Institute of Food Chemistry, Technical University of Berlin, Sekr. TIB4/3-1, Gustav-Meyer-Allee 25, 13355 Berlin, Germany
E-mail: musks@foodchemistry.de
[2] Berlin Fishery Board, Havelchaussee 149/151, 14055 Berlin, Germany
[3] Senate of Berlin, Department for Labour, Social Services and Women's Issues, Environmental Health Unit, Oranienstr. 106, 10969 Berlin, Germany

Abstract This chapter compiles some data on the occurrence and fate of synthetic musk compounds in the aquatic environment of Berlin, Germany. The city of Berlin and its suburbs represent a highly urbanized area with low surface water flows and high amounts of raw sewage produced by a population of around 4 million people. The treated sewage effluents from the municipal sewage treatment plants, containing up to 20 µg L^{-1} of synthetic musk compounds, are discharged into rivers and canals in several of Berlin's central districts. The low surface water flows result in high proportions of raw sewage and in high concentrations of musk compounds in the waterways. The concentrations of synthetic musks upstream of the Berlin waters are negligible. Thus, Berlin's surface waters can be divided into areas with no or only very low contamination and areas where synthetic musks are found at µg L^{-1} concentrations. The investigations for synthetic musk compounds carried out in Berlin are unique because of the variety of matrices (sewage effluents, sewage sludge, surface water, sediments, and fish) and the large number of samples that have been analyzed over a long time period (1991–2000). For example, more than 1500 fish samples were analyzed for musk xylene residues in the years between 1991 and 1998. The large number of samples was found to be indispensable for a reliable statistical evaluation of the occurrence data. Moreover, long-term investigations of these compounds in fish also allow to assess some timely trends of contamination by synthetic musks.

Keywords Contamination · Sewage · Sewage sludge · Surface water · Sediments · Fish tissues

1 Description of the Berlin Situation

1.1 The Berlin Area

The contamination of the surface waters by municipal waste water discharges is of great relevance in the Berlin area. Thus, high concentrations of sewage contaminants are found in the receiving waters because of the high proportions of municipal sewage effluents in the surface waters. These high contamination levels result from low surface water flows and from the large amounts of raw sewage produced by its population of around four million people. Being situated in a densely populated area and considering the associated burden of waste contaminants, the intensive use of the waterways and their high nutrient levels, Berlin's waters are subject to high ecological pressure.

Due to the high contributions of bank filtration and artificial ground water enrichment of approximately 75% in Berlin's drinking water production [1], polar organic compounds such as residues from pharmaceutically active compounds (PhACs) and pesticides may at contaminated sites also leach into the groundwater aquifers [2, 3]. Thus, several polar organic contaminants have been detected at concentrations up to the low $\mu g\ L^{-1}$-level in wells close to the contaminated surface waters [2, 3]. Nonpolar contaminants such as synthetic musk compounds should, however, be adsorbed to the sediments. Therefore, it is unlikely that they will leach through the subsoil. On the other hand, these nonpolar compounds can accumulate in sewage sludge, in sediments, and in fish and other aquatic biota. The accumulation of the musk compounds in fish is of great importance both for Berlin's fisheries and the consumers (also refer to the next section).

Figure 1 shows a map of Berlin and its important lakes, rivers and canals. This map also shows the locations and drainages of the sewage treatment plants (STPs). The municipal area of Berlin includes in total an area of 889 km^2. 59 km^2 are used for agriculture, 155 km^2 are forests and a total water area of 57 km^2 consists of several rivers, approximately 60 lakes (>1 ha) and about 500 natural ponds [1, 4, 5]. In addition, 54 km^2 (5.7% of the Berlin area) are used as freshwater fishing areas. Berlin's most important rivers Spree and Havel are typical shallow and slow flowing lowland rivers with a mean water depth of 7 m (max. depth: 16 m) [4]. The river Spree enters Berlin from the south-east, passes through the whole

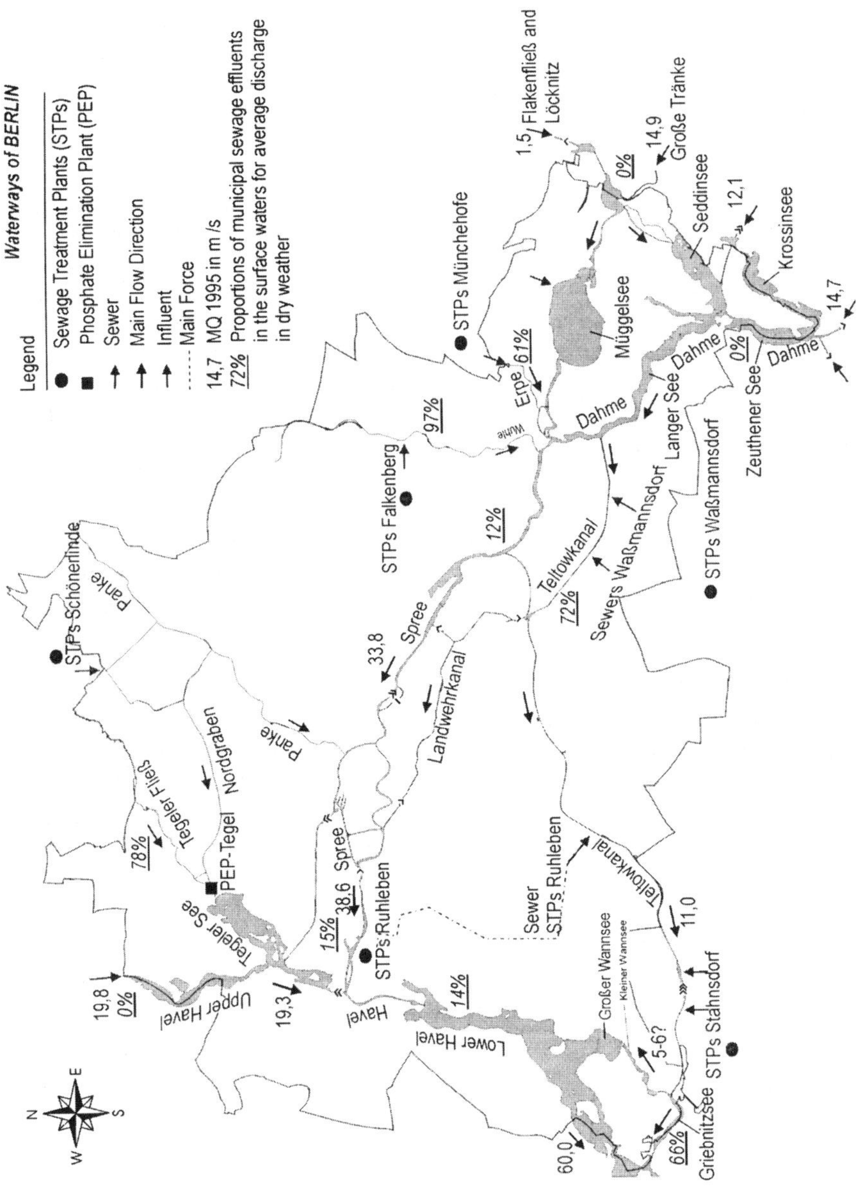

Fig. 1 Map of Berlin showing the rivers, canals and lakes of Berlin and the locations of the STPs (Schönerlinde, Falkenberg, Münchehofe, Waßmannsdorf, Marienfelde (phased out in 1999), Stahnsdorf, and Ruhleben). Reprinted with permission from the Berlin Fishery Board

city and discharges into the river Havel in the north-western districts of Berlin. The river Havel enters Berlin from the north-west and leaves the town in south-west direction.

The municipal sewage produced in the Berlin area is purified by six (formerly seven) STPs. With regard to their capacities, Berlins most important STPs are located south of Berlin in Waßmannsdorf and in the north-west of Berlin in Ruhleben. During autumn and winter time, the sewage effluents from the STP in Ruhleben are discharged into the River Spree which merges with the Upper Havel. From April to September (bathing season), the Teltowkanal, a canal in the south of Berlin (shown in Fig. 1), is fed via a waste-water disposal line (force main) by additional effluents from Berlin's largest STPs in Ruhleben. Table 1 shows the approximate and the maximum capacities for all STPs from the Berlin area. In several small brooks located in inner city districts, downstream of STPs, the mean proportions of sewage in the surface waters are very high. Thus, the brooks Erpe, Panke and Wuhle show mean proportions of 58, 64, and 93% of sewage in the surface water, respectively [5].

The above-mentioned Teltowkanal carries the highest loads of sewage discharges. The Teltowkanal was built between 1901 and 1906 [6]. The canal was used as drainage for rainwater and industrial waste water from districts formerly located outside of Berlin. Additionally, it was used as a shipping canal for industrial supply and as a short cut for the shipping routes between the rivers Oder and Elbe. The canal has a total length of approximately 35 km and connects the rivers Dahme and Havel [6]. Today, it is characterized by high proportions of sewage effluents discharged into the canal by Berlin's two largest STPs in Ruhleben (only during spring and summertime) and Waßmannsdorf. Additionally, the STPs in Marienfelde (phased out in September 1999) and Stahnsdorf discharge their effluents into this canal. At several locations in the Teltowkanal, munici-

Table 1 STPs in the Berlin area, their purification capacities (dry weather data from 1998), and the receiving waters of the sewage effluents [1, 5]

STPs	Capacity ($m^3 d^{-1}$)	Receiving waters
Stahnsdorf	40,000	Teltowkanal → Havel
Münchehofe	55,000	Neuenhagener Fließ → Spree → Havel
Marienfelde (phased out in 1999)	65,000	Teltowkanal → Havel
Schönerlinde	70,000	Nordgraben → Tegler See → Havel Panke → Spree → Havel
Falkenberg	110,000	Wuhle → Spree → Havel
Waßmannsdorf	145,000	Teltowkanal → Havel
Ruhleben[a]	240,000	Teltowkanal → Havel (April–September) Spree → Havel (October–March)

[a] From April to September the effluents from the STP in Ruhleben are discharged into the Teltowkanal. From October to March the effluents are discharged into the river Spree.

pal sewage effluents account for up to 40% of the average surface water flow [5]. Under extreme conditions (dry periods with low surface water flows), the proportions of municipal sewage may reach up to 84% [5]. With regard to contaminations from municipal sewage discharges, the canal represents a scenario that may be called a "worst-case situation".

1.2 Fishery in Berlin

In 1996, commercial and recreational fishery in Berlin was carried out by 13 full-time and 17 part-time enterprises and by approximately 30,000 sports fishermen [4]. The total natural fish production is estimated at about 100 kg fish per ha and year [4]. Fish species of commercial and other interest are the European eel (*Anguilla anguilla* L.), pike perch (*Stizostedion lucioperca* L.), perch (*Perca fluviatilis* L.), roach (*Rutilus rutilus* L.), pike (*Esox lucius* L.), common carp (*Cyprinus carpio* L.), tench (*Tinca tinca* L.), European catfish (*Silurus glanis* L.), and the American crayfish (*Orconectes limosus* Raf.). Fish species such as bream (*Abramis brama* L.) and bleach (*Blicca björkna* L.) are numerous but of lower commercial interest. In 1996, the total annual harvest in Berlin was estimated at 113 tonnes of high valued fish and about 405 tonnes of low valued fish resulting in a total production of 96.5 kg fish per ha and year [4].

Eels are of special importance for Berlin's fishery. In Berlin, juvenile European eels ("glass" eels and elvers) are released into the surface waters because natural and artificial barriers hinder the juvenile eels from reaching these waters. Originally, European eels are spawned in the Sargasso Sea, cross the Atlantic Ocean, enter rivers and lakes of Europe, feed on benthic organisms and in their later lifespan also on smaller fish. The adult eels suffer from insufficient food in the Atlantic Ocean when returning to their spawning grounds which are off the North American coast. Thus, in contrast to the American eels, the European species stores large contents of fat (on average 30–38% in adult eel) in lipid depots to succeed in returning to the spawning area located far from their living areas in Europe [4]. Immediately after arriving at the European coast the European eels start their storage of lipids. After one year in European freshwater, eels have a total length of about 12 to 16 cm. Legal fishing on eels starts at the earliest with 45 cm of total length. Such eels already have fat contents ranging from 22 to 30% in their edible muscle tissues [4].

2 Analytical Methods

All environmental samples (water samples, sewage sludges, aquatic sediments, and fish tissues) were analyzed using analytical methods and standard operation procedures as described in detail elsewhere [7–10]. Suitable internal standards and/or surrogate standards such as musk xylene-d_{15} and AHTN-d_3 (Dr. Ehrenstorfer, Augsburg, Germany) were used for quantification and quality control. In this section, the principles of the procedures applied for the analysis of the different environmental matrices are briefly described.

2.1 Water Samples

In the course of the surface water monitoring in 1996 [7], the water samples were analyzed by manual solid-phase microextraction (SPME) using 100 μm polydimethylsiloxane fibers. SPME with detection by gas chromatography-mass spectrometry (GC-MS) with selected ion monitoring (SIM) was found to be a rapid and reliable tool for the screening of synthetic musks in surface and sewage water samples [7]. Internal calibration with a suitable internal standard enabled reproducible quantification of nitro musks and polycyclic musk compounds in environmental samples down to the low ng L^{-1}-level [7].

In the other studies carried out in 1996 and 1997 [9, 10], simultaneous steam distillation-solvent extraction (SDE) was used for the extraction of five polycyclic musks from surface and sewage water samples [9, 10]. The concentrated sample extracts containing the polycyclic musks were analyzed applying GC-MS (Ion Trap) detection. In spiking experiments carried out at different concentration levels recovery rates were between 82 and 101% with standard deviations between 4 and 12% [10]. The limits of detection of the polycyclic musks were between 5 and 20 ng L^{-1} [10].

2.2 Sewage Sludge Samples

In the course of the study carried out in 2000 [11–13], the analysis of sewage sludge samples was performed using Soxhlet extraction with a mixture of cyclohexane and ethyl acetate, followed by clean-up using gel-permeation chromatography (GPC) and miniaturized silica gel columns. The whole clean-up procedure was carried out in accordance with the German standard method DFG S19 [14]. For the analysis, one aliquot of the sludge sample was not completely dried to avoid losses of the analytes. In parallel, another aliquot of the sample was used to determine the dry weight of the sludge and to calculate the content of the residues in the dry matter (dm). The extracts containing the nitro musks and the six polycyclic musk compounds were analyzed using GC-MS detection with SIM [12]. The limits of detection of the polycyclic musks were between 0.005 and 0.01 mg kg^{-1} dm.

2.3 Sediment Samples

The freeze dried sediment samples were mixed with sodium chloride and suspended in pure water. This suspension was extracted by SDE as described for the water samples [9, 10]. Sulfur was removed from the extracts using pyrogenic copper powder and ultrasonic treatment prior to analysis by GC-MS (Ion Trap) detection [9, 10]. Mean recoveries between 78 and 96% with standard deviations between 4 and 11% were determined in spiking experiments at different concentration levels. The limits of detection of the polycyclic musks were between 0.005 and 0.05 mg kg^{-1} dm [10]. In the course of a study carried out in 1999

[11–13], aquatic sediment analysis was performed using Soxhlet extraction, sample clean-up and detection by GC-MS as described for the analysis of the sewage sludge samples.

2.4 Fish Tissue Samples

The analysis of the fish tissue samples was performed using Soxhlet extraction with a mixture of cyclohexane and ethyl acetate, followed by clean-up using gel-permeation chromatography (GPC) and miniaturized silica gel columns. The whole clean-up procedure was carried out according to the German standard method DFG S19 [14]. The extracts containing the synthetic musk compounds were analyzed applying GC-Quadrupole MS or GC-Ion Trap detection [8–12]. In spiking experiments with eel samples, mean recoveries between 78 and 95% were determined for the polycyclic musk compounds [9, 10]. The limits of detection determined for the synthetic musks in spiking experiments with eel tissue samples were between 0.001 and 0.03 mg kg^{-1} wet weight.

3 Synthetic Musks in Surface Water and Sewage Effluents

In 1996, a surface water monitoring for several organic and inorganic sewage contaminants was carried out in Berlin by Heberer et al. [7]. Thirty surface water samples, representative of the Berlin area with respect to possible contaminations by sewage discharges, were analyzed for several environmental pollutants, namely phenols, synthetic musks, some polar pesticides, pharmaceutical residues and other polar organic compounds [2, 7, 15]. In fact, this was the first monitoring for synthetic musk residues in Berlin surface waters and it provided an overall picture of the degree of contamination of the Berlin waters by these and several other organic contaminants such as pharmaceuticals also originating from municipal sewage discharges. Figure 2 shows a map of Berlin's waterways, the locations of the STPs (A–G), their sewers, and the sampling sites of the surface water monitoring [7].

As shown in Table 2, musk ketone was the only nitro musk compound which was detected in most of the water samples. However, the concentrations of musk ketone were on average 20 times lower than those of the polycyclic musk compound HHCB shown in Table 3. Musk ambrette and musk tibetene were not found in any of the water samples. Musk moskene and musk xylene were only identified in a single water sample which was collected at sampling site no. 7 (Fig. 2) from a brook named Erpe. This sampling site was upstream from the municipal STP Münchehofe. Therefore, the polycyclic musks and musk ketone were only found at trace-level concentrations in this water sample. The occurrence of musk moskene (170 ng L^{-1}) and musk xylene (180 ng L^{-1}) only in this water sample was unusual and could not be explained by discharges of municipal sewage effluents. It has been suggested that there was another source for this contamination which could not be identified [7].

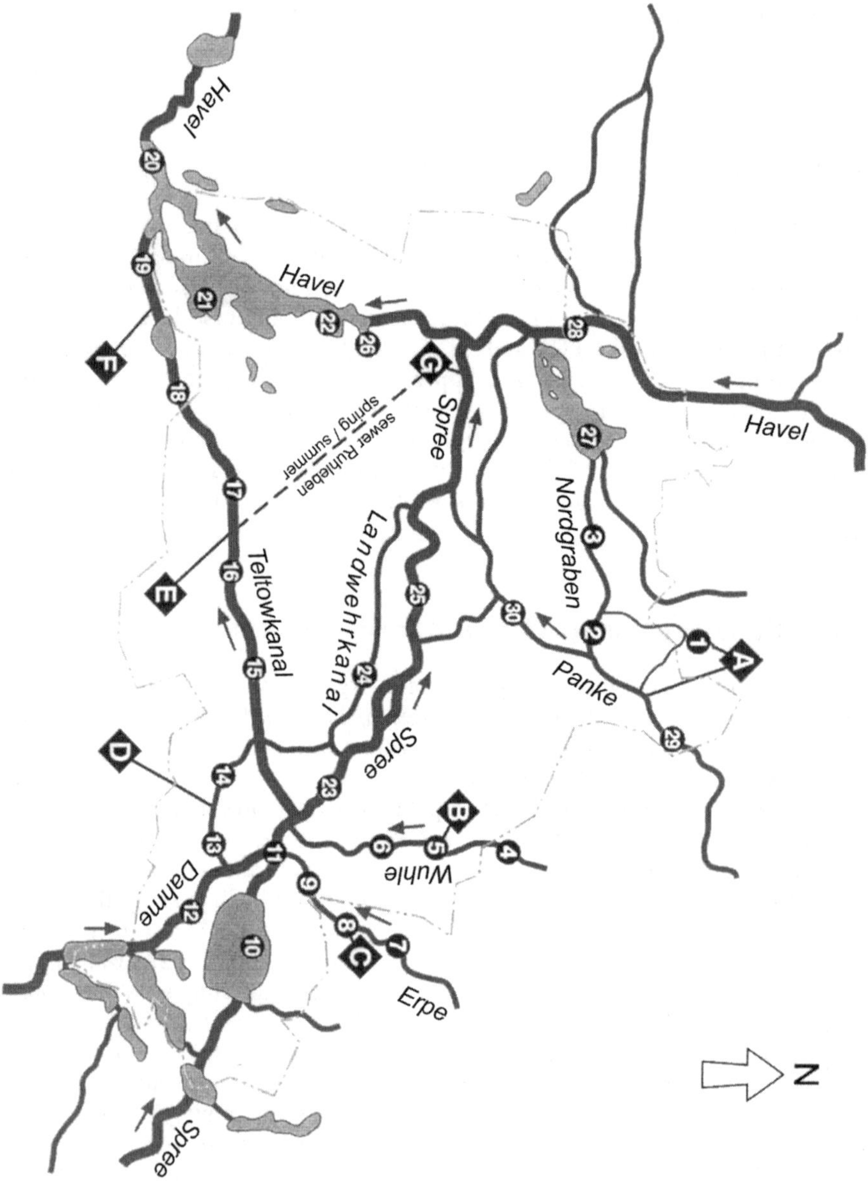

Fig. 2 Map of Berlin showing the locations of the STPs (A: Schönerlinde, B: Falkenberg, C: Münchehofe, D: Waßmannsdorf, E: Marienfelde, F: Stahnsdorf and G: Ruhleben) and the sampling locations of the surface water monitoring in Berlin carried out in September 1996 [7]. Adapted with permission from [7]. Copyright 1998, Wiley

Table 2 Concentrations of nitro musks (in ng L^{-1}) found in a screening analysis of 30 representative surface water samples collected from the Berlin area (n = 3)

No.	Sample	Musk ambrette	Musk ketone	Musk moskene	Musk tibetene	Musk xylene
1	Sewer from Schönerlinde	<10	320	<10	<10	<10
2	Nordgraben, Buchholz	<10	120	<10	<10	<10
3	Nordgraben, Rosenthal	<10	360	<10	<10	<10
4	Wuhle, Ahrensfelde	<10	<5	<10	<10	<10
5	Sewer from Falkenberg	<10	220	<10	<10	<10
6	Wuhle, Marzahn	<10	390	<10	<10	<10
7	Erpe, Neuenhagen	<10	70	170	<10	180
8	Sewer from Münchehofe	<10	160	<10	<10	<10
9	Erpe, Friedrichshagen	<10	130	<10	<10	<10
10	Müggelsee	<10	<5	<10	<10	<10
11	Spree, Köpenick	<10	20	<10	<10	<10
12	Dahme, Grünau	<10	<5	<10	<10	<10
13	Teltowkanal, Treptow	<10	<5	<10	<10	<10
14	Teltowkanal, Rudow	<10	120	<10	<10	<10
15	Teltowkanal, Tempelhof	<10	20	<10	<10	<10
16	Teltowkanal, Lankwitz	<10	40	<10	<10	<10
17	Teltowkanal, Lichterfelde	<10	90	<10	<10	<10
18	Teltowkanal, Kleinmachnow	<10	70	<10	<10	<10
19	Teltowkanal, Dreilinden	<10	70	<10	<10	<10
20	Havel, Potsdam	<10	<5	<10	<10	<10
21	Havel, Wannsee	<10	40	<10	<10	<10
22	Stößensee, Spandau	<10	<5	<10	<10	<10
23	Spree, Treptow	<10	<5	<10	<10	<10
24	Landwehrkanal, Kreuzberg	<10	<5	<10	<10	<10
25	Spree, Tiergarten	<10	<5	<10	<10	<10
26	Havel, Spandau	<10	70	<10	<10	<10
27	Tegeler See	<10	20	<10	<10	<10
28	Havel, Tegelort	<10	<5	<10	<10	<10
29	Panke, Buch	<10	<5	<10	<10	<10
30	Panke, Wedding	<10	30	<10	<10	<10

n.d.: not detected.
Data taken from Heberer et al. [7].

As compiled in Table 3, the polycyclic musks HHCB, AHTN, and ADBI were present in all of the water samples of the surface water monitoring at concentrations up to the μg L^{-1} level. In all samples, the concentrations of HHCB were on average about twice as high as those measured for AHTN, whereas the concentrations of ADBI were on average approximately 20 times lower than those of HHCB [7].

Although polycyclic musk compounds were detected in all 30 water samples, it was evident that peak concentrations were observed at those sampling sites located downstream of STPs discharges [7]. In these samples the concentration levels of HHCB and AHTN reached the μg L^{-1} level. Nevertheless, HHCB, AHTN, and ADBI were also found at very low concentrations at sampling sites located upstream from the municipal STPs. This observation has been explained by the

Table 3 Concentrations of polycyclic musks (in ng L^{-1}) found in a screening analysis of 30 representative surface water samples collected from the Berlin area (n = 3)

No.	Sample	ADBI	AHTN	HHCB
1	Sewer from Schönerlinde	410	5800	10,800
2	Nordgraben, Buchholz	50	1200	2400
3	Nordgraben, Rosenthal	260	3950	6450
4	Wuhle, Ahrensfelde	80	200	150
5	Sewer from Falkenberg	430	1950	6300
6	Wuhle, Marzahn	520	6800	12,500
7	Erpe, Neuenhagen	60	90	70
8	Sewer from Münchehofe	110	5800	10,100
9	Erpe, Friedrichshagen	120	4700	8300
10	Müggelsee	40	40	30
11	Spree, Köpenick	40	390	600
12	Dahme, Grünau	10	70	50
13	Teltowkanal, Treptow	10	30	20
14	Teltowkanal, Rudow	50	2400	4300
15	Teltowkanal, Tempelhof	20	120	550
16	Teltowkanal, Lankwitz	40	170	800
17	Teltowkanal, Lichterfelde	50	800	1400
18	Teltowkanal, Kleinmachnow	30	750	1300
19	Teltowkanal, Dreilinden	80	1050	1650
20	Havel, Potsdam	20	170	800
21	Havel, Wannsee	60	650	750
22	Stößensee, Spandau	60	550	850
23	Spree, Treptow	10	60	50
24	Landwehrkanal, Kreuzberg	20	110	90
25	Spree, Tiergarten	60	160	1000
26	Havel, Spandau	90	1000	1700
27	Tegeler See	30	70	50
28	Havel, Tegelort	300	450	300
29	Panke, Buch	30	60	50
30	Panke, Wedding	40	750	1450

Data taken from Heberer et al. [7].

ubiquitous occurrence of these contaminants in the urban environment that leads to a certain background contamination even at those sites not influenced by municipal sewage discharges [7].

At those three sampling sites (nos. 1, 5, and 8) where the samples were collected directly from the sewers of the STPs, polycyclic musks were detected at concentrations up to 15 μg L^{-1}. However, the highest concentrations were measured in the Wuhle, a brook which consists by more than 90% of sewage effluents. Thus, at sampling location no. 6 concentrations of 12.5 μg HHCB L^{-1} and 6.8 μg AHTN L^{-1} were found. These concentrations are among the highest amounts that have ever been reported for in surface or sewage water [11, 16].

In another surface water survey on the occurrence of these contaminants in the Berlin surface waters, Fromme et al. [9, 10] analyzed 102 water samples for polycyclic musk compounds.

These investigations, carried out in 1996 and 1997, confirmed the results of the above-mentioned monitoring study. During the study, the mean flow values varied from 0.002 to 0.1 m s^{-1} in the investigated surface waters. For the evaluation of the contamination data Fromme et al. [9, 10] divided the Berlin waters in three areas bearing low (A), moderate (B) and high (C) proportions of sewage effluents. The area of low contamination (A) comprises waters such as the lakes Langer See, Müggelsee, Zeuthener See, Seddinsee, and Dämeritzsee, all located in the south-east of Berlin (Fig. 1). Moderate proportions of sewage effluents are found in the river Havel, in lake Tegel and in lake "Großer Wannsee" (area B). The Teltowkanal in the south of Berlin and it's receiving waters lake Griebnitzsee and lake "Kleiner Wannsee" located in the south-western districts of Berlin are highly contaminated by sewage effluents (area C). The mean concentrations of HHCB in the surface waters were found to be 0.07 µg L^{-1} in the low contaminated areas (A), 0.23 µg L^{-1} in the moderately contaminated areas (B) and 1.59 µg L^{-1} in the highly contaminated areas (C). The results of this study are presented in detail in Table 4.

Despite the differences in sampling dates and methodologies used in both studies (SPME [7] and SDE [9, 10]), the results of surface water samples collected from identical areas showed good correlation of the contamination data. This holds also true for the results measured for sewage effluent samples. In the scope of the surface water monitoring in 1996 [7], random samples were taken from three different sewers (Schönerlinde, Falkenberg, and Münchehofe). Among the

Table 4 Concentrations (in µg L^{-1}) of polycyclic musks in Berlin surface water samples collected from areas with low (A), moderate (B) and high (C) proportions of sewage effluents [9, 10]

	Area[a]	HHCB	AHTN	ADBI	ATII	AHDI
Mean value	A	0.07	0.02	–	–	–
	B	0.23	0.07	–	–	-
	C	1.59	0.53	0.02	0.07	0.07
Standard deviation	A	0.06	0.01	–	–	–
	B	0.20	0.06	–	–	–
	C	0.72	0.25	0.01	0.03	0.04
Median	A	0.05	0.02	–	–	–
	B	0.15	0.05	–	–	–
	C	1.48	0.47	0.02	0.06	0.07
90 percentile	A	0.14	0.03	–	–	–
	B	0.49	0.14	–	–	–
	C	2.73	0.91	0.04	0.09	0.11
Maximum value	A	0.32	0.06	0.01	0.01	–
	B	0.81	0.27	0.01	0.05	0.03
	C	3.15	1.10	0.06	0.13	0.17
N > LOD/N[b]	A	31/34	21/34	1/34	1/11	0/9
	B	40/40	39/40	3/40	8/15	4/15
	C	28/28	28/28	25/28	8/8	8/8

[a] Areas of low (A), moderate (B) and high (C) contamination by municipal sewage effluents.
[b] Number of samples > limit of detection (LOD)/total number of samples.

nitro musk compounds (Table 2), only musk ketone was found in these samples at considerable concentrations of 0.32, 0.22, and 0.16 μg L^{-1}, respectively. In earlier investigations of former sewage irrigation farms (SIFs) located south of Berlin [17] similar concentrations have been reported for the nitro musk compounds. Thus, in 1994, 130 ng musk ketone L^{-1}, 30 ng musk xylene L^{-1} and 10 ng musk ambrette L^{-1} were detected in a sample of municipal sewage water (from the STP in Stahnsdorf) to be treated at these SIFs [17].

In terms of the monitoring study in 1996, the concentrations of the polycyclic musk compounds in the municipal sewage effluents were found to be significantly higher than those of the nitro musks (Table 3). Thus, HHCB was detected at concentrations of 10.8, 6.3, and 10.1 μg L^{-1}, respectively. These concentrations were about twice as high than those measured for AHTN (5.8, 1.95, and 5.8 μg L^{-1}). ADBI was found to be less important than HHCB or AHTN, but the concentrations detected for ADBI (0.41, 0.43, and 0.11 μg L^{-1}) were on average still higher than those of the most important nitro musk compound musk ketone.

In a subsequent study in 1996 and 1997, 30 proportional daily composite samples of treated sewage effluents from five municipal treatment facilities were analyzed for polycyclic musk residues [9, 10]. The results of these investigations are compiled in Table 5. Again, the mean value determined for HHCB (6.85 μg L^{-1}) was much higher than that of AHTN (2.24 μg L^{-1}) and the mean concentrations of ADBI, ATII, and AHDI (0.11, 0.31, and 0.27 μg L^{-1}) were significantly lower.

Figure 3 shows the seasonal fluctuations of the concentrations measured for HHCB and AHTN in sewage effluents from the STP in Ruhleben between August 1996 and September 1997. The highest concentrations were detected in March, April, and in early May of 1997. No relationship was observed between the concentrations of the polycyclic musks and the corresponding proportions of rainwater in the sewage effluents. These proportions were, however, very low (1.7 to 13%) for this particular STP.

The maximum values measured in the purified sewage effluents in Berlin are comparable to those measured by Simonich et al. [18] in sewage influents of an activated sludge STP (13.7 μg HHCB L^{-1} and 10.7 μg AHTN L^{-1}) and may suggest poor removal of synthetic musks by the Berlin STPs. The different studies are, however, not directly comparable because several parameters have not been

Table 5 Concentrations (in μg L^{-1}) of polycyclic musks in sewage effluent samples collected from five municipal STPs in Berlin [9, 10]

	HHCB	AHTN	ADBI	ATII	AHDI
Mean value	6.85	2.24	0.11	0.31	0.27
Standard deviation	2.64	0.86	0.04	0.20	0.13
Median	6.65	2.16	0.12	0.26	0.23
90 percentile	10.80	3.36	0.17	0.62	0.36
Maximum value	13.33	4.36	0.21	0.70	0.58
N>LOD/N[a]	30/30	30/30	30/30	24/25	12/12

[a] Number of samples > limit of detection (LOD)/total number of samples.

reported or considered, e.g., the composition of the sewage waters (percentages of municipal or industrial sewage and rain drainage), dry weather and rainy periods or the total drinking water consumption. The total drinking water usage per inhabitant in Germany of 127 L per day is also relatively low when compared to approximately 300 L per day in the USA [19].

In the course of the surface water monitoring carried out in 1996, the impact of the municipal sewage effluents on the surface water quality was demonstrated by Heberer et al. [7]. The diagram in Fig. 4 exemplarily shows the concentration profiles of HHCB and AHTN in the Teltowkanal [7].

Heberer et al. [7] described the polycyclic musks as being much more significant indicators for contaminations by municipal sewage effluents than any of the conventional chemical parameters, such as DOC (dissolved organic carbon) or COD (chemical oxygen demand). Both HHCB and AHTN were detected in the surface water samples of the Teltowkanal at maximum concentration levels of 4.3 and 2.4 $\mu g\ L^{-1}$, respectively. For both compounds, peak concentrations were observed at those sites where sewage effluents were released into the canal by the three municipal STPs. One week before the surface water monitoring was carried out at the end of September in 1996, the force main that discharges sewage from Berlin's largest STP in Ruhleben into the Teltowkanal was switched off. At that time, the STPs in Ruhleben already discharged their effluents into the river Spree (from September to April). From May to the mid of September, the force main is used to direct the municipal sewage effluents into the Teltowkanal to avoid

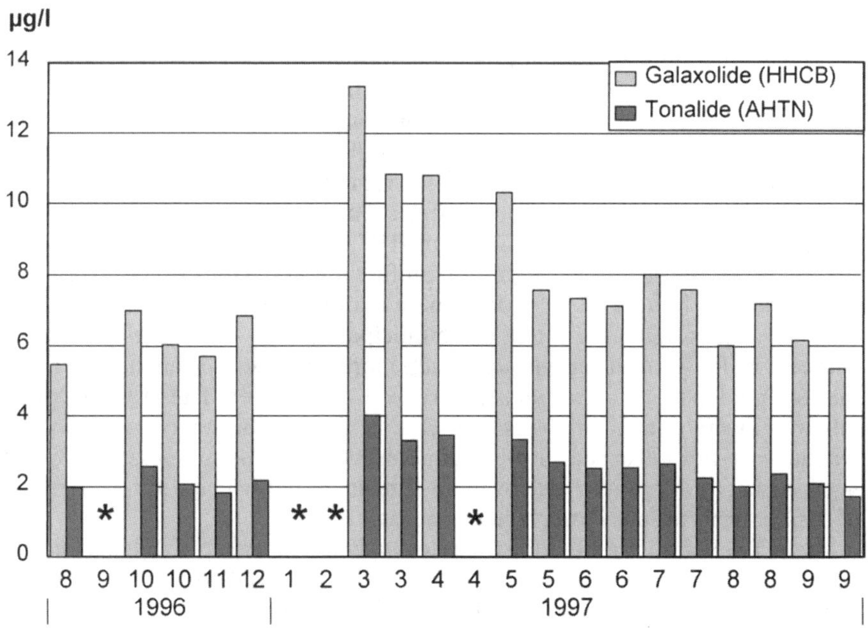

Fig. 3 Concentration of HHCB and AHTN in effluents from the STPs in Ruhleben during the period between August 1996 and September 1997. * No sample collection. Reprinted from Fromme et al. [10] with permission from Elsevier (2001)

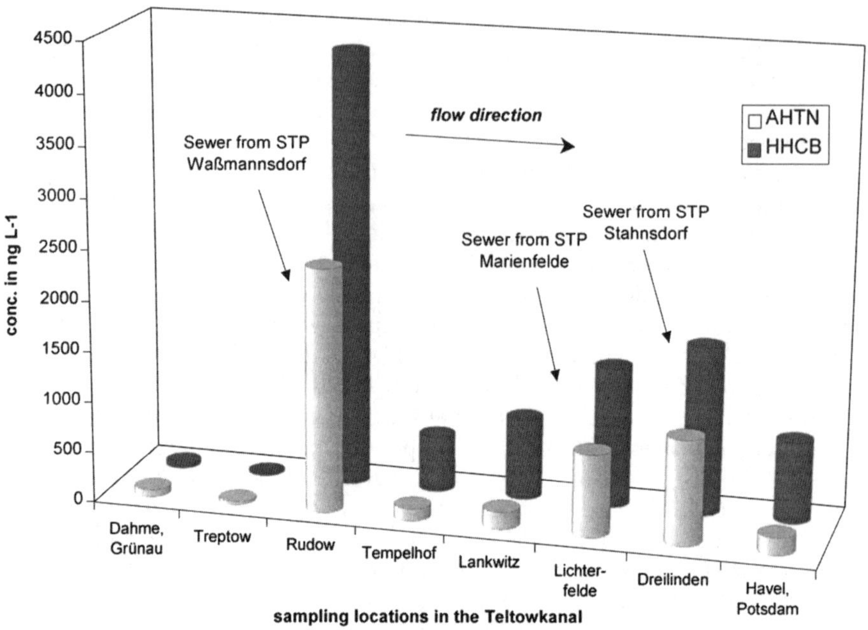

Fig. 4 HHCB and AHTN concentration profiles in the Teltowkanal. Selected results from a surface water monitoring carried out in 1996 in Berlin, Germany. STPs: sewage treatment plants. Adapted with permission from Heberer et al. [7]. Copyright 1999, Wiley

microbial problems at the bathing areas located in the Havel river and lake "großer Wannsee". However, during the sampling period only three STPs (Waßmannsdorf, Stahnsdorf, and Marienfelde) discharged their effluents into the canal.

Downstream from the discharges of the STPs the concentrations of the polycyclic musks were found to decrease very rapidly [7]. This could be explained both by dilution but also by sorption effects. The importance of sorption effects for the decrease of the concentrations of HHCB and AHTN has also been confirmed by comparing the concentration profiles measured for the nonpolar musk compounds with those of the polar contaminants such as pharmaceutical residues [2] also originating from the municipal sewage discharges. Thus, the decrease of the concentrations was much more pronounced for the polycyclic musks than for the drug residues and reflects the different fate of these contaminants in the aquatic environment. Due to their polar structures, the drug residues are not adsorped in the sediments, whereas the nonpolar polycyclic musk compounds are adsorbed by the sediments resulting in a stronger decrease of their concentrations in the aqueous phase.

At the final sampling site, which was located outside of Berlin in Potsdam (location no. 20 in Fig. 2) where the canal has already merged with the river Havel, HHCB, AHTN, and ADBI were still detected at concentrations of 800, 170 and 20 ng L^{-1}, respectively [7]. Here, the residues did not only originate from the Teltowkanal, they also resulted from effluents of other municipal STPs located upstream from it's confluence with the river Havel. At this site, the river Havel carries all municipal sewage inputs from the Berlin area and its approximately four million inhabitants. The discharges from the river Havel flow into the river Elbe and, finally, reach the North Sea, where residues of synthetic musks have also been detected at very low trace-level concentrations [20, 21].

4 Synthetic Musks in Sewage Sludge

In a recent study of sewage sludges from three municipal STPs (Ruhleben, Stahnsdorf, and Waßmannsdorf) in Berlin, 14 day collective samples were analyzed for residues of polycyclic musk compounds [11–13]. HHCB and AHTN were detected in the sewage sludges at average concentrations of 8.26 and 3.56 mg kg^{-1} dm, respectively. The average concentrations determined for AHDI (0.5 mg kg^{-1} dm), ADBI (0.26 mg kg^{-1} dm) and ATII (0.13 mg kg^{-1} dm) were significantly lower. DPMI was not detected in any of these samples. The results of these investigations are shown in detail in Table 6.

The investigations carried out in Berlin confirm recent results from Switzerland published by Herren and Berset [22]. They detected HHCB and AHTN at concentrations between 0.7 and 12.1 mg kg^{-1} dm. ADBI and DPMI were only found at concentrations below 1 mg kg^{-1} dm [22].

5 Synthetic Musks in Sediments

The investigations of sediment from the Berlin area are unique both in quantity and quality. In most of the other studies only a few, random samples have been investigated. Thus, no precise conclusions concerning the average levels and the distribution of polycyclic musk compounds in aquatic sediments could be drawn from these investigations. The Berlin results allow an accurate evaluation of the contamination of aquatic sediments from urban areas by polycyclic musks.

In a comprehensive study on the occurrence of polycyclic musk compounds in sediment from the Berlin area [9, 10], 59 surface sediment samples, collected from a depth of 10 cm, were taken from the central bed of the Berlin waters using

Table 6 Concentrations (in mg kg^{-1} dm) of polycyclic musks in sewage sludge from three STPs located in the Berlin area: 14 days collective samples (February 2000; n = 2), the limits of detection varied between 0.005 and 0.01 mg kg^{-1} dm [13]

Analyte	Ruhleben	Stahnsdorf	Waßmannsdorf	Average
HHCB	7.30	11.45	6.03	8.26
AHTN	3.08	5.07	2.52	3.56
AHDI	0.46	0.65	0.39	0.50
ATII	0.20	0.03	0.17	0.13
ADBI	0.26	0.33	0.19	0.26
DPMI	nd	nd	nd	nd

nd = not detected.

Table 7 Concentrations (in mg kg^{-1} dm) of polycyclic musks in aquatic sediments of Berlin, collected from areas with low (A), moderate (B) and high (C) proportions of sewage effluents [9, 10]

	Area[a]	HHCB	AHTN	ADBI	ATII	AHDI
Mean value	A	–	0.02	–	–	–
	B	0.22	0.26	0.010	0.021	0.011
	C	0.92	1.10	0.025	0.101	0.036
Standard deviation	A	–	0.01	–	–	–
	B	0.16	0.21	0.009	0.019	0.010
	C	0.70	0.85	0.031	0.076	0.028
Median	A	–	0.02	–	–	–
	B	0.23	0.24	0.008	0.021	0.009
	C	0.91	0.93	0.005	0.100	0.031
90 percentile	A	–	0.03	–	–	–
	B	0.38	0.52	0.021	0.044	0.022
	C	1.90	2.21	0.066	0.202	0.068
Maximum value	A	0.03	0.04	–	–	–
	B	0.52	0.61	0.023	0.051	0.026
	C	2.20	2.60	0.068	0.220	0.093
N > LOD/N[b]	A	1/19	12/19	0/19	0/19	0/19
	B	17/20	19/20	11/20	11/20	10/20
	C	18/20	18/20	16/20	17/20	16/20

[a] Areas of low (A), moderate (B) and high (C) contamination by municipal sewage effluents.
[b] Number of samples > limit of detection (LOD)/total number of samples.

a grab sampler. The results of this study are compiled in Table 7. Again, the results were evaluated by Fromme et al. [10] separating the samples by their origin from areas with low (A), moderate (B), and high (C) proportions of sewage effluents in the corresponding surface waters.

As can be seen in Table 7, the contamination of sediment correlates with the burden of sewage effluents in the aqueous phase [10]. Large variations of the polycyclic musk concentrations were observed with samples collected from highly contaminated areas (C). Fromme et al. [10] proposed that these variations may be caused by inhomogeneous sample materials, containing varying proportions of organic substances, which were, however, not determined in this study. In another study carried out in 1999 [11–13], several aquatic sediment samples were collected from different sites with low and high percentages of municipal sewage. Up to 3.6 mg HHCB kg^{-1} dm and up to 2.5 mg AHTN kg^{-1} dm were found at contaminated sites. The concentrations measured for AHDI (up to 0.14 mg kg^{-1} dm), ADBI (up to 0.05 mg kg^{-1} dm) and ATII (up to 0.03 mg kg^{-1} dm) were significantly lower. DPMI was not found in the samples [11].

Comparing the results from surface water with those of the sediment samples from the same waters, it becomes evident that the ratios of HHCB and AHTN are different in both environmental compartments. In the surface waters, the concentrations of HHCB were on average twice as high as those measured for AHTN.

In sediment, the concentrations are shifted in favor of AHTN. The concentrations of AHTN are equal to or even higher than those measured for HHCB. Data published by Winkler et al. [23] indicated that AHTN is bound stronger to solid particular matter (SPM) than HHCB. Thus, in many SPM samples, AHTN is found at concentrations similar to or even higher than those of HHCB. This observation is also confirmed by the results obtained for sediment samples collected from the Berlin waters.

6
Bioaccumulation of Synthetic Musks in Fish from Canals, Rivers, and Lakes in Berlin

Between 1991 and 1998, the world's most comprehensive study on the occurrence of residues of synthetic musk compounds in fish was commissioned by the Berlin Fishery Board [11]. More than 1500 fish samples from different species such as eel, pike, pike perch, perch, bream, white bream, rudd, roach, tench, carp, and crucian carp caught in rivers, lakes, and canals in Berlin were analyzed for residues of nitro musks and polycyclic musks. In this section, the results of these investigations will be discussed in detail. Several statistical analyses have been carried out to evaluate the large amount of data. The huge data pool allowed a selective evaluation of the results, e.g., to assess the contamination of different fish species in restricted (contaminated or non-contaminated) habitats. Thus, the Berlin data are well suited for a statistical evaluation of the contamination of fish from urbanized areas by synthetic musk compounds. It was even possible to establish temporal trends showing the decrease or constancy of the contamination by individual musk compounds.

The investigations have shown that in the Berlin waters fish live within very limited habitats. Even migrating fish species such as the European eel (*Anguilla anguilla* L.) could be included in this evaluation. The European eel is an excellent indicator to monitor contamination by nonpolar compounds. Due to their high fat content, eels strongly bioaccumulate these compounds in their tissues. Thus, the average concentrations of musk compounds per kilogram wet weight are much higher compared to other fish species. Nevertheless, the statistical evaluation of contamination data from eel tissue samples is often very difficult, especially, in rivers where eels migrate naturally from less or non-contaminated areas into contaminated areas and back into non- or less contaminated areas. In Berlin, the situation is different because several natural and artificial barriers hinder the eels from reaching the Berlin waters by natural migration. Juvenile eels ("glass" eels) are released into the Berlin waters by the Berlin Fishery Board. They live for several years in very small habitats. On their way back to their spawning areas the adult eels leave Berlin via the lower Havel. Due to the geographical situation of Berlin, fish from highly contaminated areas always migrate downstream not passing areas with low or no sewage contamination before leaving the city of Berlin. Eels from non-contaminated areas only have to pass the contaminated areas once in their lifetime when they leave their habitats towards the spawning areas. These eels can, however, easily be identified by their characteristic pigmentation.

6.1 Contamination by Nitro Musk Compounds

In 1991, musk xylene was the first musk compound included in the fish monitoring program of the Berlin Fishery Board. In the years between 1991 and 1996, 1471 fishes from different species were analyzed for residues of musk xylene. The results of these investigations are compiled in Table 8. Musk xylene was detected in approximately 56% of these samples. In 26% of the samples the concentrations exceeded the limit of quantification (LOQ) of 0.001 mg kg^{-1} ww (wet weight).

In 1995 and 1996, the investigated spectrum of synthetic musk compounds was extended. Musk ketone and several polycyclic musk compounds were included as important parameters in this study. Table 9 shows the results for musk xylene and musk ketone analyzed in 476 fishes from different species caught in Berlin between 1996 and 1998. For data evaluation the results were divided into two parts: 1. results for fish caught from contaminated areas and 2. results for fish caught from non-contaminated areas. Thus, waters having an impact from municipal sewage discharges will be called "contaminated areas". The investigations of Berlin's surface waters have already shown that the highest concentrations of musk compounds were found in the Teltowkanal, several minor brooks (e.g., Wuhle, Panke and Erpe), several lakes (e.g., Griebnitzsee, "Kleiner Wannsee", and "Großer Wannsee"), and in several (inner city) parts of the rivers Havel and Spree. The small brooks are not important for Berlins fishery but the Teltowkanal, the river Spree and especially the river Havel are among the most important fishing areas in Berlin. Thus, the term "fish from contaminated areas in Berlin" comprises fish caught in these areas and in several other waters which are expected to be contaminated by sewage effluents such as the Griebnitzsee which connects the Teltowkanal and the Havel river. The term "fish from non-contaminated areas in Berlin" comprises fish caught in waters such as lake Müggelsee, river Dahme and several other waters almost located in the south-eastern districts of Berlin. These waters are located upstream from Berlins municipal STPs.

As shown by the results in Table 9, musk xylene was detected at concentrations above the LOQ in approximately 43% of all 324 fishes caught from contaminated areas. About 48% of these samples contained musk ketone at concentrations above the LOQ. On the other hand, all 152 samples of fish caught in the non-contaminated areas were below the LOQ both for musk xylene and musk ketone. The results also show that fish from the Berlin waters stay in their ancestral habitats,

Table 8 Concentrations (in mg kg^{-1} ww or lw) of musk xylene in fish from different species caught in Berlin between 1991 and 1996

Compound/basis	N	< LOD[a]	DL < x < LOQ[b]	Maximum value	Average	Median	95 percentile
Musk xylene/ww	1471	648	438	1.16	0.018	<LOQ	0.062
Musk xylene/lw	1471	648	438	11.5	0.158	<LOQ	0.772

[a] Number of samples below the LOD (limit of detection).
[b] Positive samples with concentrations below the limit of quantification (LOQ = 0.001 mg kg^{-1} ww).

Table 9 Concentrations (in mg kg^{-1} ww or lw) of musk xylene and musk ketone in fish from different species caught in Berlin between 1996 and 1998. Differentiation between fish caught from contaminated and non-contaminated areas

Compound/basis	N	< LOQ	Maximum value	Average	Median	95 percentile
Fish from contaminated areas in Berlin (e.g., Teltowkanal, Griebnitzsee, "Kleiner Wannsee" and "Großer Wannsee", large parts of rivers Havel and the Spree)						
Musk xylene/ww	324	186	0.08	0.006	<LOQ	0.043
Musk xylene/lw	324	186	0.49	0.031	<LOQ	0.178
Musk ketone/ww	324	168	0.38	0.022	<LOQ	0.132
Musk ketone/lw	324	168	1.56	0.125	<LOQ	0.66
Fish from non-contaminated areas (e.g., Müggelsee, river Dahme, Langer See)						
Musk xylene/ww	152	152	<LOQ	<LOQ	<LOQ	<LOQ
Musk xylene/lw	152	152	<LOQ	<LOQ	<LOQ	<LOQ
Musk ketone/ww	152	152	<LOQ	<LOQ	<LOQ	<LOQ
Musk ketone/lw	152	152	<LOQ	<LOQ	<LOQ	<LOQ

Table 10 Percentage of samples below the limit of quantification (LOQ) for musk xylene and musk ketone in fish from different species caught in Berlin in 1995 and 1996 [24]

Species	Number of samples N	Musk xylene % < LOQ[a]	Musk ketone % < LOQ[a]
Eel (1995)	82	45	40
Eel (1996)	122	41	41
Perch	22	95	91
Bream	30	93	78
Pike	13	100	92
Roach	44	95	95
Pike perch	27	100	96

[a] Percentage of samples below the limit of quantification.

because no significant contamination was found in fish from non-contaminated areas.

In 1995 and 1996, 340 fish samples from different species caught from the Berlin area were analyzed for residues of musk ketone and musk xylene. As shown in Table 10, most positive results above the limits of quantification were found in fish tissue samples from eels. Thus, musk xylene and musk ketone were found in more than 50% of all eel samples. With one exception (22% of all bream samples contained musk ketone above the LOQ), nitro musk residues were only found in less than 10% of all tissue samples from other fish species.

In the early 1990s, Rimkus et al. [25–28] reported frequent findings of nitro musks in biota samples. Due to these and other findings, there was a voluntary phasing out of musk xylene use by the fragrance industry in Germany. The investigations of fish on musk xylene residues over a longer time period enabled a long-term evaluation of the data and made the recognition of any temporary trends in musk xylene concentrations possible. For this purpose, data were selected and compared for only one fish species (eel) caught in a limited, highly contaminated area (the Teltowkanal). This could only be done because of the huge data pool which, despite the limitations of species and location, still allowed a reliable statistical evaluation (n=102). The eel (*Anguilla anguilla*) was selected because most data was available for this species and because eels are excellent indicators for nonpolar contaminants. Due to their high lipid content they accumulate high concentrations of these compounds in their tissues. Thus, the synthetic musk concentrations on wet weight basis are on average much higher compared to other fish species as already demonstrated in Table 10. The limitation to only one species also increases the reliability of the contamination data and allows the comparison of the individual concentrations on lipid basis. Thus, species dependent factors such as differences in behavior and nutrition or differences in the metabolism of contaminants that might affect the bioaccumulation of the musk compounds do not disturb the interpretation of the results.

The Teltowkanal was selected as indicative location because of the high and constant inputs of municipal sewage effluents discharged into this canal. In this respect, the living conditions are very constant in this restricted area. Figure 5 shows a box plot diagram which demonstrates the decrease of the contents of musk xylene between 1991 and 1998 in eels caught from the Teltowkanal. The decrease of the concentrations can be explained by the voluntary phasing out of musk xylene use by the fragrance industry in Germany [11, 12]. The general

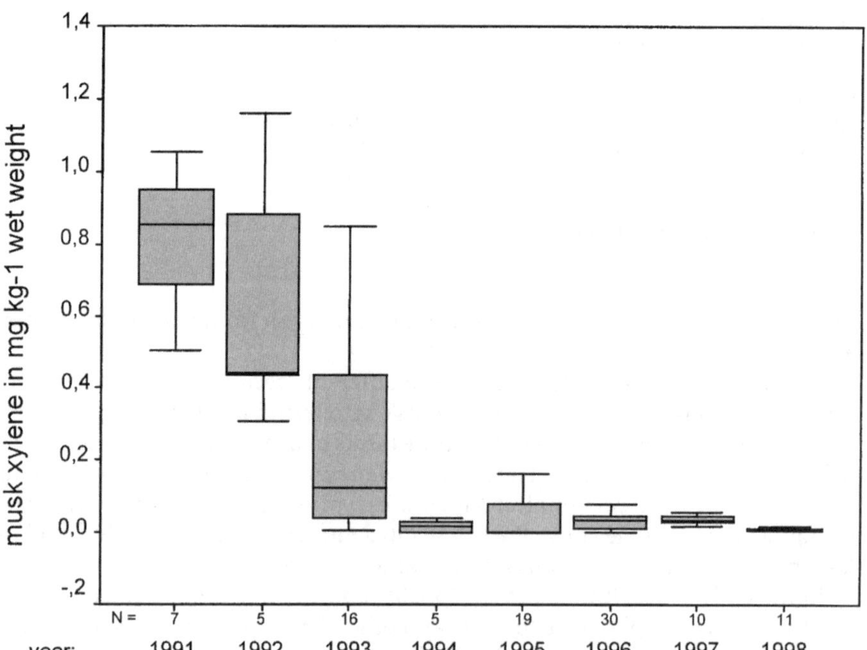

Fig. 5 Box plot diagram of musk xylene concentrations in eels caught between 1991 and 1998 in the Teltowkanal in Berlin, Germany. Reprinted with permission from Heberer et al. [11]

concern about the environmental fate and the toxicology of nitro musk compounds resulted in a decrease of their use as fragrances whereas the importance of the polycyclic musks was increasing [29, 30]. The Berlin data, also presented in more detail in Table 11, show a significant decrease of musk xylene concentrations as early as 1993. As early as 1994, the median of the concentrations detected in eel samples was close to the analytical limit of quantification (LOQ). This trend was established in the following years. In 1998, the concentrations of musk xylene were very close to the LOQ of only 0.01 mg kg^{-1} lipid weight (lw) in almost all of the analyzed samples. Due to the delay in the decrease of the concentrations of persistent organic compounds that is usually observed with environmental biota samples, it seems to be evident that the use and the discharges of musk xylene had already dramatically decreased before 1993 [11, 12].

Table 12 summarizes the concentrations of musk ketone and musk xylene detected in 51 eels caught in the Teltowkanal between 1996 and 1998. The average concentrations measured for musk ketone were almost constant and about three times higher than those for musk xylene. This trend was also confirmed by the results obtained in the investigations of 204 eel samples caught from the Berlin area in 1995 and 1996 presented in Table 13. In general, the concentrations of the

Table 11 Concentrations (in mg kg^{-1} lw) of musk xylene in eels from the Teltowkanal caught between 1991 and 1998

Year	N	< LOD[a]	LOD < x < LOQ[b]	Maximum value	Average	Median	Standard deviation
1991	7	0	0	4.17	3.36	3.20	0.56
1992	5	0	0	4.11	2.99	3.00	0.85
1993	16	0	0	3.36	1.02	0.41	1.09
1994	4	1	1	0.24	0.139	0.158	0.103
1995	19	0	10	0.67	0.173	<LOQ	–
1996	30	0	0	0.49	0.144	0.150	0.104
1997	10	0	0	0.21	0.140	0.130	0.043
1998	11	0	0	0.07	0.044	0.040	0.023
Lipid %[c]	102	–	–	36.0	22.8	24.1	6.62

[a] Number of samples below the LOD (limit of detection).
[b] Number of positive samples with concentrations below the limit of quantification.
[c] Lipid content in percent.

Table 12 Concentrations (in mg kg^{-1} ww) of nitro musks in eels from the Teltowkanal 1996–1998

	N	Samples < LOQ[a]	Maximum value	Average	Median	95 percentile
Musk ketone	51	0	0.38	0.10	0.07	0.24
Musk xylene	51	0	0.08	0.03	0.03	0.07

[a] Number of samples below the limit of quantification (between 0.001 and 0.01 mg kg^{-1} ww).

Table 13 Concentrations (in mg kg^{-1} ww) of musk xylene and musk ketone in eels caught from the Berlin area in 1995 and 1996 (n = 82 (1995) and n = 122 (1996)) [24]

	Minimum value		Maximum value		Average value		Median	
	1995	1996	1995	1996	1995	1996	1995	1996
Musk xylene	<LOQ	<LOQ	0.160	0.079	0.015	0.012	0.002	0.003
Musk ketone	<LOQ	<LOQ	0.260	0.380	0.026	0.039	0.035	0.005
Weight (g)	117	60	580	724	245	214	217	192
Length (cm)	39	32	65	69	50	48	49	47
Lipid (%)	3.1	1.8	40.0	40.8	23.2	21.1	25	23

nitro musk residues were clearly below those measured for the most important polycyclic musks HHCB and AHTN in the same fish tissue samples (next section).

6.2 Contamination by Polycyclic Musk Compounds

In the mid of the 1990s, Eschke et al. [31–33] reported the first findings of polycyclic musk compounds in the aquatic environment and their accumulation in fish from areas contaminated by municipal sewage effluents. Therefore, samples of 476 fishes caught from different surface waters in Berlin were analyzed for six important polycyclic musk compounds [11] to observe the degree of contamination of fish from the Berlin area. These investigations were again commissioned by the Berlin Fishery Board and carried out in the years between 1996 and 1998. Those fish samples (n=324) which were collected from surface waters contaminated by municipal sewage effluents contained HHCB and AHTN at average wet weight concentrations of 0.51 and 0.20 mg kg^{-1} (median 0.14 and 0.04 mg kg^{-1}), respectively. The results for the tissue samples of fish caught from contaminated waters are presented in detail in Table 14. Maximum values of up to 384 mg HHCB kg^{-1} lw and 88 mg AHTN kg^{-1} lw were detected. ADBI, AHDI, and ATII were also detected in several samples, but at concentrations considerably lower than those of HHCB and AHTN. DPMI concentrations were always below the LOQ.

Table 15 compiles the results measured for those samples (n=152) obtained for fish collected from Berlin surface waters not affected by municipal sewage effluents. In about half of these samples, synthetic musks were found at trace-level concentrations. As discussed earlier, these trace-level concentrations may originate from the ubiquitous distribution of these compounds in the urban environment. Trace amounts of synthetic musks may also originate from sources upstream of Berlin or may be caused by very limited migration of fish into contaminated areas. In general, the observed concentrations were very low and for all compounds the median values were below the LOQ. With the exception of HHCB even the average values and the 95 percentile values were below the LOQ. The maximum value of 0.26 mg kg^{-1} ww and the average value of 0.02 mg kg^{-1} ww measured for HHCB were about 20 times lower than those values determined for fishes from contaminated areas (Table 14).

Table 14 Concentrations (in mg kg^{-1} ww or lw) of polycyclic musks in fish samples from different species. Results for fish caught between 1996 and 1998 in contaminated areas in Berlin (e.g., Teltowkanal, Griebnitzsee, large parts of the Havel and the Spree river)

Compound/ basis	N	Samples < LOQ	Maximum value	Average	Median	Standard deviation	95 percentile
HHCB/ww	324	36	4.80	0.51	0.14	0.77	2.20
HHCB/lw	324	36	383.9	11.2	3.36	29.2	54.9
AHTN/ww	324	97	2.30	0.20	0.04	0.33	0.90
AHTN/lw	324	97	88.3	2.69	0.55	6.90	10.8
ADBI/ww	324	222	0.20	0.003	<LOQ	–	0.014
ADBI/lw	324	222	5.61	0.096	<LOQ	–	0.77
AHDI/ww	152	68	0.21	0.016	0.004	0.031	0.079
AHDI/lw	152	68	0.89	0.114	0.017	0.191	0.58
ATII/ww	152	47	0.19	0.019	0.006	0.034	0.103
ATII/lw	152	47	1.42	0.171	0.053	0.258	0.76
DPMI/ww	78	78	<LOQ	<LOQ	<LOQ	–	<LOQ
DPMI/lw	78	78	<LOQ	<LOQ	<LOQ	–	<LOQ

Table 15 Concentrations (in mg kg^{-1} ww or lw) of polycyclic musks in fish samples from different species. Results for fish caught between 1996 and 1998 in non-contaminated areas in Berlin (e.g., Müggelsee, Dahme river, Langer See)

Compound/ basis	N	Samples < LOQ	Maximum value	Average	Median	95 percentile
HHCB/ww	152	87	0.26	0.02	<LOQ	0.08
HHCB/lw	152	86[a]	7.5[a]	0.53	<LOQ	3.33
AHTN/ww	152	150	0.06	<LOQ	<LOQ	<LOQ
AHTN/lw	152	150	0.50	<LOQ	<LOQ	<LOQ
ADBI/ww	152	141	0.004	<LOQ	<LOQ	<LOQ
ADBI/lw	152	141	0.087	<LOQ	<LOQ	<LOQ
AHDI/ww	63	62	0.004	<LOQ	<LOQ	<LOQ
AHDI/lw	63	62	0.016	<LOQ	<LOQ	<LOQ
ATII/ww	63	54	0.014	<LOQ	<LOQ	0.004
ATII/lw	63	54	0.31	<LOQ	<LOQ	0.035
DPMI/ww	26	26	<LOQ	<LOQ	<LOQ	<LOQ
DPMI/lw	26	26	<LOQ	<LOQ	<LOQ	<LOQ

[a] One pike perch was excluded from this calculation. This particular sample was detected with a moderate HHCB wet weight concentration of 0.11 mg kg^{-1}. Due to its very low fat content of only 0.1% the calculated value of 113 mg kg^{-1} lw was unusually high.

These results also confirm that fish from the Berlin waters remain in their ancestral habitats, because only low contamination (almost by HHCB) was found in fish from non-contaminated areas. The habitat is often very small but due to the high amounts of nutrients in the Berlin waters there is no need for the fish to leave the closer areas of their individual habitats. On the other hand, even the high degree of contamination by municipal sewage effluents does not seem to have any impacts on their migration behavior. Thus, the occurrence of inorganic and organic contaminants originating from the municipal STPs does not stress the fish and does also not force them to move into non-contaminated areas. The fish do not recognize these contaminants as hazards and, thus, the areas contaminated by municipal sewage effluents are only recognized as sources for additional nutrients.

A "worst-case scenario" for the accumulation of the polycyclic musk compounds in eels is again demonstrated by the example of the Teltowkanal, a canal highly contaminated by municipal sewage effluents. Table 16 compiles the results measured for synthetic musks in 51 tissue samples from eels caught from this canal between 1996 and 1998 [11, 12]. The most recent results are presented in Table 17 showing the contamination data of 13 eels caught in the Teltowkanal in 1999 [13].

HHCB, AHTN, AHDI, and ATII were detected above the LOQ in all of the analyzed samples. It was found that 70% of all samples contained ADBI above the

Table 16 Concentrations (in mg kg^{-1} ww or lw) of synthetic musks in eels from the Teltowkanal 1996–1998 [11, 12]

Compound/basis	n	Samples < LOQ[a]	Maximum value	Average	Median	95% percentile
HHCB/ww	51	0	4.80	1.48	1.44	3.7
AHTN/ww	51	0	2.30	0.70	0.57	1.40
AHDI/ww	21	0	0.21	0.07	0.05	0.20
ATII/ww	21	0	0.19	0.07	0.05	0.19
ADBI/ww	51	19	0.02	0.004	0.004	0.014
DPMI/ww	11	11	<LOQ	<LOQ	<LOQ	<LOQ
Musk ketone/ww	51	0	0.38	0.10	0.07	0.24
Musk xylene/ww	51	0	0.08	0.03	0.03	0.07
HHCB/lw	51	0	18.5	6.39	6.17	14.7
AHTN/lw	51	0	8.85	2.97	2.69	6.1
Lipid %	51	–	36.0	23.9	25.2	33.8

[a] Number of samples below the limit of quantification (between 0.001 and 0.01 mg kg^{-1} ww).

Table 17 Concentrations (in mg kg^{-1} ww) of polycyclic musks in eels from the Teltowkanal 1999 [13]

Analyte	n	Samples < LOQ[a]	Maximum value	Average	Median	Standard deviation
HHCB	13	0	1.75	1.40	1.54	0.39
AHTN	13	0	1.23	0.94	0.99	0.29
AHDI	13	0	0.19	0.14	0.19	0.04
ATII	13	0	0.21	0.09	0.075	0.06
ADBI	13	0	0.065	0.037	0.029	0.018
DPMI	13	13	<LOQ	<LOQ	<LOQ	–

[a] Number of samples below the limit of quantification (between 0.001 and 0.01 mg kg^{-1} ww).

LOQ, whereas DPMI was not detected in any of the samples. Again, HHCB is the main contaminant in these samples found at concentrations about twice as high as those of AHTN. The concentrations of AHDI, ATII, and ADBI were found to be more than one order of magnitude lower than those measured for HHCB. The values for the average and the median concentrations of the polycyclic musks were very similar. In particular, the average and median values of HHCB and AHTN per kg lipid showed a difference of only less than 5 and 10%, respectively. This also reflects the text book like distribution of the synthetic musk compounds for one species (eel) in a surface water that is constantly polluted by municipal sewage effluents containing these compounds at considerable concentrations (Teltowkanal). The eel population in the Teltowkanal is intact which is also reflected by the uniform distribution of the fat content (Table 16) and the length (not shown) of the eels in the canal.

7 Conclusions

In 1991, the first investigation of a synthetic musk compound (musk xylene) was conducted with fish samples from the Berlin area. In the meantime, several sewage sludge samples, 30 sewage samples, 132 surface water samples, 66 sediment samples, and more than 1500 fish tissue samples have been investigated for nitro musks and polycyclic musk compounds. In all compartments, HHCB was found to be the main contaminant among the synthetic musk compounds. Nevertheless, the distribution of the polycyclic musks differs in the various compartments as also demonstrated in Fig. 6.

In sewage, sewage sludge and in contaminated surface water, HHCB was found at concentrations up to 13.3 $\mu g\ L^{-1}$, 11.5 $mg\ kg^{-1}$ dm, and up to 12.8 $\mu g\ L^{-1}$, respectively. In all of these samples, the average concentrations of HHCB were about twice as high as those measured for AHTN. In the aquatic sediments collected from sewage contaminated areas, the concentrations were shifted in favor of AHTN. The concentration of AHTN (mean value: 1.10 $mg\ kg^{-1}$ dm) was equal to or even higher than that of HHCB (mean value: 0.92 $mg\ kg^{-1}$ dm). In tissue samples of fish collected from contaminated surface waters, HHCB was the major contaminant detected at average concentrations of 0.51 $mg\ kg^{-1}$ ww (max. values up to 4.8 $mg\ kg^{-1}$ ww or 384 $mg\ kg^{-1}$ lw). As for the surface waters, the concentrations of HHCB were about twice as high as those determined for AHTN. AHTN is only slightly better accumulated in fish than HHCB. As will be shown below, this is also reflected by the bioconcentration factors (BCFs) calculated for AHTN and HHCB. In general, HHCB and AHTN were dominant in all contaminated samples from all compartments. ADBI, AHDI, ATII, musk xylene, and musk ketone were found to be less important, occurring at individual concentrations more than one order of magnitude lower than those of HHCB. With very few exceptions, DPMI, musk ambrette, musk moskene, and musk tibetene were not detected in any of the analyzed samples.

Based on these extensive contamination data, Fromme et al. [10] calculated the BCFs for HHCB and AHTN in eel samples. For these evaluations, they have chosen 15 (9 for AHTN) closely defined surface water areas where contamination

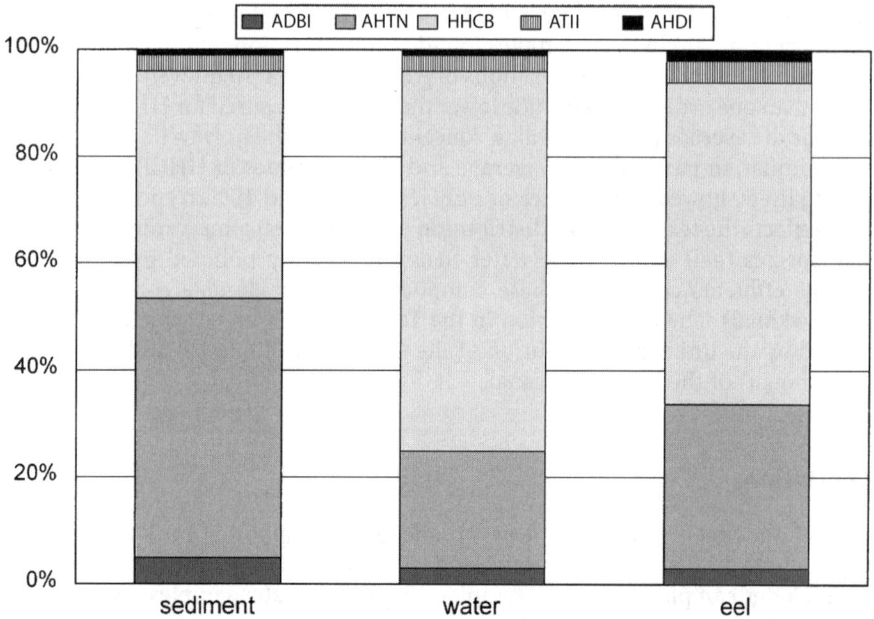

Fig. 6 Mean percentile contribution of the individual compounds to the total content of polycyclic musks in sediment, water, and eel. Reprinted with permission from Fromme et al. [9]. Copyright 2000 by the American Chemical Society

data were available for both surface water and eel samples. As an example, the mean values of HHCB for water and eel samples from these areas are shown in Fig. 7.

Applying Spearman rank correlation (r=0.74), good correlation was seen between the average concentrations in waters and those in biological samples. Based on wet weight a BCF mean value for HHCB of 862 (range: 201–1561) and for AHTN of 1069 (range: 250–1791) was calculated for eels. The corresponding BCF values based on the lipid content were calculated to be 3504 (HHCB) and 5017 (AHTN) [10]. These calculated values were, however, not based on defined experimental conditions, but were gained under natural conditions. Factors such as the content of SPM in the aqueous phase were not taken into consideration. The values obtained for eels from the Berlin area are higher than those calculated from Eschke et al. [33] under natural conditions. Based on data from the river Ruhr Fromme et al. [10] calculated mean BCF values based on wet weight of 650 for HHCB and 297 for AHTN.

Acknowledgements The authors thank the research department of the Technical University of Berlin for funding the investigations carried out in 1999 and 2000 in terms of the scientific project entitled "Investigation of the fate of polycyclic musk compounds in the aquatic system of Berlin as an example for urban waters". Special thanks go to A. These, T. Otto, and K. Pilz for preparing the samples, to M. Ricking (FU Berlin) for providing several sediment samples, and to the Berliner Wasserbetriebe for providing the sewage sludge samples.

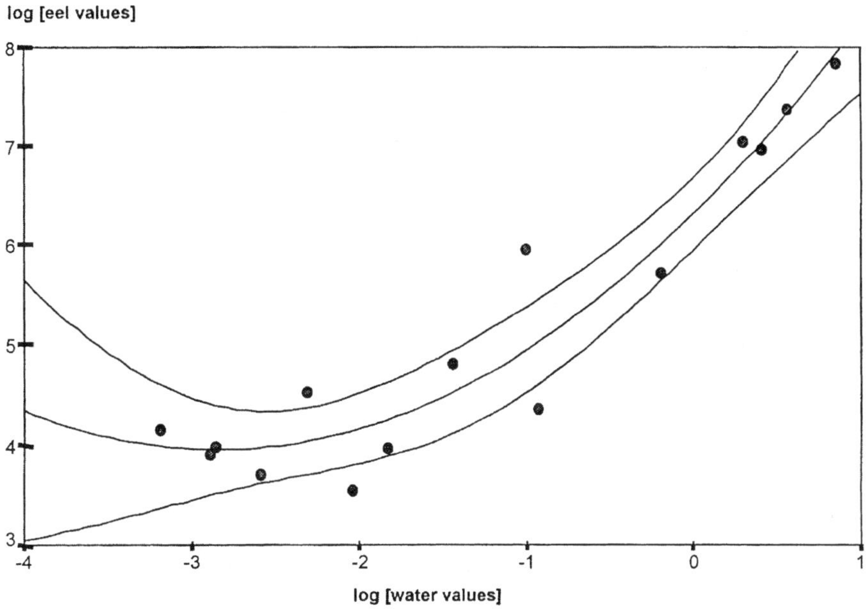

Fig. 7 Comparison of mean HHCB concentration in surface water and eel samples from 15 Berlin lakes and rivers (*inner line* represents quadratic regression, *outer lines* 95%-confidence interval). Reprinted with permission from Fromme et al. [10]. Copyright by Elsevier (2001)

References

1. Senatsverwaltung für Stadtentwicklung, Umweltschutz und Technologie (1997) Berlin digital environmental atlas, 2nd edn. Kulturbuchverlag, Berlin, free online access via: http://www.sensut.berlin.de/sensut/umwelt/uisonline/dua96/html/edua_index.shtml
2. Heberer T, Schmidt-Bäumler K, Stan H-J (1998) Acta Hydrochim Hydrobiol 26:272
3. Heberer T (2002) J Hydrol 266:176
4. Berlin Fishery Board (1997) Fishery in Berlin. Public communication, Berlin
5. Senatsverwaltung für Stadtentwicklung, Umweltschutz und Technologie (1999) Abwasserbeseitigungsplan Berlin. Kulturbuchverlag, Berlin
6. Der Senator für Stadtentwicklung und Umweltschutz (1983) Der Teltowanal – Wassermenge, Wassergüte, Sanierungskonzeptionen. In: Besondere Mitteilungen zum gewässerkundlichen Jahresbericht des Landes Berlin. Public communication, Berlin
7. Heberer T, Gramer S, Stan H-J (1999) Acta Hydrochim Hydrobiol 27:150
8. Fromme H, Otto T, Pilz K (1999) Chemosphere 39:1723
9. Fromme H, Otto T, Pilz K (2000) Polycyclic musk fragrances in the aquatic environment, chap 15. In: Lipnick RL et al. (eds) Persistent, bioaccumulative, and toxic chemicals II: assessment and new chemicals. Symposium Series 773. American Chemical Society, Washington DC, p 203
10. Fromme H, Otto T, Pilz K (2001) Water Res 35:121
11. Heberer T, These A, Grosch UA (2001) Occurrence and fate of synthetic musks in the aquatic system of urban areas – polycyclic and nitro musks as environmental pollutants in surface waters, sediments, and biota. In: Daughton CG, Jones-Lepp T (eds) Pharmaceuticals and personal care products in the environment: scientific and regulatory issues. Symposium Series 791, American Chemical Society: Washington, DC, p 142

12. Heberer T, These A, Grosch UA (2000) Occurrence and fate of synthetic musks in the aquatic system of urban areas–polycyclic and nitro musks as environmental pollutants in surface waters, sediments and biota. Invited lecture in terms of the 219th American Chemical Society (ACS) National Meeting, March 26–31, 2000, San Francisco, USA
13. Heberer T (2002) Acta Hydrochim Hydrobiol 30:227
14. DFG (Deutsche Forschungsgemeinschaft) (1991) Arbeitsgruppe Analytik: Methodensammlung Rückstandsanalytik von Pflanzenschutzmitteln, Multimethode S 19. 11. Lieferung, VCH, Weinheim
15. Schmidt-Bäumler K, Heberer Th, Stan H-J (1999) Acta Hydrochim Hydrobiol 27:143
16. Rimkus GG (1999) Toxicol Lett 111:37
17. Heberer T (1995) Identification and quantification of pesticide residues and environmental contaminants in ground and surface water applying capillary gas chromatography – mass spectrometry (in German). Wissenschaft Technik, Berlin, p 272
18. Simonich SL, Begley WM, Debaere G, Eckhoff WS (2000) Environ Sci Technol 34:959
19. Engelhard T (2000) Geo Magazin Feb 2000:42
20. Gatermann R, Hühnerfuss H, Rimkus G, Wolf M, Franke S (1995) Mar Pollut Bull 30:221
21. Bester K, Hühnerfuss H, Lange W, Rimkus GG, Theobald N (1998) Chemosphere 32:1857
22. Herren D, Berset JD (2000) Chemosphere 40:565
23. Winkler M, Kopf G, Hauptvogel C, Neu T (1998) Chemosphere 37:1139
24. Fromme H (1999) Nitromoschusverbindungen und Bromocyclen in Fischen aus Berliner Gewässern. In: Institute of Environmental Analysis and Human Toxicology (ITox) im BBGes. Beiträge zum Umwelt- und Gesundheitsschutz. Public communication, Berlin
25. Rimkus G, Wolf M (1993) Lebensmittelchemie 47:26
26. Rimkus G, Wolf M (1993) Dtsch Lebensm-Rundsch 89:171
27. Geyer HJ, Rimkus G, Wolf M, Attar A, Steinberg C, Kettrup A (1994) Z Umweltchem Ökotox 6:9
28. Rimkus G, Wolf M (1995) Chemosphere 30:641
29. Rimkus G, Brunn H (1996) Ernährungs-Umschau 43:442
30. Rebmann A, Wauschkuhn C, Waizenegger W (1997) Dtsch Lebensm-Rundsch 93:251
31. Eschke HD, Traud J, Dibowski HJ (1994) Vom Wasser 83:373
32. Eschke HD, Traud D, Dibowski HJ (1994) UWSF-Z Umweltchem Ökotox 6:183
33. Eschke HD, Dibowski HJ, Traud D (1995) UWSF-Z Umweltchem Ökotox 7:131

The Handbook of Environmental Chemistry Vol. 3, Part X (2004): 151–188
DOI 10.1007/b14120

Synthetic Musks in Bioindicators: Monitoring Data of Fish and Human Milk Samples from the Czech Republic

Jana Hajšlová · Lucie Šetková

Institute of Chemical Technology, Technická 3, Department of Food Chemistry and Analysis, 16628 Prague 6, Czech Republic
E-mail: jana.hajslova@vscht.cz

Abstract Until the middle of the 1990s, no attention was paid to the pollution of the environment in Czech Republic by synthetic musk fragrances and to the exposure of Czech population to these xenobiotics. However, in response to increasing information on high levels of synthetic musks in freshwater ecosystem reported by many industrial countries in the last decade, extensive monitoring study was conducted in Czech Republic in the years 1997–2000. More than 800 samples were collected at 11 sampling sites and examined for the presence of both polycyclic and nitro musk compounds. Their occurrence in various fish employed as bioindicator of river pollution as well as investigation of human milk documented omnipresent character of these widely used synthetic chemicals. High levels of synthetic musks were typically found downstream large urban areas. The relative concentrations in almost all analysed fish samples (perch, bream, chub, barbel and trout) decreased in following order: HHCB>AHTN⩾MX ≥MK≈AHDI>ATII≈ADBI. Polycyclic musks represented mainly by HHCB and AHTN accounted typically for more than 80% of total musk content; nevertheless, the bioaccumulation phenomena are distinctly species-dependent. Besides fish, application of semipermeable membrane device (SPMD) was shown to be a conceivable tool for the integral monitoring of the presence of musk containing sewage water in aquatic ecosystems. As fish represents a minor component of the Czech food basket, general occurrence of both groups of synthetic musk compounds in milk samples collected from 59 nursing mothers documented the significance of non-dietary routes of human exposure. The levels as well as relative abundance of synthetic musks varied in a wide range with polycyclic compounds prevailing in most of samples. Critical assessment of our results together with their comparison with data generated abroad in similar studies is presented in this chapter.

Keywords Fish · Human milk · Bioindicators · Polycyclic musks · Nitro musks · Aquatic ecosystem · River Elbe · River Moldau · River Tichá orlice

1 Introduction

Findings of the occurrence of synthetic musk fragrances in biota from German rivers reported by several authors [1–3] in the first part of the last decade caused in 1996 the establishment of a pilot study concerned with the examination of fish collected from the river Elbe at Hřensko (locality on the Czech-German border) for the presence of these new contaminants. As anticipated, the ubiquitous character of synthetic musk compounds was confirmed. Relatively extensive contamination of examined fish samples was documented; both representatives of polycyclic musk fragrances and nitro musks were found. With regard to the growing concern about aquatic ecosystem pollution, these compounds have involved since 1997 with the Czech monitoring program (established by the Ministry of Environment in 1994). Passive biological monitoring (PBM) approach utilizing fish for this purpose has been applied.

Musk compounds were also analysed in a set of human milk samples which were collected within another project focusing on the occurrence of persistent organic pollutants in this kind of bioindicator. Results obtained within both these studies are reported and discussed in this chapter.

2 Fish as Bioindicator: Set-up of Monitoring Study

In accordance with the common practice [4] used for the monitoring of persistent organochlorine micro-contaminants in freshwater and marine ecosystems, fish served as the main bioindicator of river pollution. Within four monitoring years (1997–2000) over 800 freshwater fish samples were collected using electro fishing at 11 sampling sites located along river Elbe and its tributaries, the rivers Moldau and Tichá Orlice in Czech Republic (Fig. 1); a few samples were also collected in the locality Podolí, upstream of Prague (very old specimen or those of extreme weight within the particular set of fish species were excluded).

Since eel, which has been widely used as a very suitable bioindicator organism of water pollution [4], did not occur in sufficient quantities in these localities, we employed several other species representing different categories as regards the typical feeding habits, wandering behaviour etc. for this purpose. Chub (*Leuciscus cephalus*), bream (*Abramis brama*), barbel (*Barbus barbus*) and perch (*Perca fluviatilis*) collected from rivers Elbe (Labe) and Moldau (Vltava) represent abundant fish species available in all localities each year. In the river Tichá Orlice, trout (*Salmo trutta*) is the only available fish. It should be emphasized that the migration of fish along the rivers was very restricted because of dams and/or weirs surrounding sampling localities. Sampling of fish was carried out regularly in all localities in the period between July and September by the University of Southern Bohemia, Research Institute of Fish Culture and Hydrobiology (Vodňany, Czech Republic).

Frozen samples (fillets, skin removed) that were delivered to our laboratory were held at −18 °C until analysis. The sets of examined fish from individual rivers are characterized in detail (Tables 1, 2 and 3).

Fig. 1 Sampling sites at rivers Elbe (Labe), Moldau (Vltava) and Tichá Orlice

Table 1 Characteristics of fish samples from river Elbe (Labe) examined within monitoring program

Fish species	Characteristics of samples collected during monitoring program in particular localities (1 – Kunětice, 2 – Srnojedy, 3 – Štětí, 4 – Hřensko)															
	1997				1998				1999				2000			
	1	2	3	4	1	2	3	4	1	2	3	4	1	2	3	4
Chub																
No. of pooled/ individual samples	2 (6)[a]	1 (5)	4 (17)		3 (10)	3 (10)	3 (10)	3 (11)	4 (20)	4 (17)	4 (20)	4	11	10	10	27
Mean length (mm)	320	304	284		334	300	385	383	394	342	445	443	289	274	368	210
RSD (%)	12	12	9		7	14	8	8	8	13	7	7	13	20	19	58
Mean fat (wt%)	1.1	1.5	1.8		1.0	1.0	1.4	2.2	2.2	1.6	1.5	3.0	1.3	2.5	1.8	4.0
RSD (%)	9		33		20	40	7	14	45	19	20	33	38	68	44	28
Bream																
No. of pooled/ individual samples	1 (4)[a]	2 (8)	1 (2)	20	2 (5)	3	3 (4)	3 (10)	3 (7)	4 (15)		21	5		10	6
Mean length (mm)	201	248	252	302	335	412	212	336	392	482		401	282		266	326
RSD (%)	8	27	4	10	14	4	16	12	4	7		11	11		17	16
Mean fat (wt%)	0.7	0.8	1.8	1.7	0.7	6.4	1.6	1.9	1.1	5.0		2.9	1.2		1.9	0.7
RSD (%)		13		41	14	30	6	25	16	14		10	33		26	43

Table 1 (continued)

Fish species	Characteristics of samples collected during monitoring program in particular localities (1 – Kunětice, 2 – Srnojedy, 3 – Štětí, 4 – Hřensko)															
	1997				1998				1999				2000			
	1	2	3	4	1	2	3	4	1	2	3	4	1	2	3	4
Barbel																
No. of pooled/ individual samples	1	1	1	1		3			2 (3)	3		3	8			3
Mean length (mm)	468	472	455	460		353			347	408		424	433			456
RSD (%)						1			31	8		13	6			17
Mean fat (wt%)	2.6	4.9	7.3	0.6		3.2			2.5	2.4		9.2	2.7			6.5
RSD (%)						22			36	8		15	44			32
Perch																
No. of pooled/ individual samples	1 (2)[a]	1 (5)		1			3		2 (11)	3 (15)	3 (8)		10		5	6
Mean length (mm)	163	185		205			146		148	216	188		155		161	154
RSD (%)	4	12					24		24	6	8		10		12	21
Mean fat (wt%)	0.6	0.7		0.6			0.6		0.7	0.6	0.6		0.7		0.8	0.9
RSD (%)							17		14	17	14		29		13	11

[a] In brackets: total number of fish samples composing pool samples.

Table 2 Characteristics of fish samples from river Moldau (Vltava) examined within monitoring program

Fish species	Characteristics of samples collected during monitoring program in particular localities (1 – Hluboká, 2 – Štěchovice, 3 – Klecany, 4 – Zelčín)											
	1998				1999				2000			
	1	2	3	4	1	2	3	4	1	2	3	4
Chub												
No. of pooled/ individual samples	8 (18)[a]		6 (17)		3 (10)	2 (3)	18		10	9	20	9
Mean length (mm)	258		258		296	288	300		322	316	323	300
RSD (wt%)	20		22		18	12	11		21	18	17	18
Mean fat (%)	0.8		2.0		2.7	2.7	4.2		2.6	2.5	3.0	2.2
RSD (%)	25		5		56	30	48		35	32	30	60
Bream												
No. of pooled/ individual samples	5 (12)[a]		3		2 (3)		2				14	10
Mean length (mm)	244		419		298		488				362	308
RSD (%)	13		7		3		6				7	8
Mean fat (wt%)	1.2		5.1		1.6		6.7				4.2	0.7
RSD (%)	8		16		13		24				93	43

Table 2 (continued)

Fish species	Characteristics of samples collected during monitoring program in particular localities (1 – Hluboká, 2 – Štěchovice, 3 – Klecany, 4 – Zelčín)											
	1998				1999				2000			
	1	2	3	4	1	2	3	4	1	2	3	4
Barbel												
No. of pooled/ individual samples			3 (10)[a]				3 (10)				3	
Mean length (mm)			408				471				408	
RSD (%)			10				10				11	
Mean fat (wt%)			3.2				5.3				4.3	
RSD (%)			19				19				5	
Perch												
No. of pooled/ individual samples		5 (14)	4 (11)[a]			3 (9)	18		7	13	18	
Mean length (mm)		162	166			214			204	185	227	
RSD (%)		26	17			6			39	21	25	
Mean fat (wt%)		0.3	0.7			0.7			0.9	0.8	0.9	
RSD (%)		33	14			14			11	25	33	

[a] In brackets: total number of fish samples composing pool samples.

Table 3 Characteristics of fish samples from river Tichá Orlice examined within monitoring program

Fish species	Characteristics of samples collected during monitoring program in particular localities (1 – Červená voda, 2 – Králíky, 3 – Lichkov)								
	1998			1999			2000		
	1	2	3	1	2	3	1	2	3
Trout									
No. of pooled/ individual samples	3 (13)[a]	3 (10)	3 (14)	3 (15)	3 (15)	3 (15)	3 (12)	3 (12)	3 (11)
Mean length (mm)	173	224	221	197	267	250	160	208	202
RSD (%)	12	13	12	11	9	8	11	7	10
Mean fat (wt%)	0.6	1.7	1.0	0.8	2.5	0.9	1.8	2.4	2.3
RSD (%)	33	18	59	13	12	22	6	13	9

[a] In brackets: total number of fish samples composing pool samples.

With the exception of the year 2000, in which each fish was analysed individually, pooling of samples was carried out to rationalize the number of analyses. To avoid the loss of information needed for an assessment of bioaccumulation, pooling strategy was based on the assumption that the length of fish is typically proportional to its age (which corresponds to the duration of the residence of fish in contaminated water). Samples from each locality were sorted into several groups according to their relative length. This parameter (alike the weight) was shown to correlate well with the relative age of the specimen within the population of collected samples. The lipid content extracted under conditions of analytical procedure applied for musk analysis was recorded for each analytical sample. As can be seen from Tables 1, 2 and 3, the among-season variation of both the mean lengths and lipid contents of fish caught in individual localities was relatively large, reflecting not only some variance due to the sampling process itself but also the influence of conditions existing in respective aquatic ecosystem, in particular sampling year (the later factor significantly determines the form of resident biota). The highest variability was found in the bream population. For instance the relative standard deviation (RSD) calculated for lengths of the whole set of these species collected along the river Elbe through the year 2000 was 25%, for the chub population having approximately the same mean length this value was about 18%, for examined barbel the RSD of lengths was the lowest (10%). A perspicuous correlation between the mean length of particular fish species and their lipid contents even in a single sampling period has not been found in any of the sampling localities. The highest correlation coefficient (0.8) was calculated for chub (n=20) and perch (n=18) from Klecany in the sampling year 2000.

Although fish was used as a main bioindicator of pollution situation at all localities (as well as fish fillets, some samples of livers were also examined for this purpose), samples of sediments were also collected here. Wet material was sieved, carefully dried and then stored in a refrigerator.

To obtain complementary information on the occurrence of synthetic musks in aquatic ecosystem and to receive complementary data needed for the assessment of bioaccumulation/metabolisation processes in fish, passive sampling utilizing semipermeable membrane devices (SPMDs) was carried out in several localities.

Regarding the target analytes, HHCB and AHTN representing polycyclic musk fragrances together with two major nitro musks – musk ketone (MK) and musk xylene (MX) – were determined in all samples throughout the whole monitoring program. In the last two years, an additional three polycyclic musks with lower production volumes and usage such as ADBI, AHDI, and ATII were analysed in fish from all localities.

Multiresidue analytical procedure used for all monitored musks was similar to that described by Rimkus et al. [5]. It consisted of following steps: (i) Soxhlet extraction of homogenized fish tissue (anhydrous sodium sulphate was added for dehydration) by hexane-dichloromethane solvent mixture (1:1, v/v), (ii) removal of bulk lipids from crude extract by automated gel permeation chromatography (GPC; Gilson 231XL, France) using Bio-Beads S-X3 gel (Bio-Rad Laboratories, USA) and cyclohexane-ethyl acetate (1:1, v/v) as a mobile phase, (iii) removal of residual matrix co-extracts by adsorption chromatography on silica gel mini-column (elution of analytes by 15% of diethyl ether in hexane, v/v), (iv) identification/quantitation of analytes by GC/MS (GC-HP 6890; DB-5MS capillary column, J&W Scientific, USA; MSD-HP 5973, Agilent Technologies, USA) operated in EI mode. The performance characteristics of this method (SIM, three characteristic ions monitored) are summarized in Table 4.

For gravimetric determination of lipids, aliquot portion of extract obtained in step (i) was used.

2.1 Synthetic Musks in Fish Samples from Selected Sites at Czech Rivers

The overview of synthetic musks levels analysed in fish samples throughout the monitoring program is given in Tables 5, 6 and 7 (results of 461 analyses are

Table 4 Performance characteristics of the GC/MS (SIM) method used for synthetic musk analysis of fish samples (isotope dilution technique using labeled D_3-AHTN and D_{15}-musk xylene as internal standards for quantitation)

Analytes	Limit of detection (LOD, μg kg^{-1} lw)	RSD (%) (at 10 LOD level, n=6)	Recovery (%) (at 10 LOD level)
HHCB (e.g. Galaxolide)	3	7	93
AHTN (e.g. Tonalide)	3	8	85
ADBI (e.g. Celestolide)	1	10	87
AHDI (e.g. Phantolide)	1	9	86
ATII (e.g. Traseolide)	1	11	85
MX (musk xylene)	5	11	82
MK (musk ketone)	5	9	91

Table 5 Musk compounds in fish from river Elbe (Labe), monitoring data 1997–2000

Fish species	Analyte	Content of analyte in fish from particular localities (μg kg^{-1} lw)															
		Kunětice				Srnojedy				Štětí				Hřensko			
		1997	1998	1999	2000	1997	1998	1999	2000	1997	1998	1999	2000	1997	1998	1999	2000
Chub	HHCB																
	Mean	695	709	631	1088	707	1169	1405	1146	840	1402	1049	849		672	739	932
	RSD (%)	14	24	10	17		26	8	21	51	18	27	23		13	26	35
	Median	695	736	632	1145		1233	1420	1175	673	1402	1059	886		708	703	791
	AHTN																
	Mean	655	735	647	957	739	950	1330	1052	888	1219	702	695		957	677	765
	RSD (%)	8	32	27	19		30	10	19	46	2	24	21		23	21	27
	Median	655	824	595	996		1152	1303	1096	740	1219	764	735		963	650	723
	MX																
	Mean	172	92	53	151	221	141	120	97	238	191	77	49		544	157	239
	RSD (%)	11	11	42	25		26	18	31	42	9	22	12		25	18	30
	Median	172	93	55	166		156	125	92	251	191	83	47		470	169	219
	MK																
	Mean	144	125	56	124	222	104	78	54	120	120	57	55		118	52	76
	RSD (%)	10	26	23	23		16	14	37	48	8	14	31		11	12	25
	Median	144	126	54	144		115	78	59	100	120	56	56		110	49	71
	ADBI																
	Mean		28	91			41	34			25	25			18	19	
	RSD (%)		30	32			32	68			16	36			17	63	
	Median		25	107			35	26			26	22			17	19	

Table 5 (continued)

Fish species	Analyte	Content of analyte in fish from particular localities (µg kg^{-1} lw)															
		Kunětice				Srnojedy				Štětí				Hřensko			
		1997	1998	1999	2000	1997	1998	1999	2000	1997	1998	1999	2000	1997	1998	1999	2000
	AHDI																
	Mean		54	221			92	60			50	56			40	49	
	RSD (%)		31	23			20	35			22	38			20	31	
	Median		48	255			86	57			52	48			36	45	
	ATII																
	Mean		58	108			95	36			25	32			20	28	
	RSD (%)		48	17			32	39			28	34			20	39	
	Median		57	116			106	34			25	30			19	29	
Bream	HHCB																
	Mean	915	3552	2001	1915	1904	2817	6364		492	4510		5417	1943	3655	2324	2816
	RSD (%)		9	6	18	13	11	5			16		27	24	15	39	25
	Median		3552	2051	1975	1904	2848	6444			4510		5526	1943	3448	2398	2687
	AHTN																
	Mean	585	1806	1210	1479	917	1162	2165		582	1194		1374	975	2135	1144	1359
	RSD (%)		3	11	24	18	13	13			57		51	19	14	25	30
	Median		1806	1260	1385	917	1083	2195			1194		1018	987	2012	1116	1305
	MX																
	Mean	149	159	99	140	279	147	86		68	304		81	129	610	230	169
	RSD (%)		10	12	30	13	7	33			25		36	91	39	20	27
	Median		159	103	129	279	143	87			304		75	105	514	232	153

Table 5 (continued)

Fish species	Analyte	Content of analyte in fish from particular localities (μg kg^{-1} lw)															
		Kunětice				Srnojedy				Štětí				Hřensko			
		1997	1998	1999	2000	1997	1998	1999	2000	1997	1998	1999	2000	1997	1998	1999	2000
	MK																
	Mean	167	149	97	96	191	95	110		167	118		69	338	129	56	72
	RSD (%)		9	14	10	1	11	12			1		41	28	17	21	19
	Median		149	98	99	191	90	106			118		70	323	121	56	70
	ADBI																
	Mean			48	25			49					34			37	34
	RSD (%)			8	12			18					50			19	41
	Median			49	23			48					24			39	31
	AHDI																
	Mean			123	78			174					84			69	85
	RSD (%)			7	22			22					44			16	33
	Median			125	82			189					78			69	78
	ATII																
	Mean			110	35			103					59			42	46
	RSD (%)			15	37			17					73			19	41
	Median			111	35			106					49			47	38
Barbel	HHCB																
	Mean	1871		2250	2373	3731	3207	3933		342				1127		2263	3792
	RSD (%)			12	8		17	12								20	
	Median			2250	2304		3279	4176								2587	

Table 5 (continued)

Fish species	Analyte	Content of analyte in fish from particular localities ($\mu g\ kg^{-1}$ lw)															
		Kunětice				Srnojedy				Štětí				Hřensko			
		1997	1998	1999	2000	1997	1998	1999	2000	1997	1998	1999	2000	1997	1998	1999	2000
	AHTN																
	Mean	1743		1319	2209	985	2715	2722		465				1206		1715	1653
	RSD (%)			32	11		15	19								18	
	Median			1319	2253		2862	3004								1908	
	MX																
	Mean	230		61	131	215	324	135		51				87	368	258	
	RSD (%)			5	11		17	18							13		
	Median			61	1239		323	125							389		
	MK																
	Mean	103		53	64	192	122	67		28				385	62	82	
	RSD (%)			4	8		11	3							10		
	Median			53	63		126	66							65		
	ADBI																
	Mean			17	14			26							17	18	
	RSD (%)			18	21			15							6		
	Median			17	15			24							18		
	AHDI																
	Mean	136	140					192							122	113	
	RSD (%)	22	11					48							14		
	Median	136	136					204							131		
	ATII																
	Mean	47	37					150							37	33	
	RSD (%)	21	5					58							11		
	Median	47	38					110							40		

Table 5 (continued)

Fish species	Analyte	Content of analyte in fish from particular localities (µg kg^{-1} lw)															
		Kunětice				Srnojedy				Štětí				Hřensko			
		1997	1998	1999	2000	1997	1998	1999	2000	1997	1998	1999	2000	1997	1998	1999	2000
Perch	HHCB																
	Mean	896		1456	1123	2437		2853			1954	5318	392				956
	RSD (%)			26	20			23			29	33					12
	Median			1294	1048			2990			2035	5847					956
	AHTN																
	Mean	980		1289	1055	967		2648			1390	1239	911				797
	RSD (%)			29	19			35			27	41					1
	Median			1133	1009			2763			1622	1025					797
	MX																
	Mean	88		109	189	195		173			162	117	33				235
	RSD (%)			4	30			15			5	55					14
	Median			107	186			190			167	107					235
	MK																
	Mean	86		96	134	296		155			129	127	170				128
	RSD (%)			6	19			18			12	68					24
	Median			92	130			165			121	103					128
	ADBI																
	Mean		68	86			91	99	67								36
	RSD (%)		10	35			18	25	63								3
	Median		70	102			101	89	70								36

Table 5 (continued)

Fish species	Analyte	Content of analyte in fish from particular localities (µg kg^{-1} lw)															
		Kunětice				Srnojedy				Štětí				Hřensko			
		1997	1998	1999	2000	1997	1998	1999	2000	1997	1998	1999	2000	1997	1998	1999	2000
	AHDI																
	Mean		141	207			242		173	153							81
	RSD (%)		21	31			21		6	63							17
	Median		151	236			262		171	156							81
	ATII																
	Mean		161	126			288		148	115							37
	RSD (%)		14	42			24		36	83							35
	Median		149	127			287		118	98							37

Table 6 Musk compounds in fish from river Moldau (Vltava), monitoring data 1998–2000

Fish species	Analyte	Content of analyte in fish from particular localities (µg kg^{-1} lw)											
		Hluboká nad Vltavou			Štěchovice			Klecany			Zelčín		
		1998	1999	2000	1998	1999	2000	1998	1999	2000	1998	1999	2000
Chub	HHCB												
	Mean	971	1351	821		368	707	2693	1274	1555			855
	RSD (%)	29	35	22		30	28	25	23	24			22
	Median	904	1607	793		368	704	2724	1240	1535			804
	AHTN												
	Mean	1019	753	520		232	407	2448	831	1283			802
	RSD (%)	34	33	25		24	22	19	18	29			21
	Median	878	794	535		232	445	2390	888	1188			759
	MX												
	Mean	160	100	65		104	32	494	160	169			146
	RSD (%)	41	40	23		22	25	14	14	26			13
	Median	146	110	58		104	33	534	160	165			136
	MK												
	Mean	121	85	37		180	24	213	121	127			87
	RSD (%)	33	32	25		16	17	14	15	18			16
	Median	107	101	39		102	25	232	103	123			86
	ADBI												
	Mean		41	61		96	16		41	22			24
	RSD (%)		24	18		29	19		22	45			83
	Median		45	63		96	18		47	21			18

Table 6 (continued)

Fish species	Analyte	Content of analyte in fish from particular localities (µg kg^{-1} lw)											
		Hluboká nad Vltavou			Štěchovice			Klecany			Zelčín		
		1998	1999	2000	1998	1999	2000	1998	1999	2000	1998	1999	2000
	AHDI												
	Mean		79	76		94	23		72	76			51
	RSD (%)		14	20		29	48		54	25			27
	Median		75	79		94	20		77	77			52
	ATII												
	Mean		45	60		83	14		34	34			34
	RSD (%)		20	22		30	29		26	29			32
	Median		39	61		83	13		36	34			42
Bream	HHCB												
	Mean	3299	1189					8654	5052	6394			2205
	RSD (%)	14	13					7	3	114			26
	Median	3299	1156					8394	5052	7971			2146
	AHTN												
	Mean	1866	1112					4764	1784	6940			1454
	RSD (%)	6	18					10	13	102			42
	Median	1866	1023					4494	1784	3470			1375
	MX												
	Mean	133	130					759	360	829			202
	RSD (%)	3	5					26	3	115			24
	Median	133	130					828	360	431			212

Table 6 (continued)

Fish species	Analyte	Content of analyte in fish from particular localities ($\mu g\ kg^{-1}$ lw)											
		Hluboká nad Vltavou			Štěchovice			Klecany			Zelčín		
		1998	1999	2000	1998	1999	2000	1998	1999	2000	1998	1999	2000
	MK												
	Mean	102	130					228	123	412			119
	RSD (%)	6	6					22	10	125			28
	Median	102	130					225	123	179			115
	ADBI												
	Mean		86						68	103			29
	RSD (%)		26						16	63			79
	Median		105						68	78			25
	AHDI												
	Mean		158						253	474			121
	RSD (%)		20						12	78			16
	Median		146						253	302			122
	ATII												
	Mean		117						88	180			62
	RSD (%)		28						6	113			27
	Median		98						88	104			57
Barbel	HHCB												
	Mean							10806	4616	2972			
	RSD (%)							17	12	11			
	Median							10844	4445	2745			

Table 6 (continued)

Fish species	Analyte	Content of analyte in fish from particular localities (µg kg^{-1} lw)											
		Hluboká nad Vltavou			Štěchovice			Klecany			Zelčín		
		1998	1999	2000	1998	1999	2000	1998	1999	2000	1998	1999	2000
	AHTN												
	Mean							12026	3049	2284			
	RSD (%)							19	27	20			
	Median							11378	2563	2143			
	MX												
	Mean							1956	524	325			
	RSD (%)							17	19	30			
	Median							1878	518	315			
	MK												
	Mean							659	180	159			
	RSD (%)							29	8	18			
	Median							584	189	143			
	ADBI												
	Mean								67	106			
	RSD (%)								10	8			
	Median								67	104			
	AHDI												
	Mean								448	233			
	RSD (%)								14	18			
	Median								412	215			

Table 6 (continued)

Fish species	Analyte	Content of analyte in fish from particular localities (µg kg^{-1} lw)											
		Hluboká nad Vltavou			Štěchovice			Klecany			Zelčín		
		1998	1999	2000	1998	1999	2000	1998	1999	2000	1998	1999	2000
	ATII												
	Mean								107	59			
	RSD (%)								18	17			
	Median								95	56			
Perch	HHCB												
	Mean			2837	784	578	2394	5446	5753	1670			
	RSD (%)			25	54	9	43	8	81	31			
	Median			2749	509	553	2329	5482	5753	1568			
	AHTN												
	Mean			1468	706	413	1895	3547	1248	1470			
	RSD (%)			21	53	8	35	11	65	32			
	Median			1521	514	402	1782	3717	1248	1443			
	MX												
	Mean			181	254	112	189	506	362	247			
	RSD (%)			40	24	4	56	10	50	35			
	Median			156	283	110	181	520	362	232			
	MK												
	Mean			180	194	145	179	143	240	193			
	RSD (%)			59	25	8	66	13	41	53			
	Median			140	223	143	165	150	255	179			

Table 6 (continued)

Fish species	Analyte	Content of analyte in fish from particular localities (µg kg^{-1} lw)											
		Hluboká nad Vltavou			Štěchovice			Klecany			Zelčín		
		1998	1999	2000	1998	1999	2000	1998	1999	2000	1998	1999	2000
	ADBI												
	Mean			98		69	127		284	192			
	RSD (%)			46		9	62		37	34			
	Median			67		65	128		284	178			
	AHDI												
	Mean			145		92	228		386	271			
	RSD (%)			8		17	56		48	27			
	Median			142		89	239		386	279			
	ATII												
	Mean			113		99	175		236	234			
	RSD (%)			38		49	56		48	27			
	Median			103		78	183		236	230			

Table 7 Musk compounds in fish from river Tichá Orlice, monitoring data 1998–2000

Fish species	Analyte	Content of analyte in fish from particular localities (µg kg^{-1} lw)								
		Červená Voda			Králíky			Lichkov		
		1998	1999	2000	1998	1999	2000	1998	1999	2000
Trout	HHCB									
	Mean	271	586	489	2330	1000	1787	713	1371	1628
	RSD (%)	11	63	12	16	34	7	42	61	3
	Median	266	343	475	2233	1096	1335	819	876	1606
	AHTN									
	Mean	282	399	333	3125	642	1022	1181	858	614
	RSD (%)	14	25	5	9	15	8	41	23	52
	Median	286	363	334	3156	702	970	1264	825	793
	MX									
	Mean	187	162	80	1792	224	446	294	226	384
	RSD (%)	13	20	3	8	16	9	50	31	65
	Median	181	162	79	1740	216	452	349	201	149
	MK									
	Mean	105	111	57	188	112	84	200	149	102
	RSD (%)	13	5	2	12	29	11	43	25	18
	Median	98	112	58	184	91	78	219	128	89
	ADBI									
	Mean		116	33		55	82		121	37
	RSD (%)		16	6		15	5		26	5
	Median		106	32		57	81		101	37
	AHDI									
	Mean		140	62		119	111		213	79
	RSD (%)		17	3		16	26		14	3
	Median		130	63		130	97		201	79
	ATII									
	Mean		67	40		45	50		93	40
	RSD (%)		19	3		22	6		33	3
	Median		62	40		46	49		74	41

summarized here). As documented throughout this chapter, the pertinence of information mediated by generated data is dependent on a comprehensive characterization of actual fish samples submitted for analysis as well as on the knowledge of conditions existing in a particular locality in respective sampling season (water flow, its temperature etc.).

In agreement with most of similar monitoring studies concerned with occurrence of lipophilic persistent chemicals in biota, the content of musks in examined samples was normalized to the lipid contents. However, the extent of sampling site contamination was sometimes more cogently illustrated by the expression of contaminants content in muscle/liver tissue. The total body burden of respective bioindicator can be better appraised in this way. Moreover, for the risk assessment of the dietary intake of hazardous chemicals contained in fish, information on their concentration in edible portions (fish fillets were examined in this study) is more relevant. As mentioned earlier, large differences in mean lipid content were found in population of particular fish species along monitored rivers. The reason for uneven adiposity of fish is undoubtedly the variable availability of food; its amount depends on trophic value of the water. The presence of ample macrozoobenthos is typical for mesotrophic waters. In this context, communal sewage might be a good source of nitrogen and phosphorus needed for its development. As long as the amount of oxygen remains sufficient for biota, the presence of abundant food is reflected by the rise of fish fatness in such (contaminated) locality. Under these conditions, bioconcentration factors on a wet weight basis are expected to increase due to a higher lipid content in organisms, although this does not necessarily result in elevated concentrations of contaminants in lipid [6]. In Fig. 2 various expressions of synthetic musk concentrations in bream obtained from two sampling sites of river Elbe in the years 1997–1999 (unfortunately no bream were available in the sampling year 2000 at these localities) are shown.

While, compared to samples from Kunětice, increased levels of both musk groups in tissues of bream from Srnojedy were apparent throughout the entire monitoring period indicating the closeness of pollution source (urban area Pardubice is located upstream of this sampling site), normalization of musk levels to lipid content rather obscured (see data from years 1998 and 1999) the differences between these two localities as regards the extent of their contamination. Although higher musk levels in locality Srnojedy were also documented by analysis of bottom sediments, corresponding concentrations of musks in lipid might be levelled due to the "diluting effect" occurring in fattier fish. Rather surprisingly, contamination of bream was very low in both localities close to the city Pardubice in the first monitoring year 1997. According to our records, long-term rains in respective summer season caused unusually high water flows in particular section of river Elbe what might have caused dilution of wastewater containing musk compounds and, consequently, lowered the extent of their uptake by fish. Another factors which are probably responsible for apparently low body burden of musks in bream from Srnojedy in 1997 is their small mean size (respectively age) as well as lower content of lipids compared to fish population collected in subsequent years.

Overall, polycyclic musks HHCB and AHTN were clearly dominating compounds in all fish samples as shown in a recent survey by Rimkus [7a]. Similar findings are reported in all studies carried out in Europe. The presence of other compounds of this group (ADBI, AHDI and ATII) could be unequivocally analysed in our samples too; nevertheless, their levels were lower by at least one order of magnitude. The nitro musk fragrances MX and MK were also proved ubiquitous in all monitored aquatic ecosystems. When comparing their typical abundance with polycyclic musks, the following concentration order was found in most of fish samples: HHCB>AHTN≫MX≥MK≈AHDI>ATII≈ADBI. It is worth noting that, in contrast to our data, significantly higher residue levels of MK compared to MX were reported in several German studies [2c, 7b,c, 8] for various freshwater fish, both from rivers (including river Elbe) and sewage ponds. In their comprehensive

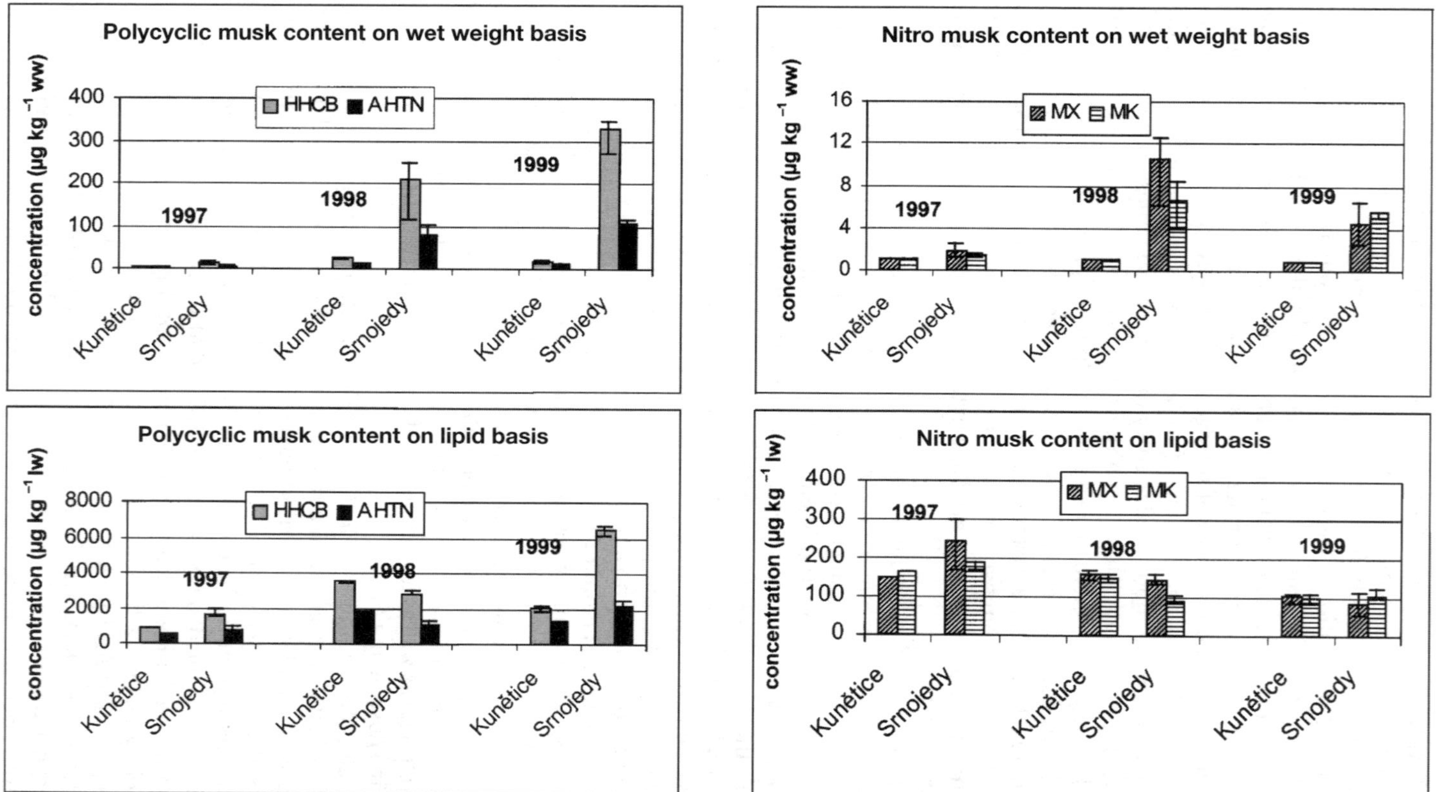

Fig. 2 Concentrations (medians) of synthetic musk compounds in breams from sampling sites located upstream (Kunětice) and downstream (Srnojedy) of urban area Pardubice

review on the fate of nitro musks in the aquatic environment, Rimkus et al. [9] referred to an important role of metabolization pathway in sewage plants resulting in a formation of monoamino metabolites from parent compounds; their presence was proved in various compartments of aquatic environment [10a,b, 11]. Adsorption in sludge is another factor influencing concentration of synthetic musks and related compounds in effluents. Although P_{ow} value (3.8) estimated for 4-NH_2-MX (main MX metabolite) is by about one order of magnitude lower than that of parent compound (P_{ow}=4.9), its hydrophobicity is still high enough to allow bioaccumulation in fish (NH_2-MK, the main metabolite of MK is too polar in this context). As shown by Rimkus et al. [9], in some fish from the German part of the river Elbe and mainly in biota from sewage ponds, 4-NH_2-MX levels exceeded its precursor even by one order of magnitude. Consequently, MX/MK ratio$\leq$1 was determined in fish in spite of lower bioconcentration factor of MK. Concentration of nitro musks in water has to be considered for the complex assessment of bioconcentration; nevertheless, opposite situation could be expected due to the higher production rates of MX compared to MK [1a]. However, since the first findings of nitro musks in the environment and human samples at the beginning of the 1990s, the industrial production and use of MX has been significantly reduced [7c]. Rather distinct nitro musks pattern recognized in fish from Czech rivers may be a result of differences in technologies employed in sewage treatment plants; biodegradation processes leading to transformation of MX might be less extensive than those in the above-mentioned German studies. The other alternative, i.e. different pattern of use volumes of individual nitro musks in the Czech Republic compared to other countries, is not probable because practically the same cosmetic/detergent products are available in both markets. Assumption on a lower conversion rate of MX into its amino metabolite in Czech sewage treatment facilities was confirmed within a pilot study in which sets of three fish species from Klecany (Moldau, sampling year 2000, characteristics of examined set of fish is shown in Tables 1, 2 and 3 were analysed not only for MX and MK but also for their metabolites represented by 2-NH_2-MX, 4-NH_2-MX and 2-NH_2-MK. Although 4-NH_2-MX (no other amino musks above LOD 10 $\mu g\,kg^{-1}$ lw were detected) exceeded concentration of MX in all samples, the metabolite/parent compound concentration ratio varied among the fish species: the maximum value of 2.2 (RSD=10%) was found in perch; in chub it was 1.9 (RSD=12%) and in bream only 1.6 (RSD=10%). Either differences in bioconcentration mechanism or in the extent of metabolization of nitro musks in fish may explain these observations. In any case, in accordance with the lowest content of 4-NH_2-MX, the highest MX/MK ratio was recorded in the later species in locality Klecany – not only in the year 2000 but also in previous sampling periods.

Generally, high correlation coefficients between concentrations of the main polycyclic musks (HHCB and AHTN) existing for all examined fish species (typically r^2>0.9) were not surprising because only mixtures of these compounds are used for fragrance compositions. Regarding nitro musks, less distinct correlation between MK and MX levels in fish was calculated in some cases. Varying extent of their biotransformation discussed above might explain this fact.

Considering the whole set of data obtained for fish used as pollution bioindicator in monitored localities, the worst situation with respect to musk levels was recognized in the localities of Klecany at Moldau, Srnojedy at Elbe and Králíky at Tichá Orlice. In all cases the sampling sites were located only a few kilometers downstream from large urban areas at which production of large volumes of sewage waters is common. Invariably, the most serious contamination of aquatic ecosystem was recognized at Klecany located approximately 4 km downstream of the outlet of one of the biggest Prague municipal sewage treatment plant. The absolutely highest concentrations (90% percentile, mg g^{-1} lw) were found in1998 in barbel: HHCB 12.6, AHTN 13.3, MX 2.3 and MK 0.9. Unfortunately, no corresponding data for this particular species (typically living in close contact with bottom sediments) are available in the literature for comparison. Nevertheless, considering the levels of polycyclic musks in bream and perch measured throughout the monitoring program in Klecany, the extent of their contamination was approximately two times higher than "the worst case situation" reported by Eschke et al. [2] in 1995 for the same species collected from the extensively polluted river Ruhr.

In accordance with findings reported, e.g. for contamination of various freshwater fish caught in several Italian rivers [12] or more recently in the river Elbe [13], HHCB levels (both, means and medians) always exceeded those of AHTN in the Czech fish samples. On the other hand, the opposite ratio for these major polycyclic musks was recorded for instance by Eschke et al. [2d], although HHCB levels in water samples taken from river Ruhr at the same time period (1994–1996) reported by the same author [2c,e] were higher compared to AHTN. It should be emphasized that environmental behaviour of these chemicals is influenced by many factors and without the knowledge of the actual pollution situation in particular compartments (levels of musks in water, suspended particulate matter and sediments) it is difficult to discuss the reason of this phenomenon.

As a part of critical assessment of the whole set of data generated throughout the monitoring program, occurrence of temporal trends in concentrations of both musk groups measured in all localities has been checked. In contrast with some reports on increasing tendency of polycyclic musks with concurrent decrease of nitro musks discussed by Rimkus [9], no statistically significant tendency of this kind was recognized in Czech rivers for either musks group. As illustrated in Fig. 3, only small variations of musk levels in specific sampling localities (less and seriously polluted ones were selected) occurred within three monitoring years both in fish tissues and sediments, in spite of some inter-annual changes in characteristics of particular matrices.

Differences in lipid contents in case of biota or slight changes in content of organic carbon in sediments may influence concentration of target pollutants. These observations may be due to a relatively insignificant fluctuation in composition of effluents from sewage treatment plants. Limited variation of musk concentrations in SPMDs deployed in river Elbe [15] as well as in various surface water samples [16] has been recently reported by other authors.

Another example of three years monitoring data obtained in our project is shown in Fig. 4 for trout collected at three localities along the small river Tichá Orlice.

As documented here, relatively low levels of musks were consistently measured in fish from sampling site downstream of the small village Červená Voda (low lipid contents in local fish were due to a limited availability of food in this "clean"

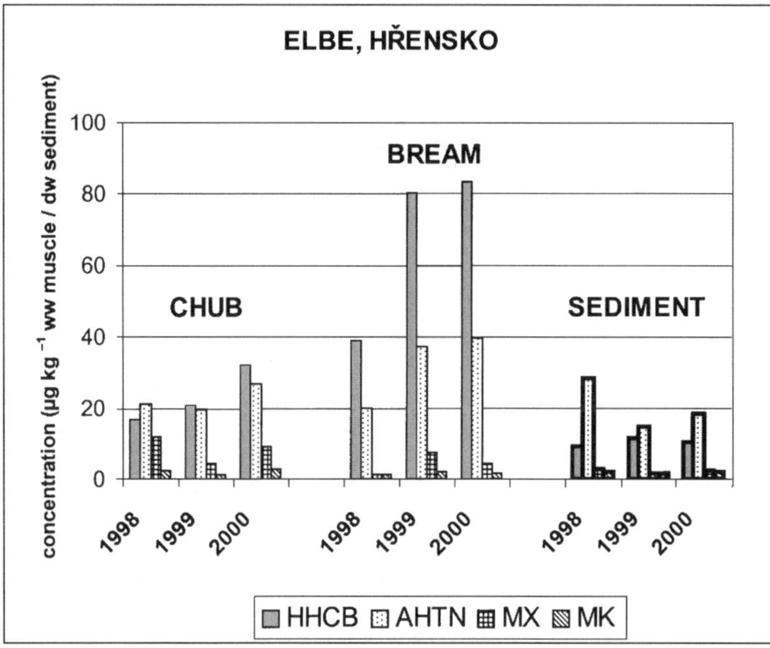

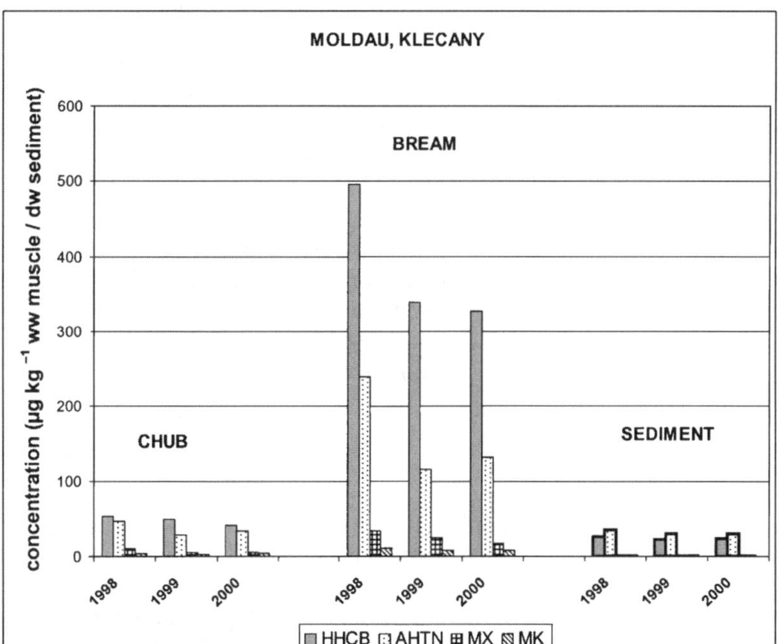

Fig. 3 Concentrations (medians) of synthetic musk compounds analysed in chub, bream and sediments collected in Hřensko (river Elbe) and Klecany (river Moldau) during monitoring program

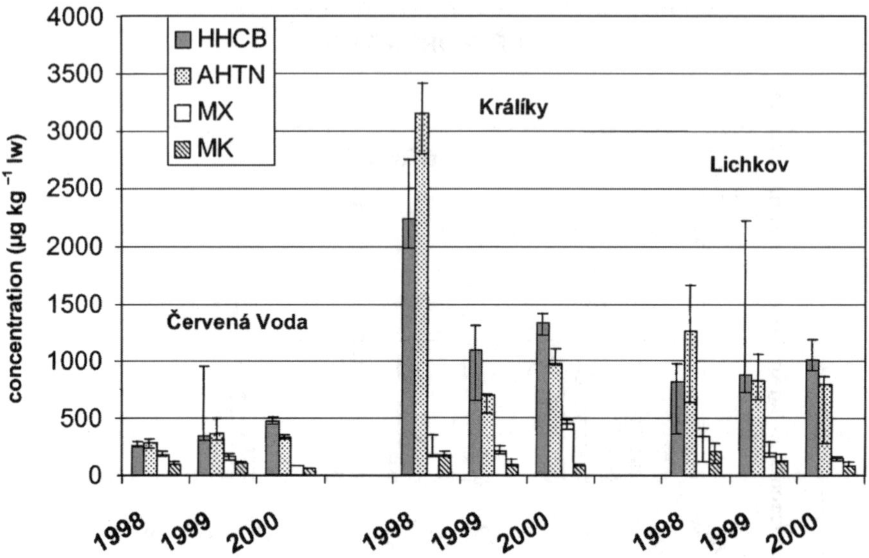

Fig. 4 Concentrations (medians) of synthetic musk compounds in trout from sampling sites at river Tichá Orlice

locality). On the other hand, samples from Králíky were in all the years distinctly more contaminated; extremely extensive pollution of trout by polycyclic musks was recorded in 1998. Interesting to note that, contrary to other years, concentrations of AHTN dominated in 1998 over HHCB. This rather untypical phenomenon was also observed in this year in the locality downstream of Lichkov. On the other hand, the levels of nitro musks in trout were at the same time relatively low. The same musks pattern was found in sediments which were, compared to other sampling sites, relatively extensively contaminated, specifically by polycyclic musks (probably not only due to high musks inputs but also because of relatively high content of organic matter which is responsible for binding of lipophilic pollutants). As one month prior to fish sampling local floods occurred at river Tichá Orlice in the year 1998, increased exposure of fish to mobilized sediments could have caused this unusual contamination pattern.

2.2
Fish as Biomonitor vs SPMD

Semipermeable membrane sampling device (SPMD) that was originally invented to study the bioavailability of hydrophobic organic chemicals such as PCBs and/or organochlorine pesticides to aquatic organisms [17] has nowadays become a widely used passive sampler finding a large range of applications including monitoring of freshwater pollution. Recently, SPMD has also been shown applicable in analysis of synthetic musks in polluted rivers and various reservoirs [10a, 13, 18]. To assess an informative/predictive value of data obtained by analysis of SPMD and to compare it with analyses of fish commonly used as biomonitor, a pilot study was conducted within our monitoring program. In Fig. 5 there are shown concentrations of musks analysed in extracts recovered from the low-density polyethylene membranes that were deployed for three weeks in the river Moldau, locality Klecany.

In Fig. 5 there are also shown data obtained for several fish species collected here at the end of SPMD sampling period. As can be seen, under conditions applied in our experiments, the concentration range of polycyclic musks accumulated in a "virtual fish" represented by SPMD corresponded approximately to chub; on the other hand, as regards relative abundance of synthetic musks, the best similarity was recorded for bream. However, the levels of nitro musks determined in triolein isolated from semipermeable membrane bag were distinctly lower than those found in any of the fish species studied; the accumulation was low specifically for MK (MX/MK ratio was significantly higher in this kind of passive sampler than in fish). Very similar observations were obtained in other localities. Considering the physico-chemical properties of particular compounds,

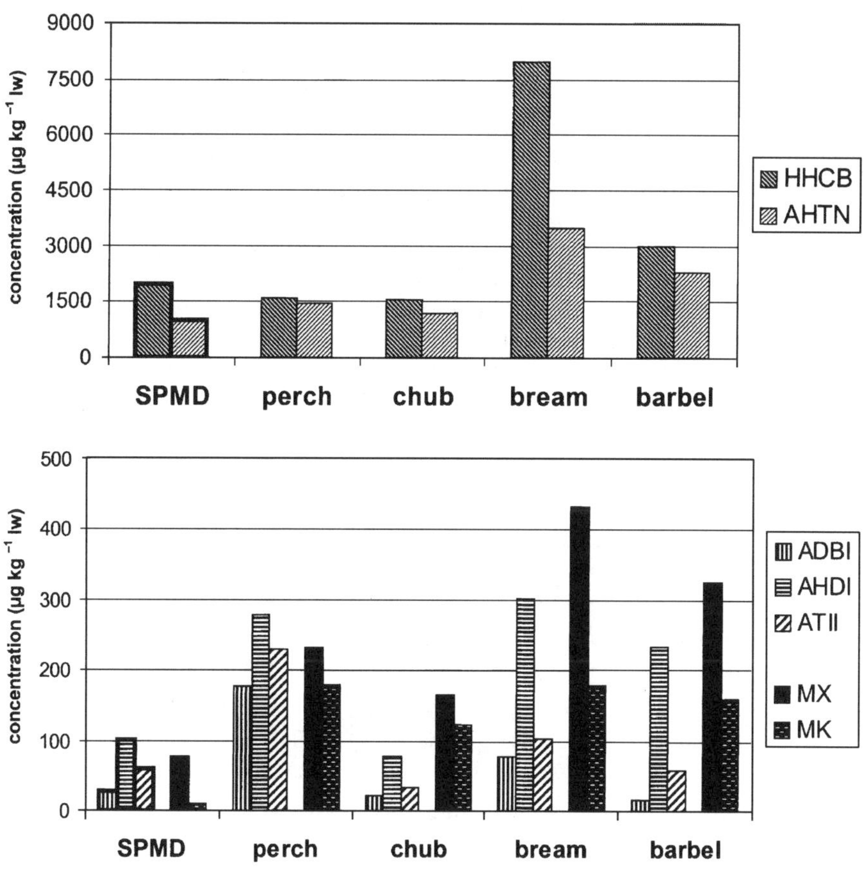

Fig. 5 Concentrations (medians) of synthetic musk compounds in perch, chub, bream and barbel, in comparison with those in SPMD (river Moldau, locality Klecany, sampling year 2000)

this phenomenon might be attributed to a lower lipophilicity of this class of musks (P_{ow} of MK and MX=4.2 and 4.9, respectively) [19] compared to polycyclic fragrances (e.g. P_{ow} of HHCB and AHTN=5.9 and 5.8, respectively) [7a]. In any case, it should be noted that musk patterns (relative abundance of individual compounds) in fish do not generally correlate with their physico-chemical properties. Significantly lower bioconcentration (uptake of chemical by absorption from water only) and bioaccumulation (combination of bioconcentration and food uptake) than expected on the basis of theory were invariably measured for polycyclic musks in several studies [7a, 16–20]. This discrepancy might be attributed to a strong metabolism taking place in these aquatic biota [10a, 13, 16]. For more details see the chapter of Biselli et al. in this monograph.

Regarding the SPMD uptake that corresponds to three weeks sampling period applied in our experiments, we assume it can be classified as a linear, integrative type of sampling, i.e. equilibrium was not approached. This assumption is in line with a recent study [8] reporting generally higher enrichment of polycyclic musk compounds in SPMDs after seven weeks of their deployment in the pond of a municipal sewage treatment plant compared to several fish species living here. Also the relative accumulation of MX and MK was higher as the result of significantly longer exposure period of SPMDs compared to our study. It should be noted that the plateau levels of AHTN and HHCB were rapidly reached (3–7 days of exposure) in bioconcentration tests employing bluegill sunfish [20]. However, as regards our study, it is rather difficult for us to speculate on the status of exposure phase of fish at the time of its catching because of different exposure conditions and wide inter-species variation as regards intake/excretion rates.

Simultaneously with synthetic musks, priority lipophilic contaminants such as PCBs and related persistent chlorinated aromatic compounds were determined in all examined samples. As illustrated in Fig. 6, the relative abundance of individual compounds composing the later group of pollutants as well as their analytical concentrations in trioleine after 21 days of SPMD deployment were

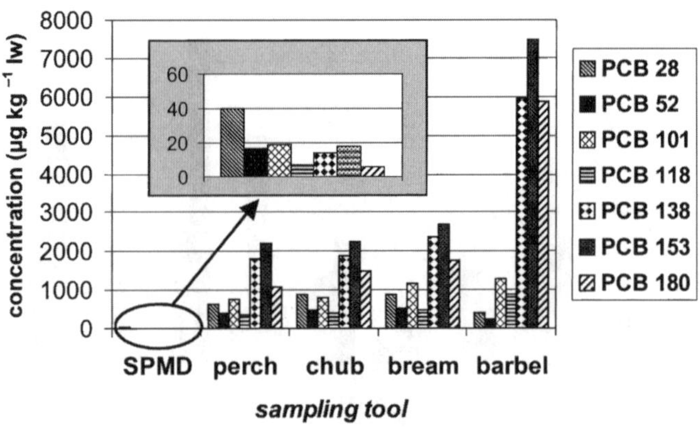

Fig. 6 Concentrations (medians) of PCBs in perch, chub, bream, and barbel in comparison with SPMD (river Moldau, locality Klecany, sampling year 2000)

mainly lower than those determined in fish. Poor agreement of the data for these chemicals obtained by this passive sampler and biomonitoring organism in spite of similar P_{ow} values to those of synthetic musks is mainly due to biomagnification occurring in fish in case of PCBs (this phenomenon obviously does not take place for musks). In any case, significantly longer deployment of SPMDs to attain equilibrium for PCBs would be needed.

2.3 Species-Dependent Pollution of Fish by Synthetic Musks

Remarkable differences in measured bioconcentration factors (BCFs) for individual musks among various fish species highlighted by several authors [7a, 8, 9, 21] have also been well documented in our study. As already shown in the foregoing Figures (see Figs. 3 and 5) and confirmed in Fig. 7, generally, the most extensive bioconcentration of musk compounds occurred in bream and/or in barbel while the lowest total musks levels were perpetually found in chub despite typically higher mean levels of lipid in this fish compared to those determined in other species.

Considering the whole data set obtained for polycyclic musks (including the minor compounds of this group) in the most abundant fish species, both the mean concentrations and median values were statistically higher in bream and/or barbel than in chub and/or perch regardless of the expression of results (either on wet weight or lipid weight). Regarding nitro musks, this trend was not so distinct. In any case, although differences in contamination of monitored localities existed, in all of them the relative amount of polycyclic musks was always higher

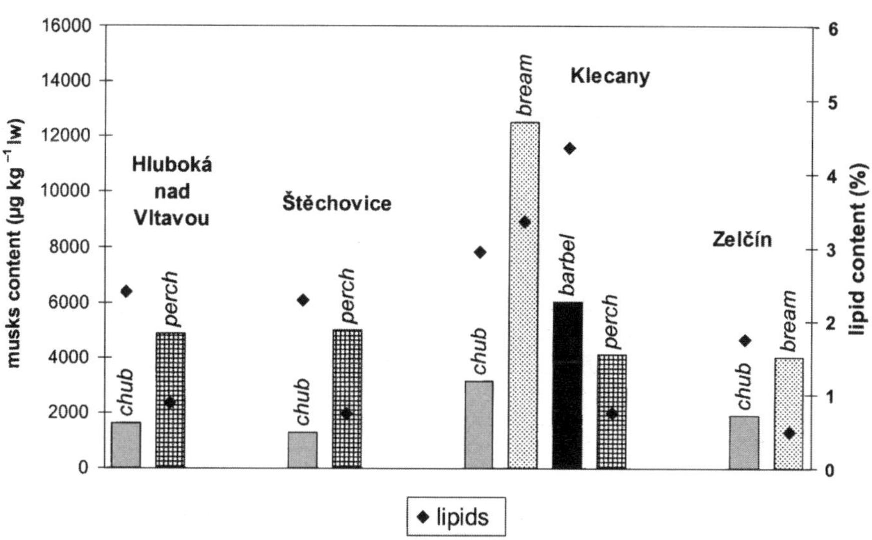

Fig. 7 Total concentrations of synthetic musk compounds (sum of examined analytes) in several fish species collected in sampling sites at river Moldau and mean values of lipid content (sampling year 2000)

in bream ranging between 90–95%; in chub their contribution to the total musks content was lower, in the range 75–86%.

Figure 8 illustrates typical patterns of analytes representing both groups of musks in fish from river Elbe, locality Hřensko.

In agreement with literature data [7a], the pattern of musks in muscle and liver of particular species were practically identical, the differences in analysed levels of musks were caused only by differing content of lipids in examined tissues. However, distinctive dissimilarities in relative abundance of HHCB and AHTN obviously due to the differences in the metabolization rate were recorded. While in chub the ratio of these dominating polycyclic musk compounds varied in a relatively narrow range typically slightly exceeding the value 1, in bream HHCB was clearly prevailing over AHTN, the ratio of major polycyclic musks ranged from approximately 1.5 up to almost 4. Considering the principles of pure bioconcentration mechanism [6] (which is theoretically simulated by SPMD sampling discussed above) as well as taking into account the data generated in presented study it might be speculated that less extensive metabolization of polycyclic musks, specifically of HHCB, occurs in bream, or vice versa a faster metabolism of AHTN. Our observations are in agreement with another study [21] concerned also with the examination of several fish species for musk levels: HHCB seems to be more prone to biotransformation compared to AHTN.

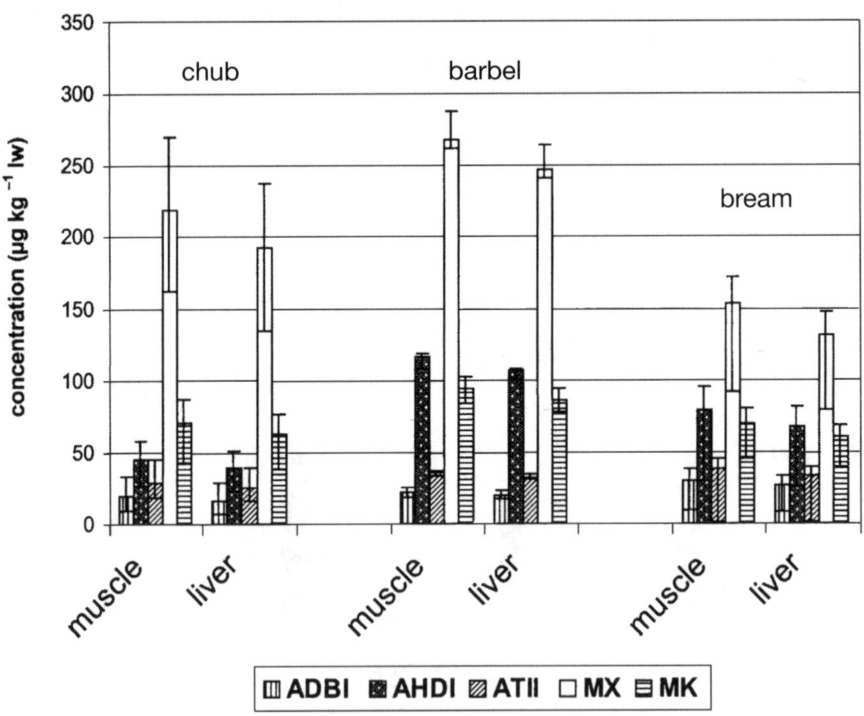

Fig. 8 Concentrations (medians) of synthetic musk compounds in muscle and liver of chub, barbel and bream (river Elbe, locality Hřensko, sampling year 2000)

3
Human Milk as Bioindicator: Set-Up of a Pilot Study

Human milk and adipose tissue is a widely employed bioindicator of the actual body burden of liphophilic and persistent compounds. While studies of a long-term exposure to organochlorine contaminants as well as studies of time-related trends in environmental contamination have already been conducted for several decades, the occurrence of musk compounds in these matrices were reported as late as at the beginning of the 1990s [22, 23a,b]. Regarding human exposure to synthetic musks, dietary intake seems to be negligible in this context, since meaningful contamination of food was found only in aquatic organisms (specifically freshwater fish) and not in other food of terrestrial animal origin. The main route of musks uptake is probably via dermal absorption due to frequent and intense dermal contact with fragrances contained in cosmetics and washed textiles [5, 23b, 25].

Before the pilot study presented here, no data were available in the Czech Republic for the assessment of human exposure by musks. In order to receive at least partial information, 59 milk samples were collected from nursing mothers (living but not necessarily born in Prague) by the Gynaecological Clinic which belongs to Medical Faculty, the Charles University, Prague. The manual sampling (milk expressed from the breast into a clean container) was conducted in accordance with WHO guidelines. Using a detailed questionnaire, relevant information on biological parameters such as age, dietary habits (specifically consumption of freshwater/marine fish), use of perfumed cosmetics, frequency of contacts with detergents, etc. were collated. Regarding the analytical procedure, isolation of analytes from milk samples was carried out according to AOAC procedure described in European norm 528 [26]. Isolated lipids (their content in milk ranged from 1.5 to 4.2 wt%) containing target analytes were processed in the same way as described above for fish samples. The limit of detection (LOD) was estimated at 10 $\mu g\ kg^{-1}$ lw for all the analytes.

3.1
Synthetic Musks in Human Milk Samples Collected in Prague

In Table 8 the obtained results are summarized; for comparison, data from several similar German studies are listed too.

In all the Czech samples, concentrations of HHCB exceeded LOD; AHTN, MX and MK were determined in 90, 92 and 32% of samples, respectively. Although, based both on mean and median values, the concentration order HHCB>AHTN>MX>MK was found in the milk set, great differences existed among examined samples (Fig. 9); the distribution of analytes was not a Gaussian one.

In some of the milk samples the opposite concentration order within the particular group of synthetic musks was determined – either AHTN slightly exceeding HHCB (in 25% of cases) or MK was higher than MX (in three cases). The ratio of sum of polycyclic musks/sum of nitro musks ranged from 1 to 20. These facts clearly document a great diversity of exposure routes of individual donators. As regards the overall extent of breast milk contamination, distinctly higher levels of polycyclic musks were found in Czech samples when compared with

Table 8 Concentrations of synthetic musks (µg kg^{-1} lw) in human milk samples from Czech Republic, Prague; comparison with data obtained in German studies (references in square parentheses)

Parameter	Origin of study, Number of samples	HHCB	AHTN	MX	MK
Minimum	Czech Rep.	13	<10	<10	<10
	[22], n[a]=391	na[b]	na[b]	10	<10
	[23b], n[a]=23	na[b]	na[b]	40	10
	[5], n=5	16	11	10	5
	[27], n=55	na	na	10	nd[c]
Maximum	Czech Rep.	720	565	156	93
	[22], n=391	na	na	1220	240
	[23b], n=23	na	na	190	90
	[5], n=5	108	58	30	15
	[27], n=55	na	na	253	103
Median	Czech Rep.	149	67	42	30
	[22], n=391	na	na	70	30
	[27], n=55	na	na	40	10
Mean	Czech Rep.	214	112	53	39
	[22], n=391	na	na	100	40
	[27], n=55	na	na	41	10
Standard deviation (RSD %)	Czech Rep.	175 (82)	114 (101)	37 (70)	24 (61)
90% percentile	Czech Rep.	509	304	106	77
	[22], n=391	na	na	210	80

[a] n = number of samples.
[b] na = not analysed.
[c] nd = not detected.

data reported by Rimkus et al. [5]. On the other hand our results were more comparable to those (not shown in Table 8) reported by Eschke [24] in another study for two milk samples. In any case, very limited set of data available at present does not allow any general conclusions.

Contrary to polycyclic musks, relatively numerous studies have been concerned with the occurrence of nitro musks in this bioindicator [22, 23a,b, 27]. Levels of nitro musks determined in Czech samples corresponded or were slightly lower than values reported by German authors (Table 8), the MX/MK ratio seems to be rather lower (relative abundance of MX higher) in our samples (however, it is worth noting that maximum levels of MX reported in extensive study carried out nine years ago in Southern Bavaria [22] were higher by almost one order of magnitude). Whether continual drop in levels of nitro musks in Czech human milk occurs (the decrease of PCBs and other persistent contaminants was unequivocally documented one of our recent studies) in accordance with downward trend described in Germany within the period 1993–1996 [28] will reveal a follow-up study planned for the year 2003.

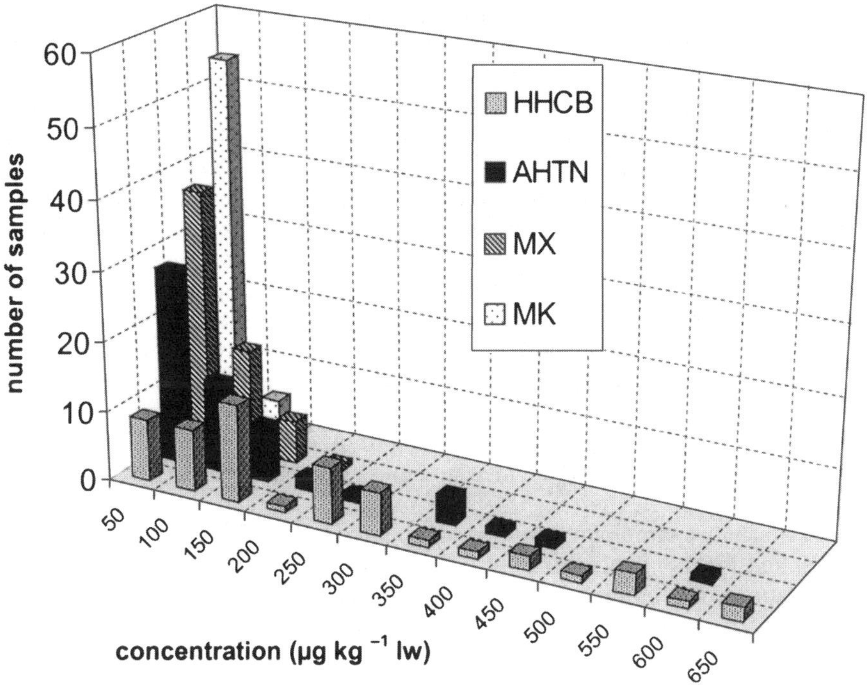

Fig. 9 Frequency distribution of synthetic musk compounds in 59 milk samples from Czech Republic (maximum value for HHCB [720 µg kg^{-1} lw] not shown here)

It should be noted, that statistical examination of all available information failed to prove any correlation between the musks levels and personal data of mothers obtained by the questionnaire. Due to generally low consumption of freshwater/marine fish (only 17% of responding mothers used to eat it each week), it might be speculated that dermal exposure is dominating way of synthetic musk intake in examined population. As declared by many other authors concerned with this issue, more research is obviously needed in this field to take effective measurements aimed at minimizing of human contamination.

4 Conclusions

Polycyclic musks as well as nitro musks were shown to be ubiquitous contaminants of aquatic ecosystems in Czech Republic. Both sediments and fish collected at monitored sampling sites of river Elbe and its tributaries Moldau and Tichá Orlice contained the same representatives of polycyclic fragrances and nitro musks as reported in other European studies. In accordance with them, the former group of musks was distinctly dominating, the maximum concentrations in fish (sampled from river Moldau downstream of Prague) were as high as 10^4 µg kg^{-1} lw (the highest levels of MX were 10^3 µg kg^{-1} lw). The relative abundance of particular synthetic musks were matrix-dependent, while in fish the typical concentration order of major musks was HHCB>AHTN≫MX>MK; in sediments the typical pattern was AHTN>HHCB≫MX=MK. In any case, significant correlation (r^2>0.8) between the contamination of sediments and content of musks in fish collected at particular locality was proved. With respect to a very similar pattern of synthetic musks in respective fish species regardless of the extent of particular locality pollution, it seems there are not distinct differences either in sewage treatment technology or the sources of waste contamination (the use of scented products).

Although any of the examined fish species may be employed as a relevant bioindicator of the presence of sewage waters in respective aquatic ecosystem, the best suited fish for this purpose seems to be bream because of its highest bioconcentration potential. This phenomenon might be attributed to low intensity of musk metabolization compared to other fish species involved in the study. As regards the sampling of fish for monitoring of musks, its age/size is not important since only bioconcentration mechanism is responsible for the uptake of these water pollutants.

An alternative sampling technique represented by SPMD was shown to be very well suited for monitoring of polycyclic musks (the levels in this "virtual fish" correspond to lipophilicity of particular substances dissolved in water). Lower sensitivity of nitro musks detection, specifically of MK, has to be taken into account due to its higher polarity.

Results obtained by analysis of breast milk samples showed extensive exposure of some donators by these synthetic fragrance compounds. However, no generalization with respect to the route of exposure could be drawn on the basis of collated information. A follow-up study focussing on clarification of this issue will be initiated.

Acknowledgements We gratefully acknowledge the financial support provided by the Ministry of Environment, project VaV 340/1/01 (fish samples) and the Ministry of Education, Youth and Sports of the Czech Republic, project MSM 223300004 (breast milk samples).

5
References

1. a) Rimkus G, Wolf M (1993) Dtsch Lebensm-Rundsch 89:171; b) Rimkus G, Wolf M (1995) Chemosphere 30:641
2. a) Eschke H-D (1994) Ruhrwassergüte, Ruhrverband, Essen, Germany, p 62; b) Eschke H-D, Traud J, Dibowski H-J (1994) UWSF-Z Umweltchem Ökotox 6:183; c) Eschke H-D, Traud J, Dibowski H-J (1994) Vom Wasser 83:373; d) Eschke H-D, Dibowski H-J, Traud J (1995) UWSF-Z Umweltchem Ökotox 7:131; e) Eschke H-D (1996) Ruhrwassergüte, Ruhrverband, Essen, Germany, p 86
3. Lagois U (1996) gwf Wasser Abwasser 137:154
4. de Boer J, Brinkman UAT (1994) Trends Anal Chem 13:397
5. Rimkus GG, Wolf M (1996) Chemosphere 33:2033
6. Geyer HJ, Rimkus GG, Scheunert I, Kaune A, Kettrup A, Zeeman M, Muir DCG, Hansen LG, Mackay D (2000) In: Beek B (ed) Bioaccumulation, new aspects and developments. Handbook of environmental chemistry, vol 2, part J. Springer, Berlin Heidelberg New York, p 1
7. a) Rimkus GG (1999) Toxicol Lett 111:37; b) Rimkus GG, Wolf M (1997) Lebensmittelchemie 51:94; c) Rimkus GG, Brunn H (1996) Ernährungs-Umschau 43:442

8. Gatermann R, Biselli S, Hühnerfuss H, Rimkus GG, Hecker M, Karbe L (2002) Arch Environ Contam Toxicol 42:437
9. Rimkus GG, Gatermann R, Hühnerfuss H (1999) Toxicol Lett 111:5
10. a) Gatermann R, Hühnerfuss H, Rimkus G, Attar A, Kettrup A (1998) Chemosphere 36:2535; b) Gatermann R, Hellou J, Hühnerfuss H, Rimkus G, Zitko V (1999) Chemosphere 38:3431
11. Herren D, Berset JD (2000) Chemosphere 40:565
12. Draisci R, Marchiafava C, Ferretti E, Palleschi L, Catellani G, Anastasio A (1998) J Chromatogr A 814:187
13. Biselli S, Gatermann R, Rimkus GG, Hühnerfuss H, Kallenborn R (2000) Third SETAC World Congress, Brighton, UK, 24–29 May 2000. Proceeding abstract 3E/p026
14. Rimkus GG, Butte W, Geyer HJ (1997) Chemosphere 35:1497
15. Winkler M, Kopf G, Hauptvogel T, Neu T (1998) Chemosphere 37:1139
16. Fromme H, Otto T, Pilz K (2001) Water Res 35:121
17. Huckins JN, Petty JD, Prest HF, Clark RC, Alvarez DA, Orazio CE, Lebo JA, Cranor WL, Johnson BT (2002) A guide for the use of semipermeable sampling devices (SPMDs) as samplers of waterborne hydrophobic organic contaminants. American Petroleum Institute, Publication No 4690
18. Schwartz S, Berding V, Matthies M (2000) Chemosphere 41:671
19. Tas JV, Balk F, Ford RA, van de Plassche EJ (1997) Chemosphere 35:2973
20. Balk F, Ford RA (1999) Toxicol Lett 111:57
21. Franke S, Meyer C, Heinzel N, Gatermann R, Hühnerfuss H, Rimkus GG, König WA, Francke W (1999) Chirality 11:795
22. Liebl B, Ehrenstorfer S (1993) Chemosphere 27:2253
23. a) Rimkus G, Wolf M (1993) Dtsch Lebensm-Rundsch 89:103; b) Rimkus G, Rimkus B, Wolf M (1994) Chemosphere 28:421; c) Rimkus GG, Wolf M (1996) Chemosphere 33:2033
24. Eschke H-D, Dibowski H-J, Traud J (1995) Dtsch Lebensm-Rundsch 91:375
25. Müller S, Schmid P, Schlatter C (1996) Chemosphere 33:17
26. CEN (1996) Fatty food - determination of pesticides and PCBs, part 2: extraction of fat, pesticides and polychlorinated biphenyls (PCBs), and determination of fat content. Paragraph 6.1.1 AOAC extraction, EN 152822. CEN, Brussels
27. Ott M, Failing K, Lang U, Schubring C, Gent H-J, Georgii S, Brunn H (1999) Chemosphere 38:13
28. Weizenegger W, Kypke K, Fetteroll B, Hahn J, Dsohnius E, Feucht M (1998) Dtsch Lebensm-Rundsch 94:103

The Handbook of Environmental Chemistry Vol. 3, Part X (2004): 189–211
DOI 10.1007/b14117

Biotic and Abiotic Transformation Pathways of Synthetic Musks in the Aquatic Environment

Scarlett Biselli[1] · Robert Gatermann[1] · Roland Kallenborn[2] · Leiv K. Sydnes[2]
Heinrich Hühnerfuss[3]

[1] Dr. Wiertz-Dipl. Chem. Eggert-Dr. Jörissen GmbH Analytical Laboratory (WEJ), Stenzelring 14b, 21107 Hamburg, Germany
E-mail: Scarlett.Biselli@wej.de

[2] The Polar Environmental Centre, Norwegian Institute for Air Research (NILU), 9296 Tromsø, Norway

[3] University of Hamburg, Institute of Organic Chemistry, Martin-Luther-King-Platz 6, 20146 Hamburg, Germany

Abstract Synthetic musks are being transformed to intermediates or relatively stable metabolites that may exhibit a strong potential for environmental harm. For example, microbial metabolisation of nitro musks in aquatic ecosystems preferentially gives rise to the formation of the amino derivatives, and, as a consequence, benzene-amine derivatives were found in environmental samples in higher concentrations than their parent compounds. Transformation of musk xylene (MX) chiefly results in 1-amino-4-*tert*-butyl-2,6-dimethyl-3,5-dinitrobenzene (4-amino-MX) and a minor product, 1-amino-2-*tert*-butyl-4,6-dimethyl-3,5-dinitrobenzene (2-amino-MX). Musk ketone (MK) is transformed by microbial processes to one main transformation product only, 1-(3-amino-4-*tert*-butyl-2,6-dimethyl-5-nitrophenyl)ethanone (2-amino-MK). In higher organisms, such as male Wistar rats the main metabolic pathway included, as a first step, reduction of the 4-nitro group of MX to the amino group, followed by acetylation of the resulting amino function and oxidation of one of the methyl groups. Alternatively, oxidation of the *tert*-butyl group appeared to be possible. The reduction of the 2-nitro group to the amino function was assumed to be less probable due to increased sterical hindrances induced by the adjacent *tert*-butyl group. Detailed investigations on the mechanism and kinetics of enzymatic MX reduction in mice and humans revealed that induction and inhibition of cytochrome P-450 2B (CYP2B) enzymes as well as the binding of 4-amino-MX to haemoglobin in blood appears to play an important role. Additional polar transformation products of the nitro musks were discussed in the literature for higher organisms, which are capable of both *N*- and *para*-hydroxylation of aniline derivatives. Recently, transformation products of polycyclic musks were also identified in considerable concentrations in environmental samples. The structure of the most abundant metabolite of HHCB was identified as a lactone (HHCB-lactone). Potential structures of photochemical transformation products of the polycyclic musks HHCB and AHTN are presently under investigation. Furthermore, in this chapter, analytical methods for the determination of the metabolites together with the parent compounds of nitro and polycyclic musks as well as environmental levels and the ecotoxicological potential of the musk metabolites are discussed.

Keywords Transformation of nitro musks · Amino metabolites of nitro musks · Transformation of polycyclic musks · Enzymatic transformation · Phototransformation · HHCB-lactone · AHTN-lactone

1 Introduction

Transformation of environmental xenobiotics is not restricted to polar compounds, but it has also been reported for various persistent substances such as pesticides of the first generation, e.g. 1,1,1-trichloro-2,2-bis(4-chlorophenyl)-ethane (*p,p′*-DDT), the chlordanes, heptachlor and γ-HCH. Enzymatic transformation by micro-organisms or in biota as well as photochemical transformation of these seemingly persistent xenobiotics gave rise to metabolites that in part turned out to exhibit a more hazardous potential than the parent compounds. Some of the best studied transformation products of this generation include 1,1-dichloro-2,2-bis(4-chlorophenyl)ethene (*p,p′*-DDE) and 1,1-dichloro-2(2-chlorophenyl)-2-(4-chlorphenyl)ethene (*o,p′*-DDE), which where shown to be responsible for the Ca^{2+}-metabolisation defects resulting in the well-known egg shell thinning of European and American bird of prey eggs and eggs from *ciconiformes* species in the early 1960s [1]. Furthermore, many reports on the impact of oxychlordane and heptachlor *exo*-epoxide on various environmental compartments and the toxicological implications can be found in the literature, e.g. in [2].

In this chapter, present knowledge on the enzymatic and photochemical transformation pathways of several synthetic musks as well as the levels and fate of the respective metabolites will be summarised. A brief description of the analytical problems and the ecotoxicological potential of the musk metabolites will also be included.

2 Transformation Pathways of Synthetic Musks

2.1 Transformation Pathways of Nitro Musk Compounds

The various mechanisms related to the reduction of nitrobenzene derivatives are well-documented in the literature and part of basic lectures of organic chemistry [3]. At low pH-values intermediates such as nitroso-benzene and *N*-phenyl-hydroxylamine may be encountered, with the final product benzeneamine (trivial name aniline). Under neutral or weakly acid conditions *N*-phenyl-hydroxylamine may dominate, while under basic conditions azoxybenzene, azobenzene and hydrazobenzene may appear or form the dominating product, respectively. This strictly pH-dependent variation of transformation mechanisms might give rise to different transformation products in limnic ($pH<7.0$) and in marine ecosystems (pH about 8 to 8.3). However, to the best of our knowledge mostly amino derivatives of nitro musks were thus far identified in environmental samples. This may be an indication that largely enzymatic transformation takes place in the environment, which in contrast to laboratory investigations, preferentially results in amino derivatives. Alternatively, it cannot be excluded that other potential metabolites as outlined above have thus far not been in the focus of ecochemists and accordingly ignored. In particular, laboratory investigations on the photochemical transformation of nitro musks revealed several stable or less stable intermediates and products that are worth being included in systematic screening studies of environmental samples [4–10].

Abiotic processes may result in intramolecular reactions as shown for MX in Fig. 1 [9]. For example, Zhao and Schwack studied the phototransformation of musk ambrette (MA) [7], musk ketone (MK) [8] and musk xylene (MX) [10]. The authors observed that upon UV irradiation in model organic solvents, MX underwent intramolecular cyclisation in the presence of cyclohexane and methanol, while photoreduction dominated in cyclohexene. On layers of cellulose stearate (as a model for transformation on clothes) the phototransformation of MX took place readily both under artificial light (sun simulator) and in natural sunlight. MX dissolved in water was also phototransformed, mainly yielding a photocyclisation product [10]. However, it is important to note that the cyclic transformation products were not detected in environmental or human samples yet. With regard to MA Zhao and Schwack [7] also included squalene in their photochemical reaction studies, i.e. a component in the lipid of the human skin surface, which was expected to be potent to affect the photochemical transformation of MA considerably. It turned out that squalene participated in the photochemical process much more actively than cyclohexene, also promoting the formation of the two conceivable monoamino derivatives of MA in addition to two photocyclisation products. The authors conjectured that nitroso- and unstable *N*-hydroxylamino intermediates are the preliminary stages of photoreduction of MA, although neither of them was detected in the reaction initiated from the sterically more hindered nitro group. Upon UV irradiation of MK [8], four photoproducts were isolated and identified, the main photoproduct was proved to be an indolinone derivative, which was previously described, by Döpp and Sailer [9] while studying the photochemistry of crystal musk ketone. Butte et al. [6], who investigated the kinetics of the photochemical transformation of MX, MK, MA, musk tibetene (MT) and musk moskene (MM) in water, discriminated two transformation pathways: nitro musks with two nitro groups next to the *tert*-alkyl group, i.e. MX, MK and MT, undergo intramolecular reactions, whereas for MA and MM intermolecular reactions were favoured giving rise to dimerization products. Kinetics of MX, MK, MT, MM and MA transformations in water by UV radiation were also studied by Neamtu et al. [4, 5] who showed that the presence of hydrogen peroxide accelerates the MK phototransformation, where the reactivity increases in the order MT<MM<MK<MX<MA, largely representing the order of the solubilities in water.

Enzymatic transformation processes in aquatic ecosystems preferentially give rise to the formation of the amino derivatives ([11] and literature cited therein; Fig. 1). It is generally assumed that microbial transformation of MX chiefly results in 1-amino-4-*tert*-butyl-2,6-dimethyl-3,5-dinitrobenzene (4-amino-MX) and a minor product, 1-amino-2-*tert*-butyl-4,6-dimethyl-3,5-dinitrobenzene (2-amino-MX). MK is transformed by microbial processes to one main transformation product only, 1-(3-amino-4-*tert*-butyl-2,6-dimethyl-5-nitrophenyl)ethanone (2-amino-MK).

In higher organisms, transformation by reductive enzymatic metabolisation is dominating. For example, Minegishi et al. [12] investigated the distribution, metabolism and excretion of MX in male Wistar rats. In particular, the 4-amino-MX transformation product was identified in the bile, faeces as well as in urine. Furthermore, using ^{3}H-MX as parent compound, the fate and occurrence of additional major transformation products were elucidated. In detail, MX, 4-amino-MX, 4-Ac-MX (1-acetylamino-4-*tert*-butyl-2,6-dimethyl-3,5-dinitrobenzene), 4-amino-2-CH_2OH-MX (1-amino-4-*tert*-butyl-2-hydroxymethyl-6-methyl-3,5-dinitrobenzene) and 4-amino-1-*tert*-BuOH-MX (1-amino-4-*tert*-hydroxybutyl-2,6-dimethyl-3,5-dinitrobenzene) were found in faeces, bile and urine, while 2-amino-MX and an unknown metabolite were detected in faeces and urine, and 2-amino-3-CH_2OH-MX in urine (1-amino-2-*tert*-butyl-6-hydroxymethyl-4-methyl-3,5-dinitrobenzene). It should be noted that the nomenclature for the MX metabolites rigorously follows the IUPAC rules using benzene as the basic hydrocarbon structure. Insofar, we did not adapt the assignments published by Minegishi et al. whose numbering is based on xylene as basic structure, which in our opinion may give rise to confusion in the literature. Furthermore, all abbreviations for MX metabolites discussed herein maintain the numbering of the pa-

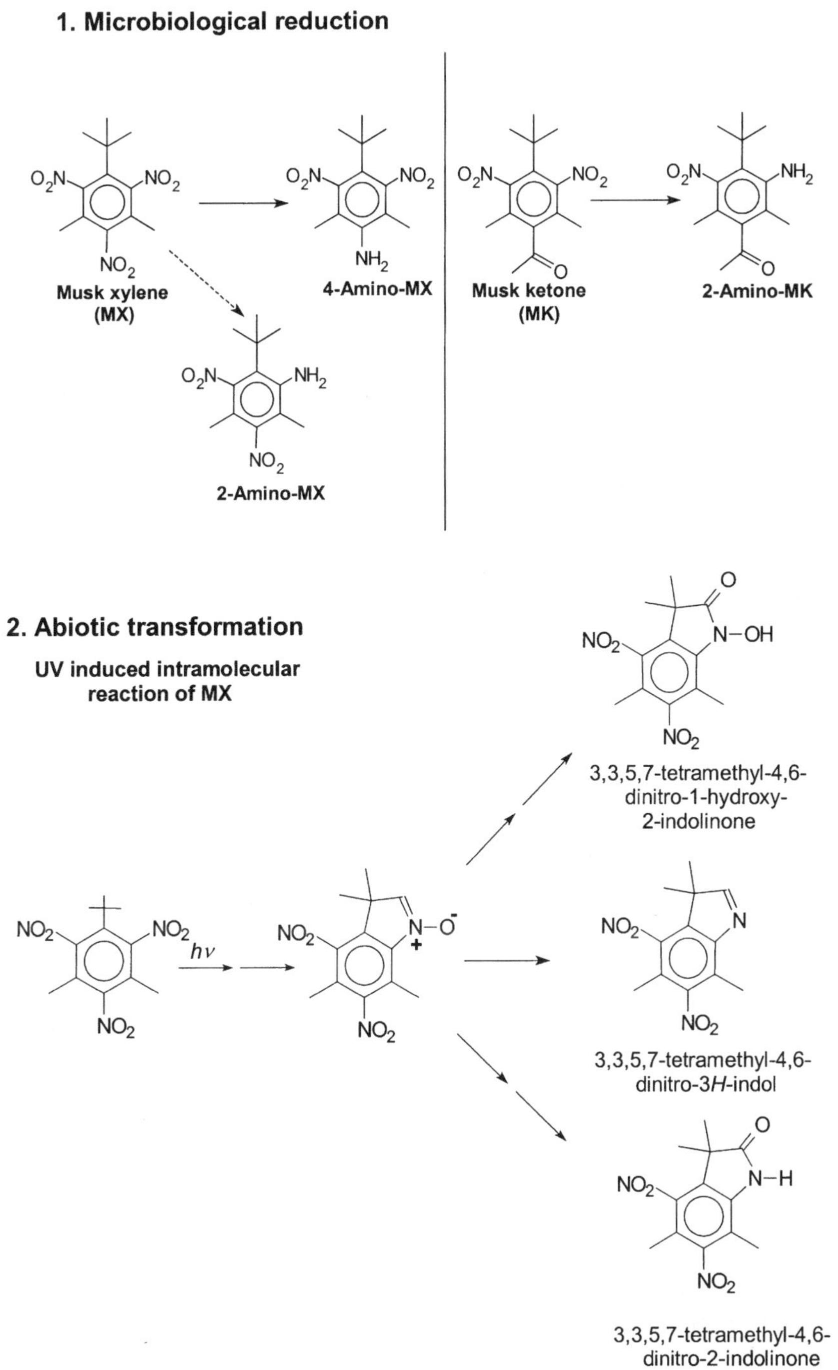

Fig. 1 Known transformation pathways for musk xylene and musk ketone with regard to microbial reduction. 2 Abiotic transformation: UV induced reaction of musk xylene [9]

rent compound which will easily allow the reader to assign the modification to the respective functional MX group.

Minegishi et al. [12] drew the following conclusions from their data set: urinary and faecal excretion accounted for 10 and 75% of the dose, respectively, on day 7 after oral administration of MX, i.e. the major route of excretion for MX was the faeces via the bile. The main metabolic pathway included, as a first step, reduction of the 4-nitro group of MX to the amino group (⇒4-amino-MX), followed by acetylation of the resulting amino function (⇒4-ac-MX). As a second step, the authors postulated oxidation of one of the methyl groups (⇒4-amino-2-CH_2OH-MX), which are more easily accessible to oxidation in the amino derivative than in the parent compound due to reduced sterical hindrances. Alternatively, oxidation of the *tert*-butyl group appeared to be possible (⇒4-amino-1-*tert*-BuOH-MX). The reduction of the 2-nitro group to the amino function (⇒2-amino-MX) was assumed to be less probable due to increased sterical hindrances induced by the adjacent *tert*-butyl group.

More detailed information on the mechanism and kinetics of enzymatic MX reduction was published by Lehmann-McKeeman et al. [13] who investigated the induction and inhibition of mouse cytochrome P-450 2B (CYP2B) enzymes by MX, 4-amino-MX and 2-amino-MX. The background of this study was formed by the observation that MX experiments with a battery of genotoxicity tests supplied negative results, but, on the other hand, a high incidence of liver tumours was produced in mice. When dosed by gavage to male B6C3F1 mice for seven days, MX increased liver weight by about 65% (dosis 200 mg kg^{-1}) and increased microsomal cytochrome P-450 content twofold as compared with the control experiment. In detail, MX increased microsomal activity for *O*-dealkylation of 7-ethoxy and 7-methoxyresorufin four- and twofold, respectively, and increased the *N*-demethylation of erythromycin approximately twofold. These results were generally consistent with increased CYP1a1, 1A2 and 3A protein levels determined by Western blotting. On the other hand, when dosed orally to phenobarbital (PB)-treated male B6C3F1 mice, MX inhibited >90% of the PB-induced *O*-dealkylation of 7-pentoxyresorufin (PROD). However, despite the in vivo inhibition, in vitro studies indicated that MX did not cause mechanism-based inactivation of CYP2B enzymes, in particular with regard to the potential for nitroreduction of MX. The latter potential, which was assumed to be catalysed by anaerobic intestinal bacteria, as well as its contribution to the inhibition of CYP2B enzyme activity was evaluated in a separate group of PB-induced mice that were dosed orally with a regimen of broad spectrum antibiotics (neomycin, tetracycline, and bacitracin), in order to reduce the gut flora prior to administration of MX. In these animals, MX did not inhibit PB-induced PROD activity. On the basis of this crucial experiment, Lehmann-McKeeman et al. suggested that:

- The parent compound MX does not inhibit the CYP2B enzyme
- Anaerobic intestinal bacteria were responsible for the nitroreduction of MX
- These amino metabolites are obligatory for the inhibition of mouse CYP2B enzymes

In subsequent studies [14, 15] Lehmann-McKeeman et al. included 2-amino-MX and 4-amino-MX, in order to discriminate their respective potential to induce

CYP2B10 and CYP1A2 mRNAs. In the in vitro studies with PB-induced mouse liver microsomes, both amines inhibited PROD activity when preincubated in the absence of NADPH. However, only 4-amino-MX caused a time- and NADPH-dependent loss of PROD activity, and the inactivation rate was a pseudo-first order process that displayed saturation kinetics. These results imply that 4-amino-MX is a mechanism-based inactivator of mouse CYP2B enzymes.

Additional polar transformation products of the nitro musks were discussed in the literature for higher organisms. For example, Debackere and Uehleke [16] prepared liver microsomes from cats, dogs and cattle, which were shown to be capable of both *N*- and *para*-hydroxylation of aniline derivatives. Furthermore, Hawkins and Ford [17] applied dermal doses of ^{14}C-labelled MA, MK or MX to male Sprague-Dawley CD rats. Most of the absorbed dose was eliminated in bile mainly in form of polar conjugated metabolites such as hydroxylated analogues formed by oxidation of the ring methyl group.

The biotransformation and toxicokinetics of MX in humans were investigated by Dekant and co-workers [18, 19]. A single dose of 0.3 mg kg^{-1} body weight of ^{15}N-labeled MX (^{15}N-MX) was administered to six volunteers (three males and three females) by the oral route and to another six volunteers (three males and three females) by the dermal route. Urine was collected for 96 h after exposure, while blood samples were taken at intervals for up to 140 days after administration. The metabolite ^{15}N-4-amino-MX in urine and ^{15}N-MX in plasma were quantified by gas chromatography/mass spectrometry with negative chemical ionisation (GC/MS-NCI). Peak plasma concentrations of ^{15}N-MX after oral administration attained values between 36 and 262 ng mL^{-1}, and 1.6 to 5.5 ng mL^{-1} after dermal exposure. The toxicokinetics of ^{15}N-MX in plasma was described by a two-compartment kinetic model with an initial decrease, due to the distribution from the blood into a second compartment (likely fat tissue) and a terminal elimination phase with an average half-life of 70 days for both routes of administration. The amount of ^{15}N-4-amino-MX in recovered urine represented 0.1–0.5% of the oral dose of ^{15}N-MX and 0.02–0.16% of the dermal dose, respectively. It is worth noting that maximum concentrations of ^{15}N-4-amino-MX were observed after a short time of invasion (18–24 h after administration of ^{15}N-MX). The subsequent elimination of the metabolite occurred by first-order kinetics with an average elimination half-life time of 11.8 h.

In addition, Dekant and co-workers [19] analysed the binding of 4-amino-MX to haemoglobin in blood samples from ten individuals not knowingly exposed to MX. The application of haemoglobin-bound metabolites as biomarkers of exposures may be used to assess cumulative exposure over a longer time range and may thus be better suited for risk assessment than quantification of urinary metabolites. Haemoglobin from the blood samples was isolated, bound metabolites were liberated as amines by alkaline hydrolysis and then determined by GC/MS-NCI. The amounts of 4-amino-MX bound to haemoglobin in the human blood samples ranged from 13 to 46 fmol mg^{-1} haemoglobin. The authors conclude that these data demonstrate the bioavailability of MX in humans. Furthermore, they note that the use of haemoglobin binding as a biomarker for nitro musk exposure warrants further studies.

2.2
Transformation of Polycyclic Musk Compounds

Anthropogenic chemicals may be subject to biotic transformation and various abiotic oxidative processes in the environment. However, little is known about the behaviour of polycyclic musk compounds under these conditions. Some studies involving HHCB have been published, and, in addition, model experiments with AHTN are currently carried out. The results thus far available will be summarised in this chapter.

2.2.1
Transformation Products of HHCB

In 1997, first evidence about the presence of more polar chemical transformation products of HHCB and AHTN was obtained. However, no suggestions for chemical structures were offered [20]. In a pioneering work Franke et al. [21] elucidated the structure of the main HHCB-transformation product, 1,3,4,6,7,8-hexahydro-4,6,6,7,8,8-hexamethylcyclopenta[*g*]-2-benzopyrane-1-one (Fig. 2; hereafter HHCB-lactone, sometimes referred to as "galaxolidone"), including its synthesis. Franke et al. [21] synthesised the racemic standard compound by oxidation of racemic HHCB using a finely powdered mixture of potassium permanganate and copper sulphate pentahydrate. Thus, they were for the first time able to verify the presence of this compound in environmental samples.

In laboratory experiments Itrich et al. [22] as well as Federle et al. [23] observed the oxidation of HHCB to the lactone in activated sewage sludge and river water (i.e. biotransformation) and in abiotic controls, using radiolabeled HHCB as standard material and high performance thin layer chromatography (HPTLC) as analytical method. The authors determined an immediate biotransformation of HHCB in the activated sewage. Several hitherto unknown transformation products were detected, the structures of which, however, were not elucidated yet. The authors determined a half-life time for HHCB in river water of 35 h and in activated sewage of 21 h, respectively. Following the principle of "Advanced Oxidation Processes", Ziesenitz [24] carried out photo- and sonochemical investigations using aqueous solutions of HHCB and AHTN in the presence of H_2O_2 aiming at potential improvements of photochemical wastewater treatment. The transformation products thus obtained included HHCB-lactone and additional "water soluble compounds". With regard to the transformation product of AHTN the author tentatively attributed a lactone structure.

Furthermore, autoxidation as a potential transformation pathway for HHCB to the lactone was proposed [11, 21]. Such an internal mechanism can be traced back both to biotic and abiotic transformation pathways. A possible scheme for the transformation is given in Fig. 2 [11]. Basically, lactone structures are known to be relatively persistent under environmental conditions. However, further transformation as proposed by Biselli (Fig. 3) cannot be fully excluded [11].

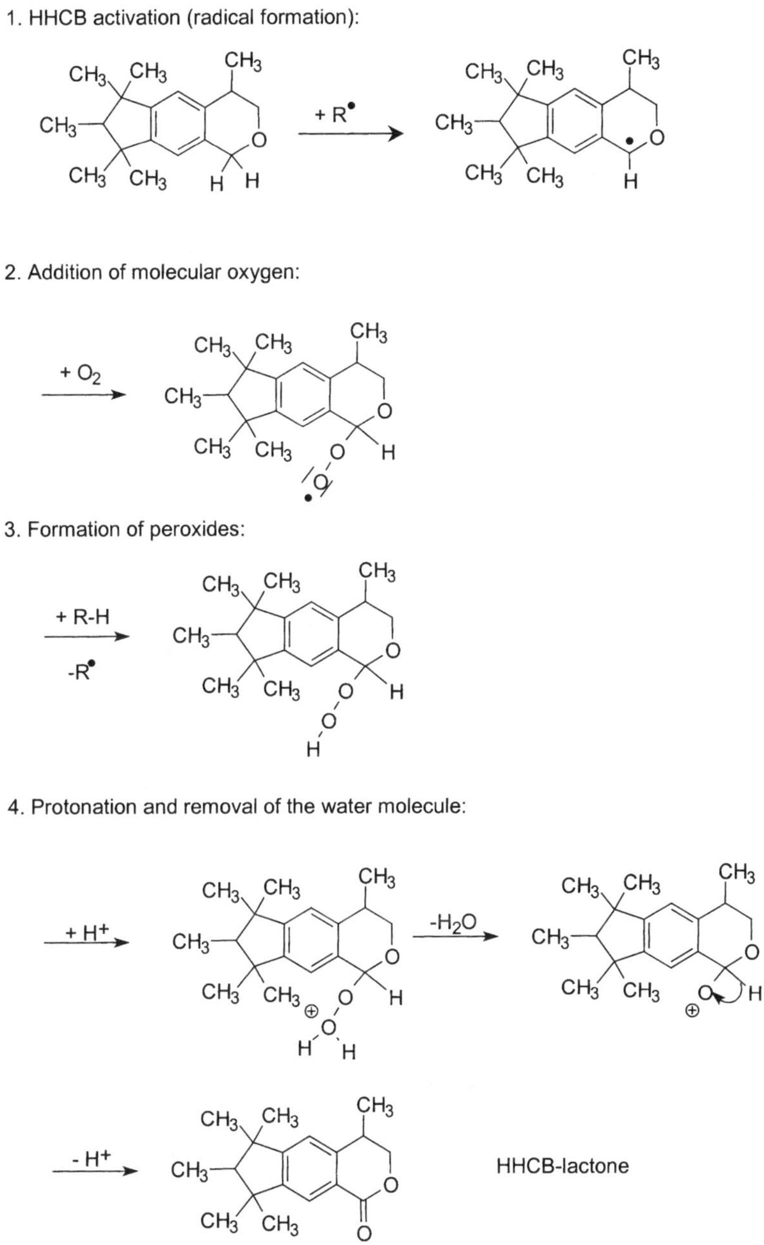

Fig. 2 Proposed formation route for HHCB-lactone in the aquatic environment

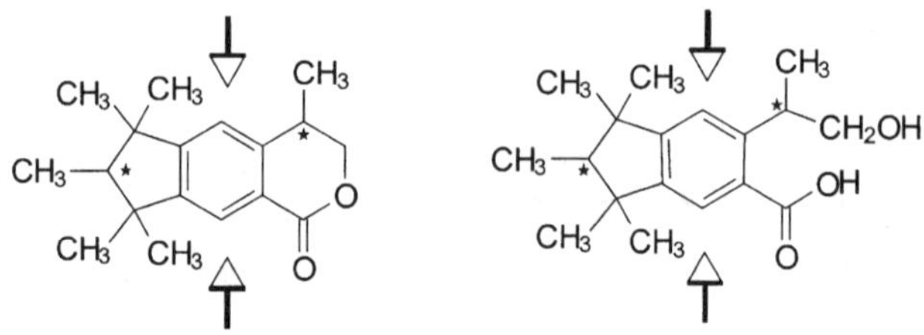

Fig. 3 Possible transformation products of HHCB-lactone and the corresponding hydroxy acid; arrows indicate the most probable points for further hydroxylation; *asterisks* mark the sterogenic centres

2.2.2
Transformation Products of AHTN

The abiotic transformation pathways of AHTN are currently being assessed by exposing the standard compound to various photooxidative conditions [25]. In order to simulate the conditions prevailing in nature [26] the light is filtered through Pyrex with cut-off at 295 nm.

Irradiation of aerated aqueous solutions of AHTN at room temperature in essence gave no products, presumably due to its low water solubility. When the temperature was increased to 308 K (35 °C), however, two products and a complex mixture of several unidentifiable compounds were isolated; the main product was 3,5,5,6,8,8-hexamethyl-5,6,7,8-tetrahydro-3*H*-naphtho[2,3-*c*]furan-1-one (**1**), the other 3-acetyl-5,5,7,8,8-pentamethyl-5,6,7,8-tetrahydronaphthalene-2-carbaldehyde (**2**) (Fig. 4).

The mechanism of the reaction apparently is both solvent and temperature dependent. When moist methanol solutions of AHTN were irradiated the conversion improved considerably. At room temperature a mixture of lactone **1** and peroxide **3** was formed, but the product ratio depended on the duration of the irradiation, because **3** appears to be photochemically unstable, and it is easily converted to **1**. As a result photolysis of AHTN at 333 K (60 °C) supplied lactone **1** only, essentially in quantitative yield.

Under natural conditions many organic compounds are also oxidised by singlet oxygen [26, 27]. Therefore, AHTN was also exposed to this oxidising agent, generated photochemically by adding sensitizers such as humic acids to the aqueous solutions and methylene blue to the methanol solutions. The product mixtures thus obtained were more complex than those formed by direct photolysis. This result implies that sensitised photooxidation in methanol leads to several new products, some of which were identified (Fig. 4C). Furthermore, the material balance is less satisfactory, because fair amounts of the products are bound to the sensitizer and can hardly be isolated. This is particularly the case, when AHTN is photolysed in water with a humic acid as sensitizer.

The results summarised above show that direct and sensitised photooxidation [28] of AHTN afford most of the products expected from reactions of a methyl-substituted acetophenone with triplet and singlet oxygen, respectively [25]. How-

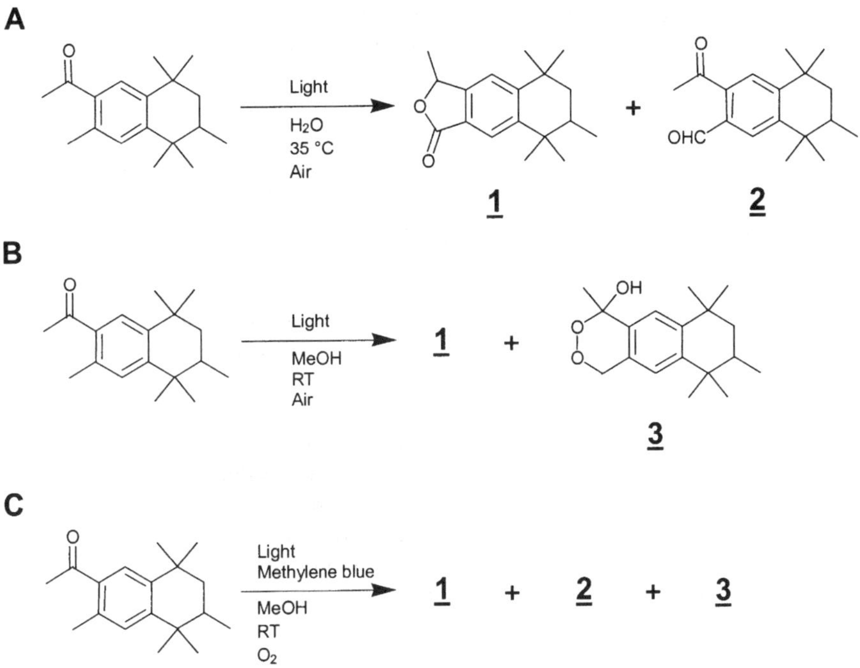

Fig. 4A–C Photochemical transformation products of AHTN: **A** in the presence of water; **B** in the presence of methanol at room temperature (RT); **C** in the presence of moist methanol, oxygen and methylene blue at room temperature (RT)

ever, the reactivity of AHTN seems to be less than that of polycyclic aromatic hydrocarbons which were also included in the study by Sydnes. Another important aspect is the fact that all the oxidation products contain functional groups, which are prone to reaction with bioorganic compounds found in biota. Therefore, it is expected that transformation products of AHTN and other polycyclic musk compounds resulting from oxidative processes are present as conjugates in biota; this may hamper their isolation and detection, and, of course, subsequent quantitative analyses. Studies in this area are currently in progress.

3 Analytical Methods

Herein, a brief survey on the methods most relevant to the determination of musk metabolites in different environmental compartments will be given [21, 29–33].

3.1 Analysis of Nitro Musks and their Metabolites

Thus far only a few selected synthetic musks have been included in official monitoring programs due to the obvious lack of adequate analytical multi-methods that allow the simultaneous determination of both the musk parent compounds and their metabolites in environmental samples. Descriptions of successful analytical approaches for the simultaneous qualitative and quantitative analyses of nitro musks and their most important amino metabolites can be found in [11, 31, 34].

3.2
Analysis of Polycyclic Musks and their Metabolites

Methods for the determination of polycyclic musks were published for air [35], water [11, 21, 34], sewage sludge [11, 34] and biota samples [34, 36, 37]. Reports on the successful identification and quantification of polycyclic musk transformation products in environmental samples are sparse and largely limited to the main metabolite of HHCB, i.e. HHCB-lactone. Due to differences in polarity and elution properties during column chromatographic separation, small adjustments must be made only, in order to recover the HHCB-lactone while applying the standard analytical methods actually developed for the parent polycyclic musks. Short descriptions of the main modifications of the respective analytical methods for the determination of HHCB-lactone in environmental water, sediment and biota samples are given below.

3.2.1
Analysis of HHCB-Lactone

3.2.1.1
Water Analysis

A liquid/liquid extraction method (*n*-hexane) followed by a silica gel clean-up as described below was successfully applied by Gatermann [34] and Biselli [11]. The HHCB-lactone elutes in fraction 5 together with all amino metabolites of the nitro musks, while the parent polycyclic musks elute both in fractions 4 and 5.

3.2.1.2
Sediment Analysis

Soxhlet extraction of moist sediments is usually carried out with acetone [11]. A subsequent cleaning with standard gel permeation chromatography (GPC) followed by a silica fractionation as described below allows the separation of the polycyclic musk HHCB and its transformation product [11].

3.2.1.3
Biota Analysis

The analysis of HHCB and its transformation products in biota samples can be carried out according to experimental approaches described for musk parent compounds in earlier studies [30, 34, 37]. After extraction by standard methods (e.g. Soxhlet, Accelerated Solvent Extraction (ASE), cold extraction) a standard

GPC clean-up is performed followed by a silica clean-up and fractionation: Fraction 1: *n*-hexane/toluene 65:35 (v/v); fraction 2: toluene (100%); fraction 3: toluene/acetone 95:5(v/v). The HHCB-lactone elutes together with MK and the monoamino metabolites of MX in fraction 2.

3.3
Multi-Methods for the Simultaneous Determination of Musk Parent Compounds and their Metabolites

Recently, Rimkus and co-workers [32] described a new method for the analysis of 17 synthetic musks and 4 transformation products of both polycyclic and nitro musks in fatty fish and human tissue samples. For the first time, standard mixtures of five macrocyclic musks were also included in this separation method. In addition to synthetic musks this multi-method also covers a large number of the organochlorine pesticides and polychlorinated biphenyls (PCB) to be quantified in fatty tissues. As a strong variability of the lipid content and lipid pattern of tissues may influence the performance of this method, an optimisation of this promising approach has to be carried out yet. For a more detailed description and evaluation of this trace analytical method, we refer to [32].

In parallel, a similar analytical concept was developed in the laboratories of the Norwegian Institute for Air Research (NILU) for the determination of nitro and polycyclic musks, the amino metabolites of MX and MK, and the main HHCB transformation product, HHCB-lactone. The latter method that also allows the parallel quantification of organochlorines and polar transformation products uses the following approach [33].

The concentrated sample extract (final sample volume 500 µL) is fractionated by a two-step GPC method consisting of a commercially available Envirogel system (Waters GmbH, Eschborn, Germany) and a tailor-made glass column (Latek, Eppelheim, Germany: 1000×15 mm) for fine fractionation, which was packed with Biobeads S-X3 (Bio-Rad Laboratories, Inc., Hercules, CA, USA). A first rough separation is carried out on the Envirogel system using a mixture of 50% cyclohexane and 50% ethyl acetate (v/v), in order to separate the main lipid fraction (1st fraction) from the synthetic musk fraction (2nd fraction). The first fraction is discarded, while the second fraction is concentrated to 0.5 mL and transferred to the glass column for fine GPC fractionation. The fraction containing all synthetic musk compounds of interest is concentrated to 300 µL for further clean-up on a small silica column (2 g precleaned silica, Baker Bond, Deventer, Holland), which is carried out using the elution sequence as follows: 1. *n*-hexane, 2. *n*-hexane/dichloromethane 9:1 (v/v), 3. *n*-hexane/dichloromethane 4:6 (v/v), 4. dichloromethane, 5. dichloromethane/ethyl acetate 1:1 (v/v). Fractions 3, 4 and 5 that contain all synthetic musk compounds and transformation products including polycyclic musk derivatives of interest are combined for the final gas chromatographic separation (GC/MS, quadrupole low resolution mass spectrometry).

4 Environmental Levels of Transformation Products of Synthetic Musks

4.1 Environmental Levels of Nitro Musk Metabolites

Nitro musk compounds are lipophilic and, thus, exhibit a limited solubility in water only. As a consequence, they are present in sewage sludge or wastewater associated with particulate matter due to strong adsorption. In a pilot study performed in a sedimentation pond of a North German municipal wastewater treatment plant, the fate of synthetic musks was investigated comprehensively with regard to their biotic transformation within the pond [11, 34]. A large number of treated water samples, sewage sludge, sediment, mussel, fish, and semipermeable membrane device (SPMD) samples was collected during the experiment. For more details see the chapter by Hühnerfuss et al. in the present monograph and [34]. The results are summarised in Table 1 for sewage sludge and sediment samples. The MX metabolite 4-amino-MX showed higher concentrations, in particular in the sediment samples as compared to 2-amino-MX and 2-amino-MK.

Recently, an additional study on nitro musks and their transformation products in sewage sludge was published by Herren and Berset [38] who also reported higher concentrations for the amino derivatives than for the parent compounds. For 4-amino-MX and 2-amino-MK, maximum concentrations of 49.1 and 15.6 ng g^{-1} dry weight, respectively, were determined. Considerably lower detection limits were achieved in the latter study as compared to the former investigation.

It is worth noting that the levels of the parent nitro musks in the effluent of the sewer were considerably lower than in the influent, while vice versa the respective amino metabolites exhibited higher levels in the effluent than in the influent (influent: MX 150 ng L^{-1}; 2-amino-MX not quantified (n.q.); 4-amino-MX n.q.; MK 550 ng L^{-1}; 2-amino-MK n.q.; effluent: MX 10 ng L^{-1}; 2-amino-MX 10 ng L^{-1}; 4-amino-MX 34 ng L^{-1}; MK 6 ng L^{-1}; 2-amino-MK 250 ng L^{-1}; [34]). Basically, these results show the efficiency of the cleaning process with regard to the parent

Table 1 Concentration levels of synthetic musks and their transformation products in sewage sludge and sediment samples (μg kg^{-1} wet weight)

Compound	Sewage sludge	Sediment 1	Sediment 2
MX	<10	0.5	0.8
MK	<10	6.3	3.6
4-Amino-MX	<10	6.3	1.3
2-Amino-MX	n.d.	n.d.	n.d.
2-Amino-MK	n.d.	n.d.	n.d.
HHCB	4800	75	160
HHCB-lactone	55	2.3	4.9
AHTN	2000	75	140

n.d.=not detectable.

Table 2 Musk xylene, musk ketone and transformation products in selected freshwater biota samples (concentrations in ng g^{-1} lipid weight according to [29])

Sample material	Lipid content [%]	MX	4-Amino-MX	2-Amino-MX	MK	2-Amino-MK
Pike-perch (river Elbe, Lauenburg, 1994)	0.3	30	90	<3	54	<3
Pike-perch (river Elbe, Lauenburg, 1994)	0.4	23	28	10	25	13
Pike-perch (river Elbe, Haseldorf, 1994)	0.5	69	46	28	64	9
Pike-perch (river Elbe, Haseldorf, 1994)	0.5	99	16	12	27	13
Pike (river Elbe, Wedel, 1993)	0.7	46	410	20	176	21
Bream (river Stör, up-stream, 1993)[a]	1.2	84	310	14	217	16
Bream (river Stör, down-stream, 1993)	1.7	10	80	4	85	10
Rainbow trout (aquaculture, Denmark, 1995)	2.5	207	205	23	111	23
Rainbow trout (aquaculture, Denmark, 1997)	2.4	11	76	<1	50	4
Rainbow trout (aquaculture, Denmark, 1997)	3.5	18	40	<2	49	7
Rainbow trout (aquaculture, Denmark, 1997)	4.2	56	48	<1	49	<1
Rainbow trout (aquaculture, Denmark, 1997)	2.8	36	60	<1	29	<1
Rainbow trout (aquaculture, Austria, 1997)	3.9	100	40	<2	22	n.a.[b]
Rainbow trout (aquaculture, Spain, 1997)	2.3	340	100	3	111	<1
Rudd (sewage pond, 1997)[c]	0.6	28	266	<2	320	7
Rudd (sewage pond, 1997)	1.1	50	300	<2	425	10
Rudd (sewage pond, 1997)	0.8	18	290	<2	300	21
Tench (sewage pond, 1997)	0.8	234	1070	30	1360	90
Tench (sewage pond, 1997)	1.3	365	1070	43	1245	90
Tench, liver (sewage pond, 1997)	4.0	100	1640	60	588	150
Tench (sewage pond, 1997)	0.8	220	1040	46	1330	80
Tench (sewage pond, 1997)	0.6	185	3600	100	1250	180
Crucian carp (sewage pond, 1997)	1.1	222	1200	60	920	90
Crucian carp (sewage pond, 1997)	1.7	280	960	70	1400	70

Table 2 (continued)

Sample material	Lipid content [%]	MX	4-Amino-MX	2-Amino-MX	MK	2-Amino-MK
Crucian carp (sewage pond, 1997)	2.0	365	380	27	1440	60
Crucian carp (sewage pond, 1997)	2.6	324	710	50	1170	50
Crucian carp (sewage pond, 1997)	2.5	355	720	30	990	40
Crucian carp (sewage pond, 1997)	1.7	305	700	40	1100	90
Crucian carp (sewage pond, 1997)	3.5	440	860	47	1800	85
Crucian carp, liver (sewage pond, 1997)	10.4	292	750	40	1090	60
Eel (sewage pond, 1997)[d]	15.7	260	60	5	420	25
Eel (sewage pond, 1997)[e]	18.1	220	55	4	360	20
Zebra mussel (sewage pond, 1997)	1.4	130	1550	77	1440	270

[a] River Stör is a tributary to the river Elbe in North Germany.
[b] n.a.=not analysed.
[c] Sewage pond of a municipal waste treatment plant.
[d] Pooled samples of six eels (189–262 g).
[e] Pooled samples of 12 eels (120–180 g).

compounds; however, they underline that metabolites are being formed during this procedure, which, therefore, have to be included in the assessment of the overall efficiency of the complete sewage treatment. In addition to transformation of nitro musk compounds into the corresponding amino products, adsorption to particles may also play an important role in the removal process which, therefore, has to be included in a comprehensive assessment.

High concentrations of the transformation products were determined in freshwater fish species (Table 2; [29]). Rimkus et al. inferred from the data set that 4-amino-MX seems to be the main transformation product accumulated in aquatic biota, whereas in water samples 2-amino-MK is dominating. Furthermore, a species dependent accumulation of musk metabolites is evident. The advantage of the fish samples taken from the sewage treatment pond is their common environmental background. As a consequence, the average concentration values of 4-amino-MX can be calculated for each species that allow a reliable assessment with regard to species dependent accumulation of this MX metabolite (rudd: 285 ng g^{-1}; tench: 1700 ng g^{-1}; crucian carp: 790 ng g^{-1}; eel: 58 ng g^{-1}; Zebra mussel 1550 ng g^{-1}). Basically, a species dependent accumulation of 4-amino-MX can also be inferred from the fishes caught in different German rivers and taken from aquacultures (pike-perch from the river Elbe: 45 ng g^{-1}; bream from river Stör:

195 ng g^{-1}; rainbow trout from aquacultures 81 ng g^{-1}). However, in this case caution has to be applied, because fishes in rivers are much more mobile, i.e. the environmental background may be different. For example, a comparison of the bream sample values (Table 2) stemming from fishes caught up-stream (310 ng g^{-1}) and down-stream (80 ng g^{-1}) the river Stör, respectively, suggests that dilution effects may bias species dependent transformation and accumulation. This stresses the importance of the unique data set obtained from fishes of the sewage treatment pond (Table 2). For more details, the interested reader is referred to the review [29].

As documented in several monitoring studies, the overall contamination levels of nitro musks have been reduced significantly within the past five years [29, 30, 36], where the levels of the amino metabolites have often been exceeding the concentrations of the parent compounds considerably at least in surface water samples (4–40 times) [31]. The latter results imply that risk assessment studies must include the determination of the transformation products, in order to estimate the environmental risk of nitro musks in a reliable way.

4.2 Environmental Levels of Polycyclic Musk Metabolites

Levels of HHCB-lactone were determined in the same water, sewage sludge, sediment, SPMD and fresh water fish samples from the sewage pond as described for nitro musk metabolites in the previous section [11, 34]. In the effluent of the sewage plant HHCB-lactone was found in concentrations around 220 ng L^{-1}, i.e. nearly one order of magnitude lower values than the parent compound HHCB (factor 8). In sewage sludge, levels of 55 µg kg^{-1} wet weight (ww) were determined for HHCB-lactone, in this case, nearly two orders of magnitude lower than the parent compound HHCB (factor 90). The results obtained for HHCB and its main metabolite HHCB-lactone in freshwater fish samples revealed an unequivocal species dependent transformation of this polycyclic musk compound (Table 3). This conclusion was further supported by enantioselective analyses of the same sample extracts, which are described elsewhere in this monograph.

The highest HHCB-lactone concentrations were found for tench (average 32 µg g^{-1} lipid weight) and zebra mussel (*Dreissena polymorpha*, average 33 µg g^{-1} lipid weight). The lowest values were determined for rudd and eel samples (Table 3), where the low lipid based concentrations in the eel samples may also be attributed to their high lipid content. The high ratios of the HHCB/HHCB-lactone concentrations in tench samples (4.5–5.0) suggest a low biotransformation of HHCB in this fish species. This conclusion is in line with the results of the enantioselective investigations, which however, indicated a strong further transformation of the *trans*-HHCB-lactone. A very detailed discussion on the stereoselective transformation of *cis*- and *trans*-HHCB as well as the enantioselective transformation of the HHCB diastereoisomers, AHTN, *trans*-ATII and AHDI can be found elsewhere in this monograph.

Based on these concentration levels in different fish species and in water, bioaccumulation factors on a lipid basis (BAF_L) were determined both for HHCB and HHCB-lactone (Table 3). Considerably higher BAF_L values were determined

Table 3 Concentration levels (μg g^{-1} lipid weight) and bioaccumulation factors BAF_L (lipid weight basis) for HHCB and HHCB-lactone determined from freshwater biota collected in a sewage pond in Northern Germany

Species	HHCB Concentr.	HHCB-lact. Concentr.	HHCB/ HHCB-lact. Ratio	HHCB BAF_L	HHCB-lact. BAF_L
Rudd (*Scardinius erythrophthalmus*)	6.3	8.9	0.7	2,300	40,000
Rudd (*Scardinius erythrophthalmus*)	7.1	8.2	0.9	2,600	37,000
Rudd (*Scardinius erythrophthalmus*)	8.0	7.0	1.1	2,800	32,000
Tench (*Tinca tinca*)	155	34	4.5	54,000	154,000
Tench (*Tinca tinca*)	160	33	4.8	57,000	151,000
Tench (*Tinca tinca*)	165	34	5.0	59,000	152,000
Tench (*Tinca tinca*)	150	30	5.0	54,000	132,000
Crucian carp (*Carassius carassius*)	42	26	1.6	15,000	116,000
Crucian carp (*Carassius carassius*)	61	22	2.8	22,000	100,000
Crucian carp (*Carassius carassius*)	95	25	3.8	34,000	114,000
Crucian carp (*Carassius carassius*)	73	21	3.5	26,000	93,000
Crucian carp (*Carassius carassius*)	53	21	2.5	19,000	96,000
Crucian carp (*Carassius carassius*)	68	28	2.4	24,000	127,000
Crucian carp (*Carassius carassius*)	85	29	3.0	31,000	130,000
Eel (*Anguilla anguilla*)	4.8	4.0	1.2	1,800	18,000
Eel (*Anguilla anguilla*)	4.6	3.9	1.2	1,700	18,000
Zebra mussel (*Dreissena polymorpha*)	120	33	3.7	44,000	148,000
Tench liver	72	21	3.5	27,000	95,000
Crucian carp liver	69	18	3.9	26,000	80,000
Average SPMDs (n=12 and n=9)	166±38	8±1.0	21	83,000[a]	36,000[a]

[a] Partition coefficient between SPMD and water (K_{SPMD}).

Table 4 Concentration levels of HHCB and HHCB-lactone in Atlantic cod liver (*Gadusmorhua*) from the Oslo Fjord region (southern Norway) in ng g^{-1} lipid weight

Sampling site	HHCB	HHCB-lactone
Lysakerfjorden	663	232
Havnebassenget; Kavringen/Hovedøya	132	154
DYNO company; Sætre (Oslofjord, NO)	337	163
Ormøya/Utøya/Bekkelagsbassenget	423	201
VEAS company (Oslofjord, NO)	301	214
Breivold/Bunnefjorden	1510	337
Larviksfjorden, Inner Harbour	27	21
Frierfjorden	31	18
Breviksfjorden	9	n.d.
Sastein	12	9

n.d. ≡ not detected.

for the HHCB-lactone as compared with the parent compound HHCB. On the other hand, the partition coefficients $K_{SPMD/W}$ obtained from SPMD samples, which are assumed to represent model bioconcentration values (for details see [39]), were much higher for HHCB than for the lactone. Accordingly, the high ratio HHCB/HHCB-lactone of 21 as compared to the respective ratios in biota from the pond (HHCB/HHCB-lactone between 0.7 and 5.0) clearly implies that only a minor part of the relatively high content of the metabolite in biota can be attributed to passive bioconcentration, but the larger part is assumed to stem from enzymatic transformation processes, which in turn are strongly species dependent [40]. This fact must be kept in mind, when interpreting the high BAF_L-values of HHCB-lactone in biota discussed above.

The concentrations of HHCB and HHCB-lactone were also determined in liver extracts from Atlantic cod (*Gadus morhua*) caught in the Oslofjord region [41]. The concentration levels determined by the NILU multi-method described above are listed in Table 4. Again, considerable levels were found for this HHCB transformation product as compared to the parent compound.

5 Toxicological Considerations

The presence of amino transformation products of the nitro musks in tissues of various organisms [29, 30] stresses the need for comprehensive investigations into the ecotoxicological potential of these metabolites. However, to date, only few studies have been dealing with the toxicological impact of nitro musk transformation products. Significant differences in the toxicity of 2-amino-MX and 4-amino-MX were determined for *Daphnia magna* [42]. However, a repetition of the experiments initiated by the Research Institute for Fragrance Materials (RIFM) resulted in considerably lower toxic effects of 4-amino-MX [43, 44]. Based on quantitative structure activity relationship (QSAR) calculations using the ECOSAR software (ECOSAR classes for Microsoft Windows; ECOWIN v0.99C, USEPA, Washington, DC), a theoretical toxicity value can be derived, which is

Table 5 Acute toxicity (EC_{50} for *Daphnia magna*, (µg L^{-1})) of the musk xylene (MX) amino transformation products performed according to the OECD guideline 202

Compound	Behechti et al. [42]	ECOSAR	RIFM study [43, 44]	
			Artificial water	Natural water
2-amino-MX	1,070	546	n.p.	n.p.
4-amino-MX	0.25	546	490	370
2,4-diamino-MX	23,300	341	n.p.	n.p.
2,4,6-triamino-MX	58,800	1,215	n.p.	n.p.

n.p.=data not published yet.

more in agreement with the corrected value of the RIFM study (Table 5). In summary, contradictory results of two similar studies are thus far available which hardly allow a general conclusion. Therefore, a more comprehensive study is overdue, aiming at an elucidation of the complex toxicology of nitro musks including their metabolites in organisms.

This situation set the stage for a comprehensive toxicological study started by the University of Hamburg (Germany) focussing on the potential mutagenicity of the nitro musk metabolites. As already known for other aromatic amines, the main toxic risk is associated with a possible DNA adduct formation by highly reactive hydroxylamine intermediates. It is assumed that the biological activation is directly related to the presence of the highly reactive and mutagenic nitrenium ion. In detail, the mutagenic and genotoxic properties of the transformation products 2-amino-MX, 4-amino-MX and 2-amino-MK were investigated using the arabinose-resistance (ARA-test) as well as umu-test and *Salmonella* microsome test, but no significant mutagenicity or genotoxicity was found [45].

Furthermore, the xenoestrogenic properties of nitro musks and their metabolites were investigated using the E-screen test according to Soto et al. [46], but the test revealed no significant xenoestrogenic effects. On the other hand, relatively high cell mortality was determined for 2-amino-MX and 4-amino-MX, however, due to the limited number of experiments performed no general conclusions can be drawn.

Chou [47] developed a new method for the determination of teratogenic and endocrine modulation activity of synthetic musk fragrances in fish and amphibians. The competitive binding to the estrogen receptor (ER-receptor) was used as a measure for the determination of the xenoestrogenic potential of synthetic musks. This method was extended to the nitro musk transformation products, in particular, the amino compounds. The detailed results of this study are presented in the chapter by Dietrich et al. in the present monograph.

Systematic investigations into the ecotoxicological potential of the metabolites of polycyclic musks, in particular, HHCB-lactone, are still urgently needed.

6
Perspectives

The impact of synthetic polycyclic musks and their metabolites on the environment is not confined to the substances thus far chiefly discussed in the literature, i.e. HHCB, AHTN, ATII and AHDI. Gas chromatograms obtained from commercially available fragrance mixtures and from SPMD sample extracts collected in the pond of a German waste water treatment plant revealed several identical signals that may be tentatively attributed to additional hitherto unknown musks contaminants, presumably by-products of the technical synthesis of polycyclic musks. The presence of these signals was also confirmed in SPMD samples of the river Elbe. It is interesting to note that enantioselective analysis of the same sample extracts showed that several derivatives of these contaminants are chiral and can be separated into their enantiomers.

As a part of a master thesis carried out at the University of Hamburg [48], different by-products of the HHCB production were isolated and identified using GC/MS methods. Therefore, it can be assumed that the same by-products are also present in the commercial fragrance mixtures, thus introduced into the environ-

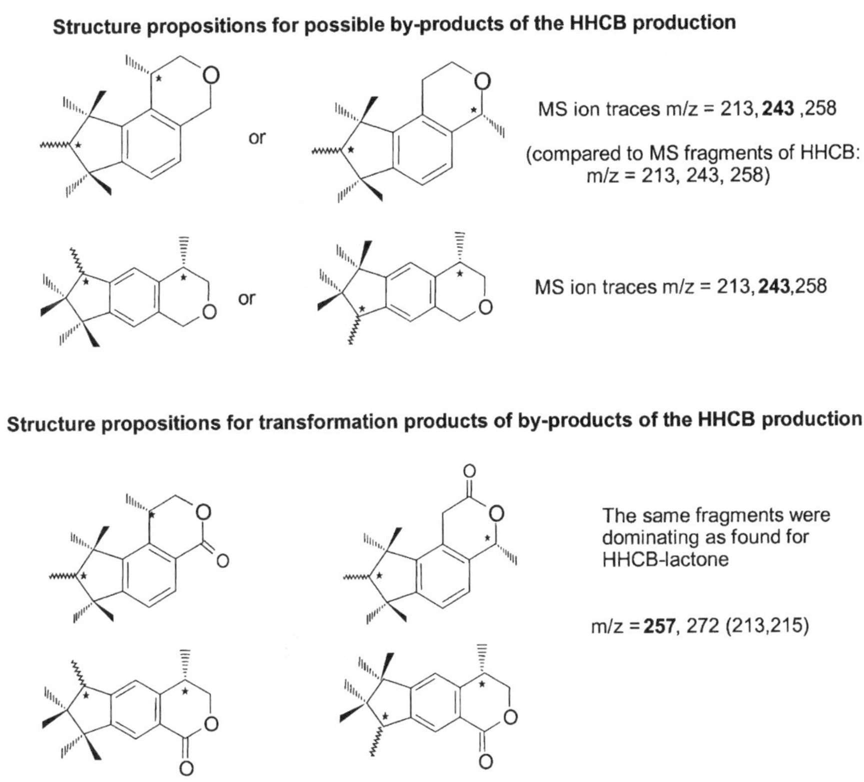

Fig. 5 Structure propositions for by-products of the HHCB production and their transformation products; *asterisks* mark the stereogenic centres

ment and subsequently transformed to more polar transformation products. Suggestions for molecular structures were made, but not confirmed yet (Fig. 5).

Based on these preliminary results, it can be safely assumed that the presence of only a few transformation products of the polycyclic musk fragrances in the environment is confirmed so far. Thus, comprehensive scientific efforts are still needed in the future to identify and elucidate the structures as well as the environmental fate and the ecotoxicological potential of the hitherto unknown fragrance components and their transformation products.

Acknowledgements We thank Dr. Martin Schlabach, Dr. Ole-Anders Braathen and Dr. I.C. Burkow for their generous support, as well as for open minded and continuous discussions about the topic. We appreciate the help of Stig Valdersnes who has synthesised and analysed the AHTN transformation products. The help of Sissel Planting, Dr. Dorte Herzke and Gro Hammerseth during sample preparation and analysis is highly appreciated. Dipl. Chem. Frank Hoffmann helped during the discussion of the data. Dr. Robert Gatermann expresses his thanks for the financial support of the Research Council of Norway.

7 References

1. Ratcliffe DA (1967) Nature 215:208
2. Kallenborn R, Hühnerfuss H (2001) Chiral environmental pollutants. Trace analysis and ecotoxicology. Springer, Berlin Heidelberg New York
3. Walter W, Francke W (1998) Beyer-Walter. Lehrbuch der Organischen Chemie. Hirzel, Stuttgart Leipzig
4. Neamtu M, Siminiceanu I (1999) Rev Chim 50:545
5. Neamtu M, Siminiceanu I, Kettrup A (2000) Chemosphere 40:1407
6. Butte W, Schmidt S, Schmidt A (1999) Chemosphere 38:1287
7. Zhao X, Schwack W (1999) Chemosphere 39:11
8. Zhao X, Schwack W (2000) Toxicol Environ Chem 74:217
9. Döpp D, Sailer KH (1975) Chem Ber 108:3483
10. Zhao X, Schwack W (1999) Intern J Environ Anal Chem 74(1/4):179
11. Biselli S (2001) Entwicklung einer analytischen Methode zum Nachweis von ökotoxikologisch relevanten organischen Problemstoffen in Sedimenten und Biota unter besonderer Berücksichtigung von Irgarol, synthetischen Moschusduftstoffen und deren Transformationsprodukten. PhD thesis, University of Hamburg, Hamburg, Germany
12. Minegishi K, Nambaru S, Fukuoka M, Tanaka A, Nishimaki-Mogami T (1991) Arch Toxicol 65:273
13. Lehmann-McKeeman LD, Johnson DR, Caudill D (1997) Toxicol Appl Pharmacol 142:169
14. Lehmann-McKeeman LD, Stuard SB, Caudill D, Johnson DR (1997) Mol Carcinog 20:308
15. Lehmann-McKeeman LD, Johnson DR, Caudill D, Stuard SB (1997) Drug Metab Dispos 25:384
16. Debackere M, Uehleke H (1964) Proc Eur Soc Study Drug Toxicity 4:40
17. Hawkins DR, Ford RA (1999) Toxicol Lett 111:95
18. Riedel J, Dekant W (1999) Toxicol Appl Pharmacol 157:145
19. Riedel J, Birner G, van Dorp C, Neumann HG, Dekant W (1999) Xenobiotika 29:573
20. Balk F, Ford RA (1999). Toxicol Lett 111:57
21. Franke S, Meyer C, Heinzel N, Gatermann R, Hühnerfuss H, Rimkus G, König WA, Francke W (1999) Chirality 11:795
22. Itrich NR, Simonich SL, Federle TW (1998) SETAC 19th Annual Meeting, 14–15 November 1998, Charlotte, NC, USA
23. Federle TW, Langworthy DE, Itrich NR, Simonich SL (2000) Third SETAC World Congress, Brighton, UK, 24–29 May 2000, Poster Presentation 4B/P003:170

24. Ziesenitz O (2000) Reaktionsverhalten polycyclischer Moschusverbindungen unter photo- und sonochemischen Bedingungen. Master thesis, University of Jena, Jena, Germany
25. Valdersnes S (2001) Diploma thesis, University of Bergen
26. Sydnes LK, Hemmingsen TH, Skare S, Hansen SH, Falk-Petersen I-B, Lønning S, Østgaard K (1985) Environ Sci Technol 19:1076
27. Sydnes LK, Hansen SH, Burkow IC (1985) Chemosphere 14:1043
28. Albini A (1981) Synthesis 4:249
29. Rimkus GG, Gatermann R, Hühnerfuss H (1999) Toxicol Letters 111:5
30. Rimkus G, Gatermann R, Biselli S, Hühnerfuss H (1999) Lebensmittelchemie 53:88
31. Gatermann R, Hühnerfuss H, Rimkus G, Attar A, Kettrup A (1998) Chemosphere 36:2535
32. Rimkus GG, Rummler M, Wolf M (2001) Lebensmittelchemie 55:60
33. Biselli S, Gatermann R, Rimkus G, Hühnerfuss H, Kallenborn R (2000) Third SETAC World Congress, Brighton, UK, 24–29 May 2000, Poster Presentation 3E/P026:156
34. Gatermann R (1999) Verteilung, Anreicherung und Transformation nitroaromatischer und polycyclischer Moschusduftstoffe sowie weiterer N- und P-haltiger Problemstoffe in der Umwelt, PhD thesis, University of Hamburg, Shaker, Aachen
35. Kallenborn R, Gatermann R, Planting S, Rimkus GG, Lund M, Schlabach M, Burkow IC (1999) J Chromatogr A 846:295
36. Rimkus G, Brunn H (1996) Ernährungs-Umschau 43:442
37. Rimkus GG (1999) Toxicol Lett 111:37
38. Herren D, Berset JD (2000) Chemosphere 40:565
39. Gatermannn R, Biselli S, Hühnerfuss H, Rimkus GG, Hecker M, Karbe L (2002) Arch Environ Contam Toxicol 42:437
40. Hühnerfuss H, Biselli S, Gatermann R, Kallenborn R, Rimkus G (2001) Organohalogen Compd 52:441
41. Gatermann R, Kallenborn R (2001) Syntetisk Musk i Torsk fra havneområder i Oslo, Porsgrunn og Larvik. Internal NILU report, pp 4
42. Behechti A, Schramm K-W, Attar A, Niederfellner J, Kettrup A (2000) Water Res 32:1704
43. Salvito D (2000) Water Res 34:2625
44. Giddings JM, Salvito D, Putt AE (2000) Water Res 34:3686
45. Vahl HH, Biselli S, Gatermann R, Hühnerfuss H, Westendorf J (2000) Third SETAC World Congress, Brighton, UK, 24–29 May 2000, Poster Presentation 5I/P018:249
46. Soto AM, Sonnenschein C, Chung KL, Fernandez MF, Olea N, Serrato FO (1995) Environ Health Perspect 103:113
47. Chou YJ (1999) Embryotoxic, teratogenic, and endocrine modulating activity of musk fragrances in fish and amphibians. PhD thesis, University of Konstanz, Konstanz, Germany
48. Heinzel N (1997) Synthese von Bis(1,3-dichlor-2-propyl)ether und (4*S*,7*RS*)-Galaxolid als Referenzsubstanzen häufiger Schadstoffe in Ökosystemen, Master thesis, University of Hamburg, Hamburg, Germany

The Handbook of Environmental Chemistry Vol. 3, Part X (2004): 213–231
DOI 10.1007/b14126

Enantioselective Analysis of Polycyclic Musks as a Versatile Tool for the Understanding of Environmental Processes

Heinrich Hühnerfuss[1] · Scarlett Biselli[2] · Robert Gatermann[2]

[1] Institute of Organic Chemistry, University of Hamburg, Martin-Luther-King-Platz 6, 20146 Hamburg, Germany
E-mail: huehnerfuss@chemie.uni-hamburg.de
[2] Dr. Wiertz – Dipl.-Chem. Eggert – Dr. Jörissen GmbH Analytical Laboratory (WEJ), Stenzelring 14b, 21107 Hamburg, Germany

Abstract In this chapter the application of enantioselective chromatography to process studies related to the microbial, enzymatic and photochemical transformation of chiral polycyclic musks will be summarised. A detailed discussion on enantiomeric ratios (ER) of HHCB, AHTN, ATII and AHDI as well as of the main HHCB metabolite HHCB-lactone analysed in tissue extracts of 18 fish samples (rudd, tench, crucian carp, eel) and one pooled zebra mussel sample from the pond of a municipal sewage treatment plant in the Federal State of Schleswig-Holstein (Germany) will be given. In addition, three water samples taken at the effluent of the sewage plant, as well as two water samples, and two series of semi-permeable membrane devices (SPMDs) consisting of six samples each from the pond were included in the study. This comprehensive data set allowed a reliable evaluation of species dependent metabolisation processes. The pattern of the polycyclic musks in the chromatograms obtained by enantioselective gas chromatography seemed to be typical of each species like a fingerprint. Very strong enantioselective metabolisation can be concluded for *trans*- and *cis*-HHCB and the respective HHCB-lactone (with a preferential metabolisation of the 4*S* enantiomers) as well as for *trans*-ATII in crucian carp. With ER values $\leq$0.1 for *trans*-HHCB and *trans*-ATII the highest enantioselectivity was observed within this class of xenobiotics. The values for the pooled mussel and SPMD samples reflect the water values. The exact assignment to the respective enantiomers was in part possible by pure enantiomeric standards.

Keywords Enantioselective GC · Enantioselective HPLC · Polycyclic musks · Polycyclic musk metabolites · Process studies · Fish · Mussel · Water · SPMD · Sewage treatment plant

1 Introduction

The application of classical achiral stationary phases in capillary gas chromatography (GC) and high-performance liquid chromatography (HPLC) allows insight into the distribution of xenobiotics in marine and terrestrial ecosystems. Furthermore, limited information about the fate of these compounds is accessible; in particular, metabolites can be determined. A deepened study of the various processes that may give rise to metabolisation in ecosystems, be it microbial, enzymatic or photochemical processes, has to account for the very property that has been assumed to be closely related to life since the basic studies of Pasteur: chirality. Homochirality appears to be a requirement for the functioning of enzymes and nucleic acids, while the specific (partial) incorporation of the respective opposite enantiomers of amino acids and carbohydrates, respectively, may block the active sites and thus give rise to potent toxins [1].

In general, it is assumed that the chiral environment of such active sites is able to select monomers of matching chirality out of a racemic mixture thus giving rise to the formation of diastereomeric complexes between the active sites of enzymes and chiral xenobiotics and their metabolites, respectively. Diastereomeric complexes and salts are known to exhibit physical properties, molecular orders (e.g. packaging) and structures that may be quite different from those of homochiral domains of the respective components [2]. As a consequence, the velocities of the metabolisation process of the enantiomers of chiral environmental pollutants may be significantly different. In some cases, only one enantiomer is being decomposed, while the second enantiomer is being accumulated in the environment.

The basic principles of chirality summarised above set the stage for various process studies about the microbial, enzymatic or photochemical transformation of chiral environmental pollutants with one or more stereogenic centres. In 1991, Faller et al. [3] raised the question "Do marine bacteria degrade α-HCH stereoselectively?", and they were the first to answer this question in the affirmative by applying GC with heptakis(3-*O*-butyryl-2,6-di-*O*-*n*-pentyl)-β-cyclodextrin as a chiral stationary phase to residual analysis of environmental samples. In a subsequent systematic investigation by Faller et al. [4], 16 water samples representing all parts of the North Sea that are interesting both from an oceanographic and chemical view were analysed with regard to enantiomeric ratios (ER) of the α-HCH enantiomers. The ER is directly obtained by peak integration of the gas chromatogram and dividing the peak area of the first eluting (+)-α-HCH through the peak area of the second eluting (–)-α-HCH enantiomer, i.e. $ER=E_1/E_2$. The results thus obtained revealed different microbial transformation pathways in the North Sea.

After the pioneering investigation by Kallenborn et al. who demonstrated the enantioselective metabolisation of α-HCH in Eider ducks (*Somateria mollissima* L.) [5] increasing attention has been paid to the chromatographic enantiomer separation of chiral xenobiotics and their metabolites in biota sample extracts. Thus, very different enzymatic transformation pathways became visible: in tissue extracts of muscle, liver and kidney of Eider ducks [5] and in liver samples of flounders [6], (+)-α-HCH was clearly enriched; almost pure (+)-α-HCH was present in liver extracts of Eider ducks [5], roe-deer [7] as well as in brain tissue extracts of Eider ducks and of harbour seals [8]. The latter results are insofar notable as they imply an enantioselective permeation of (+)-α-HCH through the Blood-Brain Barrier (BBB). The same enantioselectivity of the BBB was also observed for brain tissues of sheep [9] and humans [10], although in fat and liver tissues of sheep a preferential enrichment of the (–)-enantiomer was determined reflecting quite different enzymatic systems.

Another notable phenomenon revealed by enantioselective gas chromatography was reported by Pfaffenberger et al. [7] who observed a correlation between the concentrations of α-HCH, *cis*-heptachlorepoxide and oxychlordane, respectively, in roe-deer liver samples and the enantiomeric ratios of these compounds, which indicated that higher concentrations of these xenobiotics resulted in stronger transformation of the (+)-enantiomer of α-HCH, while higher levels of *cis*-heptachlorepoxide and oxychlordane appeared to lead to a faster decomposition of the respective (–)-enantiomer or a preferential formation of the respective (+)-enantiomer. Similar observations were made by Pfaffenberger et al. [11] for bromocyclen in fish tissue.

In general, photochemical transformation of chiral xenobiotics is assumed to be largely non-enantioselective. This hypothesis forms the basis for a discrimination between biotic (enantioselective) and abiotic processes like photodecomposition. However, Hühnerfuss et al. [12] showed that enantiomeric excesses, e.g. of β-PCCH, that may have been formed by enzymatic processes may be modified by photochemical processes provided that enzymatic processes become less important, for example due to seasonal variations of microbial activity. Another notable application of enantioselective gas chromatography was reported by Jantunen and Bidleman [13] as well as by Bethan et al. [14] who determined the enantiomers of α-HCH as tracers of air-water gas exchange.

Differential effects of the α-HCH enantiomers on cytotoxicity and growth stimulation in primary rat hepatocytes were recently observed by Möller et al. [15]. The cytotoxic effect was determined as a parameter for the acute toxicity of α-HCH, while the growth stimulation may be associated with the chronic toxicity, e.g. tumour promotion.

For a more detailed information the reader should refer to the reviews by Vetter and Schurig [16] and by Hühnerfuss [17] as well as to the monograph by Kallenborn and Hühnerfuss [18] which will allow deepened insight into the application of enantioselective analysis with chiral stationary phases to microbial, enzymatic or photochemical transformation studies of chiral environmental pollutants exhibiting one or more stereogenic centres. In this chapter the application of enantioselective chromatography to process studies related to the microbial, enzymatic and photochemical transformation of chiral polycyclic musks will be summarised [19–24].

2
Enantioselective Analyses of Polycyclic Musks

The synthetic polycyclic musks HHCB (1,3,4,6,7,8-hexahydro-4,6,6,7,8,8-hexamethylcyclopenta[*g*]-2-benzopyrane, e.g. galaxolide) and AHTN (1-(5,6,7,8-tetrahydro-3,5,5,6,8,8-hexamethyl-2-naphthalenyl)-ethanone, e.g. tonalide), which are important artificial fragrances used in a large number of perfumes, laundry detergents, fabric softeners, toiletry products, and other household products [25], as well as the polycyclic musks ATII 1-[2,3-dihydro-1,1,2,6-tetramethyl-3-(1-methyl-ethyl)-1*H*-inden-5-yl]-ethanone, e.g. traseolide), and AHDI (1-(2,3-dihydro-1,1,2,3,3,6-hexamethyl-1*H*-inden-5-yl)-ethanone, e.g. phantolide) were included in a recent study by Gatermann et al. [20]. The stereochemical structures of these four polycyclic musk compounds can be found in Fig. 1. All derivatives are chiral compounds, where HHCB and ATII exhibit two stereogenic centres and thus two diastereomeric pairs of enantiomers. However, it is important to note that the technical ATII contains more than 95% of the *trans* isomer, presumably reflecting larger differences in the physicochemical properties of the diastereomers, which in turn results in a high excess of this *trans*-diastereomer during the synthesis. As a consequence, in environmental samples largely the latter diastereomer will be encountered, which, therefore, is included in Fig. 1 only. With regard to HHCB the enantioselective syntheses and separations of all four stereoisomers was achieved, and thus it could be proved that the 4*S*,7*R*/*S*-isomers are the powerful, musky components [26]. Furthermore, the authors postulate a high three-dimensional structural similarity between 5α-androst-16-en-3-one and these two active isomers.

In the case of HHCB, the separation of the diastereomers on capillary columns commonly used in pesticide and PCB residue analysis turned out to be a challenge. Only in a few studies the diastereomers were separated, e.g. in extracts of human adipose tissue and in fish extracts using a methylpolysiloxane phase with 12–15% phenyl groups and a polyethylene glycol phase, respectively ([25] and literature cited therein).

First results about the stereoselective enrichment or transformation of HHCB and AHTN were reported by Franke et al. [19] followed by more comprehensive investigations by Gatermann et al. [20]. In the latter paper, the successful enantioselective GC separations of the four polycyclic musks shown in Fig. 1, including the diastereomeric pairs of enantiomers, were described. Furthermore, a detailed discussion on the enantiomeric ratios (ER) analysed in tissue extracts of four different fish species, a pooled mussel sample and 12 semipermeable membrane device (SPMD) samples, which were taken from a pond of a municipal sewage treatment plant in Schleswig-Holstein (Germany), was given.

HHCB
e.g. galaxolide ®

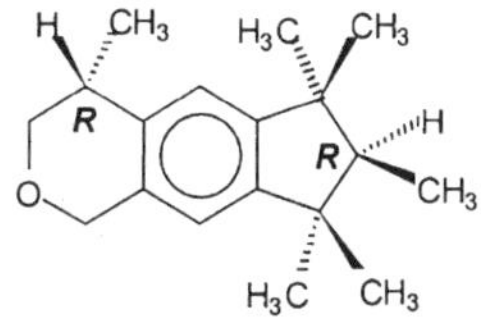

4R, 7R

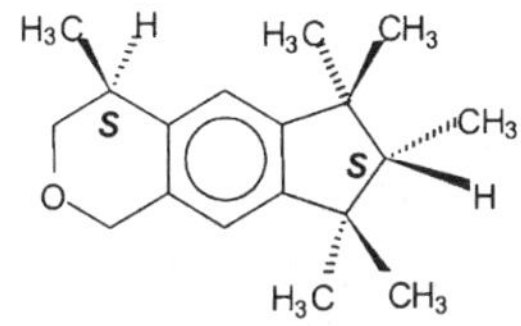

4S, 7S

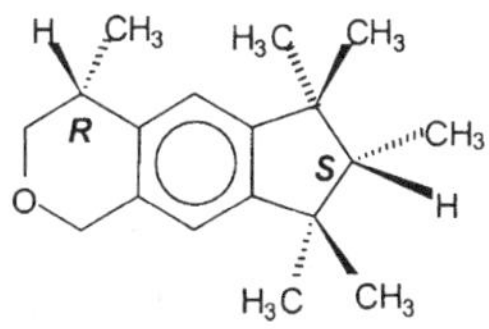

4R, 7S

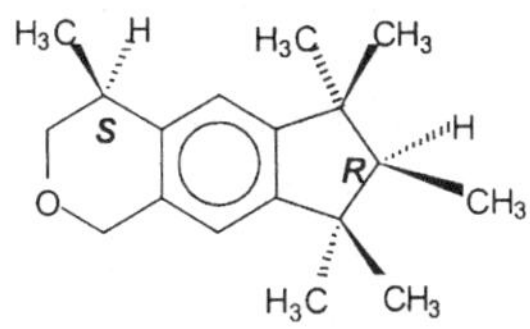

4S, 7R

AHTN
e.g. tonalide ®

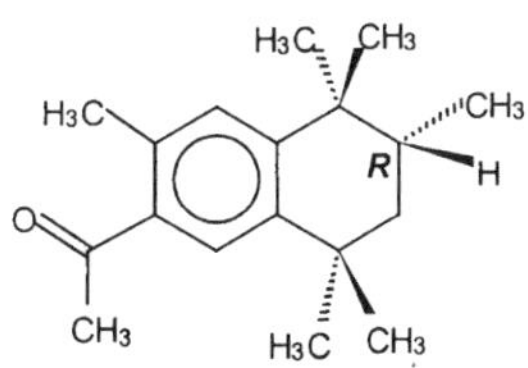

3R

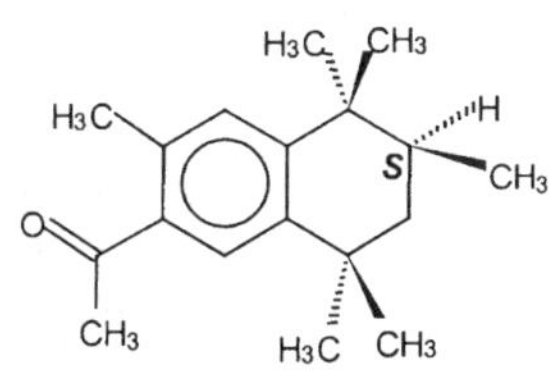

3S

***trans*-ATII**
e.g. traseolide ®

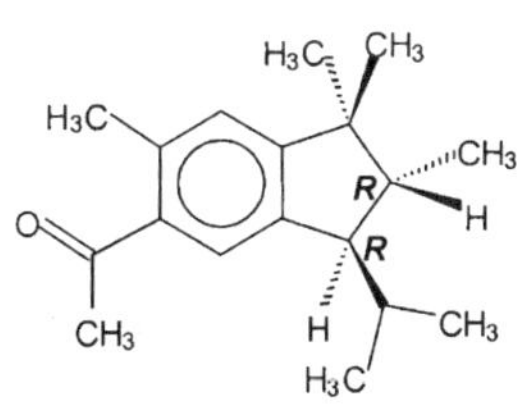

2R, 3R

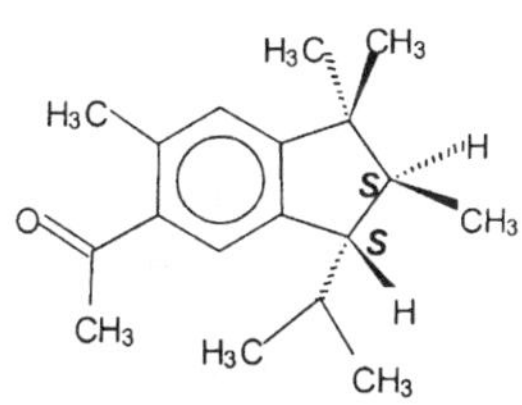

2S, 3S

AHDI
e.g. phantolide ®

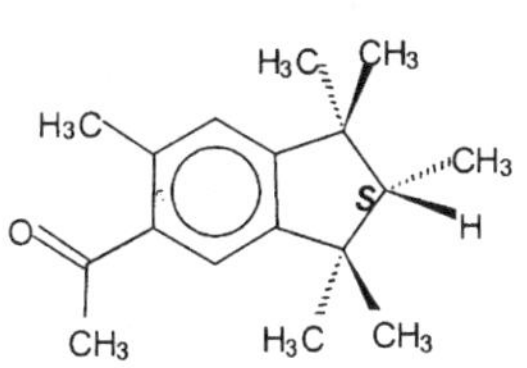

2S

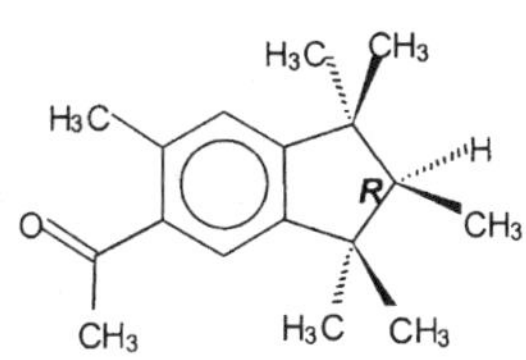

2R

Fig. 1 Stereochemical structures of *trans*-HHCB, *cis*-HHCB, AHTN, *trans*-ATII and AHDI

2.1
Enantioselective Separation of Standard Substances

The enantioselective separation of the polycyclic musk compounds including the diastereomeric pairs of enantiomers was achieved using a CE instruments 8560 Mega II gas chromatograph (Milan, Italy) equipped with a capillary column (0.25 mm i.d.; length 25 m) coated with a 1:1 mixture of OV 1701/heptakis(6-*O*-*tert*-butyldimethylsilyl-2,3-di-*O*-methyl)-β-cyclodextrin (TBDMS) including a 2-m deactivated precolumn (0.50 mm i.d.; J&W, Folsom CA) [20]. The chromatographic parameters comprised: the temperature program was started at 343 K (70 °C, 2 min isotherm), then the temperature was raised by 10 K min^{-1} to 418 K (145 °C), 0.5 K min^{-1} to 453 K (180 °C) and 10 K min^{-1} to 503 K (230 °C, 10 min isotherm); helium 5.0 as carrier gas; flow velocity 1.0 mL min^{-1}. The GC was coupled to a Finnigan MD 800 quadrupole low-resolution mass spectrometer (San Jose, USA), which was run in electron impact mode (EI) with the ion source temperature set on 453 K (180 °C) applying selected ion monitoring (SIM) mode with a dwell time of 100 ms for each ion. The sample was injected in on-column mode (2 µL injection volume). The m/z values used were: HHCB m/z=243 (m/z=213 and m/z=258 were used for further verification); AHTN m/z=243 (reference mass m/z=258); *trans*-ATII m/z=215 (reference mass m/z=258); AHDI m/z=229 and m/z=187; HHCB-lactone m/z=257.

Selected ion monitoring (SIM) fragmentograms (m/z=187, 213, 215, 229, 243 and 258) of a standard mixture containing HHCB, AHTN, *trans*-ATII and AHDI (1 ng μL^{-1} each) are shown in Fig. 2. As the technical formulation of HHCB used in fragrance formulations contains different chiral by-products with the same fragment ions m/z=258 and m/z=243 [27], possible coelution of these by-products and the polycyclic musks themselves cannot be basically excluded. However, injection of the authentic mixture confirmed that no coelution between HHCB and the by-products takes place on this column (Fig. 2). The assignment of the peaks to the diastereomeric pairs of HHCB was already successfully attained [22, 23], while the assignment to the enantiomers of the other polycyclic musks is presently attempted in a subsequent endeavour by comparing the retention times of the authentic separated stereoisomers with those of the different isomers in the racemic standards. The exact assignment to the respective enantiomers required their enantioselective separation by HPLC as pure enantiomeric standards which was successfully attained by a monofunctionalised permethyl-β-cyclodextrin stationary phase for HHCB and *trans*-ATII as well as for their chiral by-products [22, 23]. As shown in Fig. 3, good separation of the by-products and baseline separation for the third and fourth eluting isomers of HHCB was achieved. The isomers of the *trans*-ATII were nearly baseline separated, and there was also a good partition of the thus far unknown compound present in the technical product of ATII. The absolute configuration of the isomers of ATII has not been determined yet.

The two diastereomers of HHCB responsible for the significant musky odour (4*S* configuration, see Fig. 1) are eluting first on the β-cyclodextrin GC column used by Gatermann et al. [20]. Thus it turned out that the technical HHCB consists of an approximately 1:1 mixture of both diastereomers with a slight excess

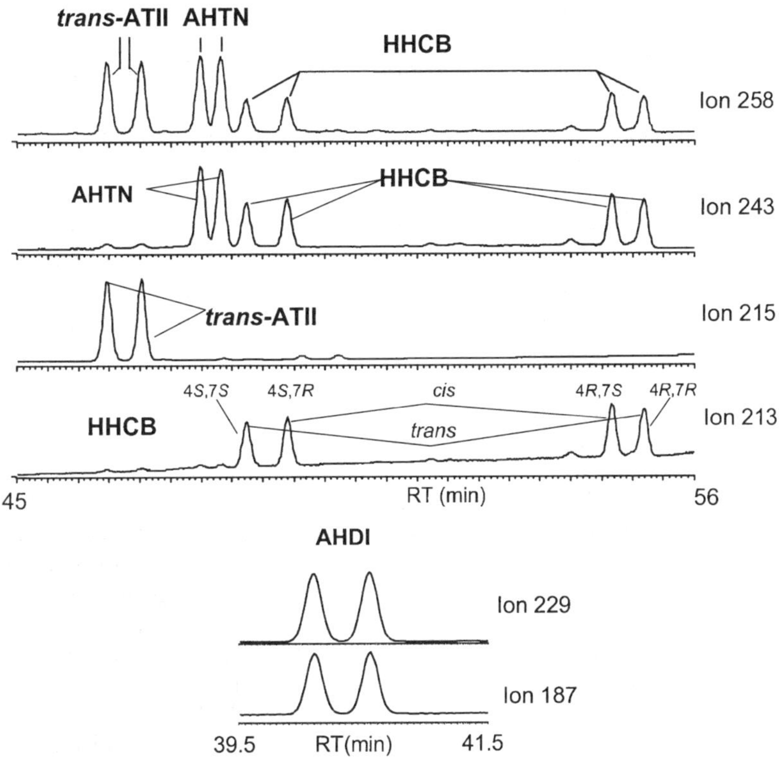

Fig. 2 SIM fragmentograms of a standard mixture containing HHCB, AHTN, *trans*-ATII, and AHDI (1 ng/µL each), stationary phase: 1:1 mixture of OV 1701/heptakis(6-*O*-*tert*-butyldimethylsilyl-2,3-di-*O*-methyl)-*β*-cyclodextrin; from [20]

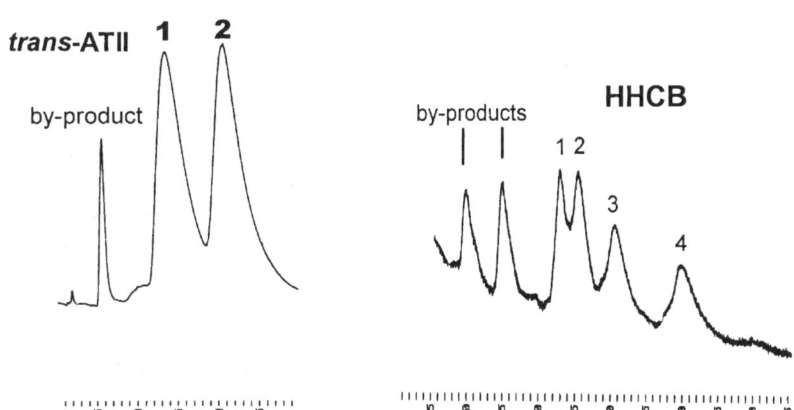

Fig. 3 Preparative enantioselective separation of *trans*-ATII and HHCB by HPLC, stationary phase permethyl-*β*-cyclodextrin on 3-aminopropyl silica gel; from [22, 23]

of the *cis* isomer. Since no significant changes in the *cis*/*trans* ratios in the water, sediment, sewage sludge and SPMD samples were found, the physicochemical properties of the isomers are expected to be relatively similar. By way of contrast, the technical ATII contains more than 95% of the *trans* isomer, as mentioned above, which may reflect larger differences in the physicochemical properties of the diastereomers. In line with these assumptions, the chromatographic separations of HHCB and ATII on the TBDMS column are different. The enantiomers of the *trans*-ATII diastereomer eluted closely together. However, for HHCB the 4*S* and the 4*R* diastereomers eluted closely together, whereas the enantiomers were extremely widely separated (see Fig. 2). In addition, the separations of the enantiomers of AHTN (m/z=258 and m/z=243) and AHDI (m/z=229 and m/z=187) are also shown in Fig. 2.

2.2 Polycyclic Musks in a Sewage Treatment Plant

Gatermann et al. [20] carried out their experiment in a sewage treatment plant in the Federal State of Schleswig-Holstein (Germany). At first, the sewage enters the treatment plant, and subsequently the treated waste water flows towards a pond, where it is allowed to remain for some weeks, assuming an average water exchange rate of about 17,000–22,000 m^3 per day and taking into account an area of 128,000 m^2 and a depth of 4 m. The effluent of the pond is closed with bars thus enabling small fishes only to enter or to escape from the pond. Therefore, equilibrium conditions can be assumed for synthetic musks in fish tissues of larger animals that have to remain within the pond. The 3 rudd, 4 tench (plus 1 tench liver), 7 crucian carp (plus 1 crucian carp liver), 2 eel, 1 pooled mussel and 12 SPMD samples investigated herein stem from this pond. For further information about experimental details concerning sample collection, extraction and clean-up, separation and quantification as well as apparatus, chemicals and reagents used in this work the reader is referred to the paper of Gatermann et al. [28].

2.2.1 *Enantiomeric Ratios of HHCB, AHTN, trans-ATII and AHDI in Water and SPMD Samples*

In the water and SPMD samples collected in the pond, only for *trans*-ATII deviations from the racemic enantiomeric ratios (ER) were found (in water samples: ER=0.7–0.8; in SPMD samples: ER=0.8), which is a clear indication of a pronounced enantioselective transformation in the water of the sewage treatment plant of this specific polycyclic musk derivative. This appears not to be the case for the other polycyclic musks investigated by Gatermann et al. [20].

2.2.2 Enantiomeric Ratios of HHCB, AHTN, *trans*-ATII and AHDI in Biota Samples

As an example for biota sample extracts, SIM fragmentograms are shown for tench and crucian carp in Fig. 4. Concentrations and enantiomeric ratios of HHCB, *cis*- and *trans*-HHCB, AHTN, *trans*-ATII and AHDI for 18 fish samples

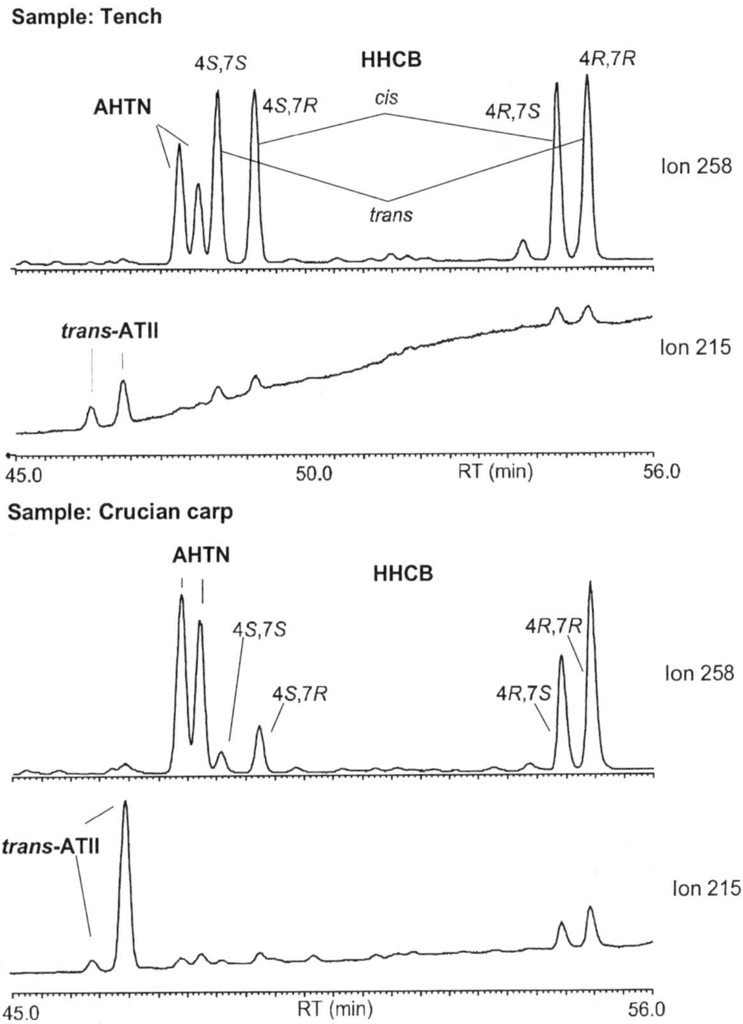

Fig. 4 SIM fragmentograms of tench and crucian carp sample extracts, stationary phase: 1:1 mixture of OV 1701/heptakis(6-*O*-*tert*-butyldimethylsilyl-2,3-di-*O*-methyl)-*β*-cyclodextrin; from [20]

and 1 pooled mussel sample from the pond of the sewage plant are summarised in Table 1. In the case of one tench and one crucian carp, liver samples were analysed in addition to the muscle tissue. The concentrations of HHCB and AHTN as inferred from non-enantioselective GC were discussed by Gatermann et al. [28].

Both the concentrations and the enantiomeric ratios appear to exhibit a species dependency. Clearly distinguishable clusters can be inferred from the data set as illustrated in Fig. 5. For example, for HHCB the following species dependent clusters can be stated: rudd (concentration=6.2–7.5 μg g^{-1} lipid;

Table 1 Concentrations (concentration; µg g^{-1} lipid) and enantiomeric ratios (ER) of HHCB, *trans*- and *cis*-HHCB, AHTN, *trans*-ATII, and AHDI in biota caught in the pond of a municipal waste water treatment plant in summer 1997 (from [20])

Species	HHCB	*trans*-HHCB	*cis*-HHCB	AHTN		*trans*-ATII		AHDI	
	conc	ER	ER	conc	ER	conc	ER	conc	ER
Rudd	6.2	0.66	1.24	5.0	0.94	0.5	0.99	0.3	0.90
Rudd	7.1	0.55	1.10	5.7	0.91	0.5	0.98	0.3	1.12
Rudd	7.5	0.57	1.07	6.1	0.80	0.5	0.97	0.3	0.95
Tench	150	1.06	0.98	30	1.65	2.3	0.60	2.2	1.31
Tench	160	1.10	0.95	32	1.97	2.0	0.38	1.9	1.47
Tench, liver	72	1.01	0.85	16	1.98	1.1	0.33	0.9	1.78
Tench	160	0.88	1.00	35	1.74	2.2	0.51	2.1	1.30
Tench	150	1.03	0.98	42	1.65	2.6	0.46	2.5	1.28
Crucian carp	39	0.12	0.44	26	1.05	1.2	0.05	1.8	0.65
Crucian carp	59	0.12	0.42	31	1.24	1.5	0.08	2.7	0.81
Crucian carp	91	0.19	0.54	31	1.19	1.9	0.20	3.1	0.94
Crucian carp	71	0.14	0.48	30	1.15	1.4	0.10	2.5	0.78
Crucian carp	50	0.10	0.41	34	1.19	1.6	0.06	2.4	0.71
Crucian carp	66	0.10	0.46	40	1.14	1.5	0.05	2.3	0.66
Crucian carp	84	0.13	0.51	34	1.15	1.9	0.07	3.0	0.64
Crucian carp, liver	69	0.10	0.48	31	1.29	1.6	0.06	2.5	0.63
Eel	4.8	0.98	1.27	2.6	0.87	0.3	1.95	0.2	0.73
Eel	4.6	0.79	1.15	2.7	n.a.	0.3	1.50	0.2	0.48
Mussel	120	0.91	1.26	45	0.90	3.4	0.67	2.3	1.02

n.a. = not analysed.

$ER_{trans=}$0.55–0.66; ER_{cis}=1.07–1.24), tench (concentration=150–160 µg g^{-1} lipid; ER_{trans}=0.88–1.10; ER_{cis}=0.95–1.00), crucian carp (concentration=39–91 µg g^{-1} lipid; ER_{trans}=0.10–0.19; ER_{cis}=0.41–0.54), eel (concentration=4.6–4.8 µg g^{-1} lipid; ER_{trans}=0.79–0.98; ER_{cis}=1.15–1.27), mussel (concentration=120 µg g^{-1} lipid; ER_{trans} 0.91; ER_{cis}=1.26). Similar clusters can be inferred from Table 1 for concentrations and enantiomeric ratios for AHTN, *trans*-ATII, and AHDI, respectively.

It is tentatively assumed that high concentrations and enantiomeric ratios close to racemic, i.e. ER≈1.0, indicate a low metabolisation potential of a species for the respective polycyclic musk derivative. On the other hand, caution has to be applied when inferring metabolisation potentials exclusively from lipid based concentrations and/or bioaccumulation factors on lipid basis (BAF_L; see Gatermann et al. [28]): low concentrations and low BAF_L may reflect a stronger metabolisation, but it cannot be excluded that specific matrix effects, e.g. high lipid contents like in eel, may pretend a stronger metabolisation than actually encountered. An unequivocal parameter for an enantioselective transformation process, however, are ER values clearly different from 1.

On the basis of these assumptions, rudd appears to exhibit a strong enantioselective metabolisation potential for *trans*-HHCB. Tench shows a low enantioselective metabolisation potential for *trans*- and *cis*-HHCB, and a moderate one

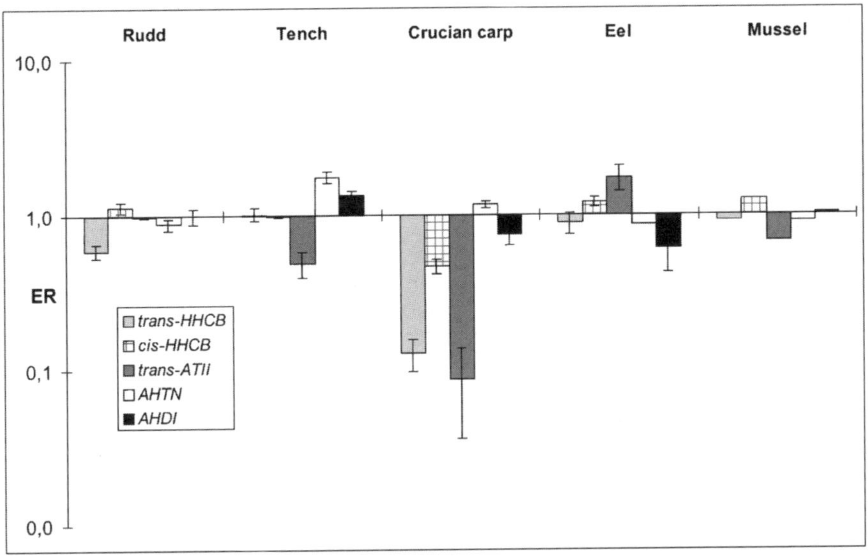

Fig. 5 Mean ER values of *cis*- and *trans*-HHCB, *trans*-ATII, AHTN, and AHDI for different aquatic species; from [20]

for AHTN, *trans*-ATII and AHDI. Very strong enantioselective metabolisation can be concluded for *trans*- and *cis*-HHCB as well as for *trans*-ATII in crucian carp, while for AHTN and AHDI a moderate metabolisation was observed. With ER values ≤0.1 for *trans*-HHCB and *trans*-ATII the highest enantioselectivity in this study was observed. In eel the high lipid content gives rise to low lipid-normalised concentrations, but the ER values indicate low to moderate enantioselective metabolisation for *trans*- and *cis*-HHCB, and AHTN, while for *trans*-ATII and AHDI a stronger enantioselective metabolisation capability was found. The values for the pooled mussel sample, known for a low metabolisation capability, reflect the water values which were also determined in the present study.

By way of contrast, in the liver the metabolic capacity is higher than in other organs. As a consequence, lower concentrations of the polycyclic musks as well as larger deviations from racemic were determined in the tench and crucian carp liver extracts (with the exception of *trans*-HHCB in the tench liver sample) as compared to the corresponding muscle tissues (Table 1).

As stated above, the lipid based concentrations and/or bioaccumulation factors may be misleading for the assignment of metabolisation potentials. However, in those cases where low concentrations, low BAF_L and significant deviations of the enantiomeric ratios from 1 are being encountered, it is justified to assume strong metabolisation capacities. This assumption is in line with previous conclusions drawn from enantioselective analyses of chlordanes in different fish species [29].

In order to support the conclusion that the lower musk concentrations encountered in crucian carp are caused by enantioselective transformation, theoretical concentrations for passive bioconcentration were calculated on the

Table 2 Experimental and calculated concentrations (μg g^{-1} lipid) of HHCB, AHTN, and *trans*-ATII in several fish species and mussels (from [20])

Species	HHCB		AHTN		*trans*-ATII	
	Experim.	Calculated	Experim.	Calculated	Experim.	Calculated
Tench	155	–	35	44	2.3	2.8
Tench, liver	72	–	16	21	1.1	1.7
Crucian carp	66	122	32	35	1.6	3.0
Crucian carp, liver	69	135	31	35	1.6	3.0
Mussel	120	–	45	–	3.4	–

following basis [20]: it was tentatively assumed that only one enantiomer is being transformed, which implies that the prevailing second enantiomer represents 50% of the original racemate, i.e. prior to the beginning of the enantioselective transformation process. In the case of *trans*-ATII, an ER of 0.8 (found in the water samples) instead of a racemic distribution was used as basic value. The calculated values for each species as well as mean experimental values are summarised in Table 2. In the case of AHTN, the measured values in crucian carp and tench do not differ significantly and are comparable with the values in mussels, which implies that the concentration levels are explainable with passive bioconcentration only and no or little metabolisation. On the other hand, in crucian carp significantly lower concentrations of HHCB and *trans*-ATII compared to tench as well as lower experimental than calculated values indicate strong enantioselective transformation of the first eluting *trans*-ATII enantiomer. In the case of HHCB, the calculation is more sophisticated because the measured concentration is the sum of two diastereomers, *cis*- and *trans*-HHCB. The comparison of the stereochemical patterns in the crucian carp sample extracts and in standard solutions shows a different behaviour of the *cis* and the *trans* isomer: For the *cis* isomer both enantiomers seem to be transformed compared to the latest eluting enantiomer of *trans*-HHCB. Therefore, it was decided to choose this stereoisomer (4*R*,7*R*) as the basic value for the calculation. The mean calculated value (122 μg g^{-1} lipid) is as high as the mussel value and only slightly lower than the mean value of the tench samples (152 μg g^{-1} lipid). Thus, it can be concluded that also for HHCB enantioselective transformation is the most important process resulting in lower concentrations for crucian carp. In addition, a higher selectivity of the *trans* compared to the *cis* isomer can be assumed.

2.2.3
Implications of the Observed Enantioselectivity for Risk Assessment Studies

The species dependent metabolisation processes as inferred from the present data set bear a notable implication: caution has to be applied if risk assessment studies have to rely on one fish species only. It may be the wrong choice, because the fish may possess an extremely low or an extremely high metabolisation potential.

Furthermore, the in part high enantioselectivity in the metabolic capacity for the polycyclic musks and the well-defined binding activities of the polycyclic musks to olfactory receptors [30–33], the known effect of natural musk as pheromones [34–36] and the structure relationship, especially of HHCB, to steroid hormones [26] raise the question whether or not synthetic musks exert an effect on the chemical communication and/or the hormone system of species in the aquatic and/or the terrestrial environment. However, this question can only be answered by systematic experiments in which tests with pure enantiomers have to be included that have to be separated by enantioselective HPLC [22, 23]. Presently, all answers on this problem must remain speculative.

3 Enantioselective Analyses of Polycyclic Musk Metabolites

A comprehensive risk assessment study should also include the metabolites of the respective xenobiotics. In the case of the polycyclic musks, recent publications by Franke et al. [19], Biselli [23] and Hühnerfuss et al. [24] presented first results on HHCB metabolites, including their syntheses and their enantioselective gas chromatographic separations. The main HHCB metabolite, 1,3,4,6,7,8-hexahydro-4,6,6,7,8,8-hexamethylcyclopenta[*g*]-2-benzopyrane-1-one (HHCB-lactone, in some references 'galaxolidone'), is shown in Fig. 6. For a detailed discussion on the preparation of the standard compound as well as its formation in the environment the reader is referred to Chap. 11 of this monograph.

Extensive enantioselective GC/MS investigations of HHCB-lactone in different environmental compartments including different fish species and sediment were recently reported by Biselli [23] and by Hühnerfuss et al. [24]. In the course of the sewage treatment plant experiment described in Sect. 2.2 the authors analysed water, SPMD and sediment samples as well as the same biota sample extracts summarised in Table 1 with regard to their HHCB-lactone concentrations and to the enantiomeric ratios of this metabolite.

3.1 Enantioselective Separation of the HHCB-Lactone Standard

As HHCB-lactone exhibits two stereogenic centres (Fig. 6), two diastereomeric pairs of enantiomers may be expected, which in turn may give rise to at maxi-

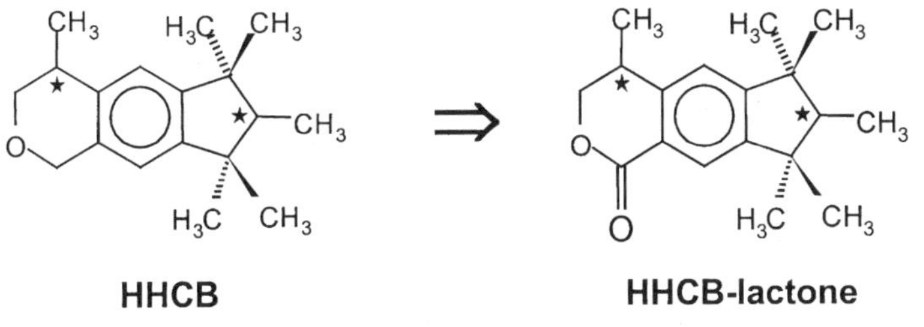

Fig. 6 The transformation of HHCB to HHCB-lactone; from [18]

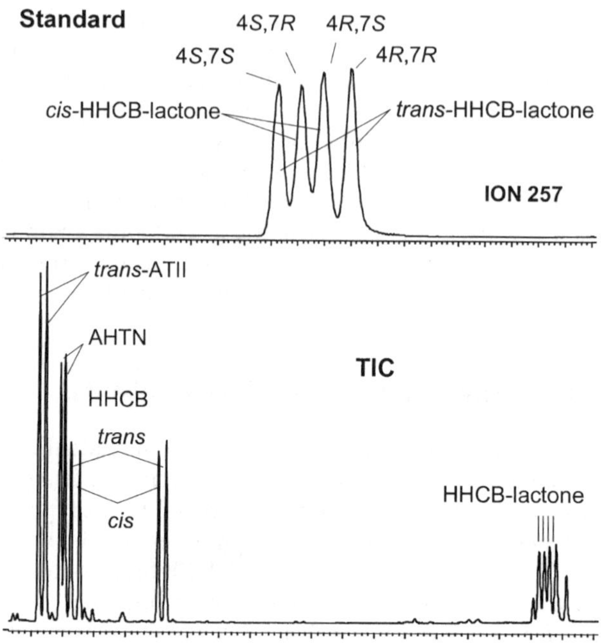

Fig. 7 SIM fragmentogram of a HHCB-lactone standard (ion 257) (*top*); standard mixture containing HHCB, AHTN, *trans*-ATII, and HHCB-lactone (TIC); stationary phase: 1:1 mixture of OV 1701/heptakis(6-*O*-*tert*-butyldimethylsilyl-2,3-di-*O*-methyl)-β-cyclodextrin (*below*); from [23, 24]

mum four peaks in the gas chromatogram. The enantioselective separation of this compound including the diastereomeric pairs of enantiomers was achieved using the same stationary phase as described for the enantioselective HHCB analysis (1:1 mixture of OV 1701/heptakis(6-*O*-*tert*-butyldimethylsilyl-2,3-di-*O*-methyl)-β-cyclodextrin (TBDMS), 2-m deactivated precolumn (0.50 mm i.d.; J&W, Folsom CA), [23, 24]). The chromatographic parameters can be found in Sect. 2.1. SIM fragmentograms (m/z=257) of the HHCB-lactone standard and total ion chromatograms (TICs) of a standard mixture containing HHCB, AHTN, *trans*-ATII and the HHCB-lactone are shown in Fig. 7. It is worth noting that the stationary phase used by Biselli [23] and by Hühnerfuss et al. [24] allowed both an enantioselective separation of HHCB-lactone, of its parent compound HHCB and of the polycyclic musks AHTN and *trans*-ATII without any significant coelutions. Detailed photochemical transformation experiments, i.e. UV irradiation of the respective HHCB enantiomers, revealed that the elution sequence of the HHCB-lactone enantiomers corresponds exactly with that of the respective parent enantiomers of HHCB [23, 24]. As indicated in Figs. 2 and 7, the peaks of the *trans*-HHCB-lactone enantiomers (4*S*,7*S* and 4*R*,7*R*) and of the *cis*-HHCB-lactone enantiomers (4*S*,7*R* and 4*R*,7*S*) exhibit the same elution sequence as those of the respective HHCB stereoisomers. Additional peaks in the TIC-region close to the HHCB-lactone peaks are presently being investigated on the basis of conjectures that are discussed in Chap. "Biotic and Abiotic Transformation Pathways of Synthetic Musks in the Aquatic Environment" of the present monograph (by-products of the HHCB-production, other transformation products?).

3.2 Enantiomeric Ratios of HHCB-Lactone Diastereomers in Environmental Samples

As an example for environmental samples, SIM fragmentograms are shown for tench, mussel, rudd, eel, crucian carp and SPMD sample extracts in Fig. 8. Average concentrations of HHCB-lactone as well as the enantiomeric ratios of both

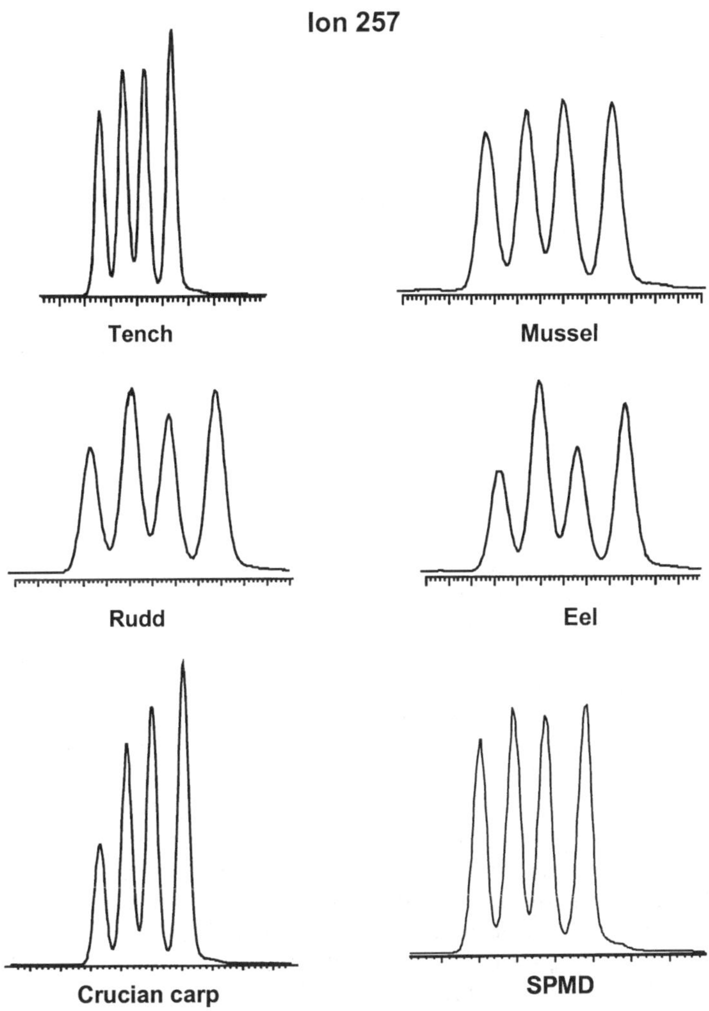

Fig. 8 SIM fragmentogram of HHCB-lactone in various fish, mussel and SPMD extracts; stationary phase: 1:1 mixture of OV 1701/heptakis(6-*O*-*tert*-butyldimethylsilyl-2,3-di-*O*-methyl)-β-cyclodextrin; from [23, 24]

Table 3 Average concentrations (μg g^{-1} lipid) of HHCB as well as average concentrations (μg g^{-1} lipid) and enantiomeric ratios (ER) of HHCB-lactone as determined in different fish species, mussel and SPMD sample extracts stemming from the pond of a sewage treatment plant (from [23, 24])

Sample	HHCB conc [μg g^{-1} lipid]	HHCB-lactone conc [μg g^{-1} lipid]	*trans*-HHCB-lactone ER	*cis*-HHCB-lactone ER
Rudd (n=3)	6.9	8	0.65	1.19
Tench (n=4)	155	33	0.68	0.99
Crucian carp (n=7)	66	24	0.39	0.83
Eel (n=2)	4.7	4	0.59	1.51
Mussel (pooled)	120	33	0.81	0.91
SPMD (n=12)	166 (n=12)	8 (n=9)	0.81	1.01

n = number of samples.

trans- and *cis*-HHCB-lactone are summarised in Table 3 for rudd, tench, crucian carp, eel, one pooled mussel sample and SPMD extracts, all stemming from the pond of the sewage plant. For comparison, average concentrations of the parent compound HHCB are also included in Table 3.

In the SPMD sample extracts which largely reflect the situation in the water no pronounced enantiomeric shifts were found both for *trans*- and *cis*-HHCB-lactone. The most intensive deviation of the enantiomeric ratios from racemic were encountered for crucian carp extracts. Basically, the latter result is in line with the corresponding result for the parent compound HHCB, though the exact ER values summarised in Table 3 indicate that the enantiomeric shifts are less pronounced for the metabolite than for the parent compound. However, it is interesting to note that both for HHCB and its transformation product an enantioselective transformation of the 4*S* enantiomers of the diastereomeric pairs was observed, i.e. the HHCB stereoisomers that are preferentially formed are being subject to preferential further transformation. In this case, an enantioselective bioconcentration from the water can be excluded because, first, the relation of HHCB and HHCB-lactone in the SPMD samples, which cannot be subject to transformation processes, as compared with the respective relations in the biota samples suggest that the uptake of the metabolite from the water plays a minor role, and, second, the enantiomeric shifts for the diasteromeric pairs of HHCB enantiomers in water were negligible. Similar coincidences of the preferential transformation of the same stereoisomers of the parent compound and the metabolite can be inferred from the results obtained for rudd and eel sample extracts. However, in these two cases, the diastereomers show a different metabolisation tendency; while for *trans*-HHCB-lactone the 4*S* enantiomer is being preferentially transformed, in the case of the *cis*-HHCB-lactone the 4*R* enantiomer is enantioselectively transformed. Summarising, the detailed analysis of the enantiomeric ratios of the polycyclic musk HHCB and its metabolite HHCB-lactone reveals the different enzymatic transformation pathways occurring in different fish species.

3.3 Conclusions for the Enantioselective HHCB Transformation

A summarising evaluation of the complete data set thus far available for the enantioselective transformation pathway of the polycyclic musk HHCB and its metabolite HHCB-lactone in various environmental compartments including different fish species reveals the following evidence (see also Tables 1 and 3):

- *Water*: with the exception of *trans*-ATII (ER=0.7–0.8), no significant transformation of the polycyclic musks was observed, i.e. no enantioselective transformation of HHCB.
- *SPMD*: with the exception of *trans*-ATII (ER=0.8), no significant transformation of the polycyclic musks was observed, i.e. no enantioselective transformation of HHCB.
- *Crucian carp*: the most intensive transformation of both HHCB and HHCB-lactone was observed, with a preferential metabolisation of the 4*S* enantiomers: medium concentrations of HHCB, high concentrations of HHCB lactone, lowest ER values for both HHCB as well as both HHCB-lactone diastereomers.
- *Rudd*: low HHCB concentrations; for *trans*-HHCB and *trans*-HHCB-lactone a preferential transformation of the 4*S* enantiomers, while for *cis*-HHCB and *cis*-HHCB-lactone a preferential transformation of the 4*R* enantiomers is encountered. At least for the *trans*-diastereomers a strong metabolisation can be inferred from the results.
- *Tench*: high HHCB concentrations; low enantioselective metabolisation potential for *trans*- and *cis*-HHCB; strong further transformation of *trans*-HHCB-lactone; no significant metabolisation of *cis*-HHCB-lactone.
- *Eel*: low HHCB concentrations (high lipid content!); low metabolisation of *trans*- and *cis*-HHCB; by contrast, a strong further transformation of both *trans*- and *cis*-HHCB-lactone was observed; for *trans*-HHCB and *trans*-HHCB-lactone a preferential transformation of the 4*S* enantiomers, while for *cis*-HHCB and *cis*-HHCB-lactone a preferential transformation of the 4*R* enantiomers was encountered.
- *Mussel*: high HHCB concentrations; low transformation both for HHCB and HHCB-lactone.

These results clearly show the potential of enantioselective chromatography for process studies related to the microbial, enzymatic and photochemical transformation of chiral polycyclic musks in environmental samples, be it water or biota. This experimental approach allows a reliable evaluation of species dependent metabolisation processes. The pattern of the polycyclic musks in the chromatograms obtained by enantioselective gas chromatography appears to be typical of each species like a fingerprint. Complemented by enantioselective HPLC, the exact assignment to the respective enantiomers may be attained. While the interpretation of concentrations of polycyclic musks in biota as determined by non-enantioselective GC may be misleading, because high lipid contents may pretend high metabolisation rates, enantiomeric ratios determined by enantioselective chromatography necessarily will reveal differential transformation processes. It is important to note that process studies by enantioselective chromatography are not confined to chiral polycyclic musks, but this experimental approach may be applied to all chiral and prochiral xenobiotics in different environmental compartments [18].

4 References

1. Jung G (1992) Angew Chem 104:1484
2. Hühnerfuss H, Neumann V, Stine KJ (1996) Langmuir 12:2561
3. Faller J, Hühnerfuss H, König WA, Krebber R, Ludwig P (1991) Environ Sci Technol 25:676
4. Faller J, Hühnerfuss H, König WA, Ludwig P (1991) Mar Pollut Bull 22:82
5. Kallenborn R, Hühnerfuss H, König WA (1991) Angew Chem 103:328
6. Pfaffenberger B, Hühnerfuss H, Kallenborn R, Köhler-Günther A, König WA, Krüner G (1992) Chemosphere 25:719
7. Pfaffenberger B, Hardt I, Hühnerfuss H, König WA, Rimkus G, Glausch A, Schurig V, Hahn J (1994) Chemosphere 29:1543
8. König WA, Hardt IH, Gehrcke B, Hochmuth DH, Hühnerfuss H, Pfaffenberger B, Rimkus G (1994) Angew Chem 106:2175
9. Möller K, Hühnerfuss H, Rimkus G (1993) J High Res Chromatogr 16:672
10. Möller K (1998) Verteilung, Schicksal und toxische Wirkung chlorierter organische Problemstoffe in der Umwelt und deren ökotoxikologische Bewertung. PhD Thesis, University of Hamburg, Hamburg, Germany, 154 pp
11. Pfaffenberger B, Hühnerfuss H, Gehrcke B, Hardt I, König WA, Rimkus G (1994) Chemosphere 29:1385
12. Hühnerfuss H, Faller J, König WA, Ludwig P (1992) Environ Sci Technol 26:2127
13. Jantunen LM, Bidleman T (1996) J Geophys Res 101:28,837; Jantunen LM, Bidleman T (1997) J Geophys Res 102:19,279
14. Bethan B, Dannecker W, Gerwig H, Hühnerfuss H, Schulz M (2001) Chemosphere 44:591
15. Möller K, Hühnerfuss H, Wölfle D (1996) Organohal Compd 29:357
16. Vetter W, Schurig V (1997) J Chromatogr A 774:143
17. Hühnerfuss H (2000) Chemosphere 40:913
18. Kallenborn R, Hühnerfuss H (2001) Chiral environmental pollutants. Trace analysis and ecotoxicology. Springer, Berlin Heidelberg New York, 209 pp
19. Franke S, Meyer C, Heinzel N, Gatermann R, Hühnerfuss H, Rimkus G, König WA, Francke W (1999) Chirality 11:795
20. Gatermann R, Biselli S, Hühnerfuss H, Rimkus GG, Franke S, Hecker M, Kallenborn R, Karbe L, König WA (2002) Arch Environ Contam Toxicol 42:447
21. Hühnerfuss H, Gatermann R, Biselli S, Rimkus GG, Hecker M, Kallenborn R, Karbe K (1999) Organohal Compd 40:401
22. Biselli S, Dittmann H, Gatermann R, Kallenborn R, König WA, Hühnerfuss H (1999) Organohal Compd 40:599
23. Biselli S (2001) Entwicklung einer analytischen Methode zum Nachweis von ökotoxikologisch relevanten organischen Problemstoffen in Sedimenten und Biota unter besonderer Berücksichtigung von Irgarol, synthetischen Moschusduftstoffen und deren Transformationsprodukten. PhD Thesis, University of Hamburg, Hamburg, Germany, 271 pp
24. Hühnerfuss H, Biselli S, Gatermann R, Kallenborn R, Rimkus G (2001) Organohal Compd 52:441
25. Rimkus GG (1999) Toxicol Lett 111:37
26. Frater G, Müller U, Kraft P (1999) Helv Chim Acta 82:1656
27. Müller S, Schmid P, Schlatter C (1996) Chemosphere 33:17
28. Gatermann R, Biselli S, Hühnerfuss H, Rimkus GG, Hecker M, Karbe L (2002) Arch Environ Contam Toxicol 42:437

29. Wiberg K, Oehme M, Haglund P, Karlsson H, Olsson M, Rappe C (1998) Mar Pollut Bull 36:345
30. Baydar A, Petrzilka M, Schott MP (1993) Chem Senses 18:661
31. Gilbert AN, Wysocki CJ (1991) Psychosom Med 53:693
32. Gower DB, Ruparelia BA (1993) J Endocrinol 137:167
33. König WA (2004) In: Wainer IW, Lough WJ (eds) Chirality in the natural and applied sciences. Kluwer Academic Publishers, Amsterdam (in press)
34. Bjerselius R, Olsén KH (1993) Chem Senses 18:427
35. Buck L, Axel R (1991) Cell 65:175
36. Stacey NE, Sorensen PW, Cardwell JR (1993) Hormonal pheromones: recent developments and potential applications in aquaculture. Coastal and estuarine studies, Berlin, pp 227–239

The Handbook of Environmental Chemistry Vol. 3, Part X (2004): 233–244
DOI 10.1007/b14125

Bioaccumulation and Ecotoxicity of Synthetic Musks in the Aquatic Environment

Daniel R. Dietrich[1] · Bettina C. Hitzfeld[2]

[1] University of Konstanz, Department of Environmental Toxicology, 78457 Konstanz, Germany
E-mail: Daniel.Dietrich@uni-konstanz.de
[2] Swiss Agency for the Environment, Forests and Landscape, 3003 Berne, Switzerland

Abstract Due to the fact that both nitro and polycyclic musk fragrances and their metabolites are not readily biodegradable in most sewage treatment plants and thus appear in the aquatic environment, the emphasis in this chapter is laid on understanding and evaluating their potential for bioaccumulation and hence their possible adverse impact (acute and chronic toxicity) on aquatic ecosystems. The bioaccumulation of these fragrances in aquatic organisms is principally governed by their inherent structure (parent compounds as well metabolites produced by microbial degradation in sewage treatment plants) and hence by their bioavailability, lipophilicity and the species specific capability of aquatic organisms to metabolize these compounds to readily excretable forms. Consequently, all potential adverse effects, whether acute, subchronic or chronic must be seen primarily as the result of the latter compound specific characteristics. Generally speaking nitro musk fragrances, due to their high bioaccumulation potential and higher resilience toward metabolic conversion appear to have potentially a greater environmental influence than their polycyclic counterparts. However, all presently available data, although primarily based on mammalian studies and limited aquatic toxicological assessments and despite their restriction in breadth and depth of detailed mechanistic understanding, suggest that neither of the musk fragrance classes pose an immediate or long-term hazard to the aquatic ecosystem despite their presence in environmental samples. However, prudence dictates that their accumulation in various organisms of several trophic levels in the aquatic ecosystems is unacceptable and that this situation should be ameliorated.

Keywords Nitro musks · Polycyclic musks · Bioaccumulation · Bioconcentration factors · Aquatic toxicity · Toxicity mechanisms

1
Introduction

In view of the yearly global production of nitro and polycyclic musk fragrances, estimated at approximately 2000 and 5600 tons, respectively (for the year 1996) [1–3], and the use of musks as fragrances and fragrance fixatives in a wide array of personal care products (e.g., washing detergents, detergents in general, perfumes, lotions, soaps and shampoos, cosmetics, etc.) it follows that most of these compounds will appear in municipal sewage treatment plants (STP). The removal of nitro musks (NMs) and polycyclic musks (PCMs) during municipal sewage treatment processes has been estimated at approximately 60–80% and 40–60%, respectively. The higher retention of NMs in the STP are explained by the presence of the aromatic ring and thus higher affinity for particles, a rather low water solubility due to the limited ring substitution with polar groups, and a moderately high lipophilicity (Table 1). In contrast, PCMs have a high water solubility, despite their inherently high lipophilicity (Table 1) and biological stability [4]. In view of the lipophilicity of NMs and PCMs and their broad form of application, it is not surprising to find these compounds as contaminants in the aquatic environment. Indeed, the concentrations detected in environmental samples range from ng L^{-1} to μg L^{-1} in effluent and surface waters [5]. The fact that NMs and PCMs can also be detected from μg kg^{-1} to mg kg^{-1} lipid weight in aquatic organisms [3, 6–9] raised serious concern as to their potential adverse effects on the aquatic ecosystem. Moreover, most recent analyses point to NM and PCM metabolites as being of greater environmental concern, due to greater metabolic stability and environmental persistence and consequently higher concentrations present in biological samples, e.g., in fish muscle, than the respective parent compounds [10–13].

The presence of not readily biodegradable compounds, i.e., musk fragrances in the aquatic environment raised major concerns, primarily in Europe, where the

Table 1 Common/tradename, abbreviation, CAS number and octanol-water partition coefficient (P_{OW}) for the most commonly used musk fragrances and their metabolites

Common/trade name	Abbreviation	CAS No.	P_{OW}	Reference
Musk xylene	MX	81-15-2	4.9	[2, 17]
2,4-di-aminomusk xylene	2,4-di-NH_2-MX	–	2.7–3.0	[25]
4-Amino musk xylene	4-NH_2-MX	107342-55-2	3.6–4.3	[2, 25]
2-Amino musk xylene	2-NH_2-MX	107342-67-6	2.7–4.3	[2, 25]
Musk ketone	MK	81-14-1	4.2–4.3	[2, 17]
2-Amino musk ketone	2-NH_2-MK	–	–	
Musk moskene	MM	116-66-5	4.4	[27]
Galaxolide	HHCB	1222-05-5	5.9–6.26	[13]
Tonalide	AHTN	1506-02-1; 21145-7-7	5.7–6.35	[13]
Celestolide	ADBI	13171-00-1	5.4	[3]
Phantolide	AHDI	15323-35-0	5.8	[3]
Cashmeran	DPMI	33704-61-9	4.5	[3]
Traseolide	ATII	68140-48-7	–	
Versalide	ATTN	88-29-9	–	

horrendous impact of polychlorinated biphenyls (PCBs), especially on fish otter (*Luttra luttra*) populations, is vivid in the memories of scientists, health officials, and the public. However, moderate to high lipophilicity, slow biodegradation, and the presence in water, sediments and edible fish tissues does not automatically warrant the comparison of musk fragrances with PCBs. Indeed, in order to assume adverse effects of any given compound in the environment, some information about its toxicity must be available. In most cases ecotoxicological data, although mostly restricted to acute standardized tests, is available and allows a rough judgment of potentially imminent acute adverse environmental effects. In those cases where no data to the acute or chronic environmental toxicity is at hand, a tentative risk assessment can be based on available mammalian toxicological data. In the case of musk fragrances, for which only a very limited number of ecotoxicological data are presently published, the broad mammalian toxicological dataset allows for an impression as to how environmentally dangerous these compounds potentially could or could not be [14, 15].

2 Bioaccumulation

All available analytical data, while showing the capability of musk fragrances to bioconcentrate in various aquatic species, do not demonstrate any capacity of these compounds for biomagnification in the aquatic ecosystem. Despite the fact that especially the nitro musk fragrance parent compounds demonstrate a moderate to high lipophilicity and potential for bioaccumulation/bioconcentration (Table 2), the capacity for bioconcentration/bioaccumulation must be differentiated in that for these compounds this appears more likely to be a function of momentary exposure of the species in question, rather than that of a lifetime upconcentration from a chronically contaminated environment. Indeed, age class analyses of fish taken from the Elbe river demonstrated no significant differences in tissue levels of NMs and PCMs from younger and older fish of the same species [16]. The concentrations of musk fragrances in the aquatic environment, including species, e.g., fish, are highly related to the distance from the STP [13]. In consequence and contrary to the situation with PCBs, the potential for toxicological effects resulting from musk parent compound exposure stems largely from the actual concentrations the species are exposed to via the ambient water in situ [17]. On the other hand, most recent analyses point to NM and PCM metabolites as being of greater environmental concern, due to greater metabolic stability and environmental persistence and thus higher concentrations present in biological samples, e.g., in fish muscle, than the respective parent compounds [2, 3, 10–12, 18–20]. In addition, primary degradation of PCM to more polar metabolites may lead to a different sorptive behavior and possibly to a higher bioavailability than the parent compounds [13, 21, 22]. However, despite the latter findings, which clearly highlight the necessity of additional clarifying experiments, there is a paucity of literature data with respect to the bioaccumulation and kinetics of musk metabolites in aquatic species. Indeed, such data would allow a better assessment of the risks posed by these metabolites for aquatic species, especially when considering the potential subacute-subchronic/chronic

Table 2 Bioconcentration factors (BCF_W) and bioaccumulation factors (BAF_W) (both based on a wet weight basis) of musk fragrances in various aquatic species

Abb.	Species (Latin name)	Species (common name)	BCF_W	BAF_W	Reference
MX	*Scardinus erythrophthalmus*	Rudd		290	[17, 18, 27, 28, 48–51]
	Tinca tinca	Tench		2400	
	Carassius carassius/C. auratus	Crucian carp		7500	
	Anguilla anguilla	Eel		40,000	
	Dreissena polymorpha	Zebra mussel		1800	
	Cyprinus carpio	Common carp	640–6740		
	Oncorhynchus mykiss	Rainbow trout	1600, 4400	10–60	
	Lepomis macrochirus	Bluegill sunfish	1600–1700		
	–	Fish unspecified	1300		
	Xenopus laevis	S.A. Clawed frog	5030		
MK	*S. erythrophthalmus*	Rudd		60	[17, 18, 27, 28, 49]
	T. tinca	Tench		230	
	C. carassius/C. auratus	Crucian carp		570	
	A. anguilla	Eel		1300	
	Dreissena polymorpha	Zebra mussel		390	
	Lepomis macrochirus	Bluegill sunfish	1380		
	Danio rerio	Zebrafish	455		
	–	Fish unspecified	2143		
	–	Fish unspecified	1100		
Musk Moskene	–	Fish unspecified	1300		[27, 28]
	X. laevis	S.A. Clawed frog	5830		
HHCB	*S. erythrophthalmus*	Rudd		20	[13, 18, 52]
	T. tinca	Tench		510	
	C. carassius	Crucian carp		580	
	A. anguilla	Eel	862 (BCF_L 3504)	290	
	Dreissena polymorpha	Zebra mussel		620	
	L. macrochirus	Bluegill sunfish	1584		
	D. rerio	Zebrafish	620		
AHTN	*S. erythrophtalmus*	Rudd		40	[13, 18, 52]
	T. tinca	Tench		280	
	C. carassius	Crucian carp		670	
	A. anguilla	Eel	1069 (BCF_L 5017)	400	
	Dreissena polymorpha	Zebra mussel		570	
	L. macrochirus	Bluegill sunfish	597		
	D. rerio	Zebrafish	600		

effects, e.g., endocrine modulation in fish and amphibians as discussed below or potential immuno-modulatory effects which have yet to be investigated.

3 Toxicity

The following paragraphs represent primarily a compilation of data for the acute, subacute, and potential for subacute-subchronic/chronic toxicity of musk fragrances and their metabolites in target species (algae, daphnia, fish, and amphibians).

3.1 Acute Toxicity

The acute toxicity and potential environmental effects of NMs and PCMs were summarized in several publications either using the EU-Technical Guidance Documents as a basis for environmental risk assessment [4, 13, 23, 24], test procedures in conformity with OECD guideline 201 and 202 for testing of chemicals [25–27], or test procedures identical or analogous to ASTM guideline E 1439-91 [28]. These publications include studies with algae (*Pseudokirchneriella subcapitata*), *Daphnia magna*, bluegill sunfish (*Lepomis macrochirus*), rainbow trout (*Oncorhynchus mykiss*), zebra fish (*Danio rerio*), fathead minnow (*Pimephales promelas*), and the South African clawed frog (*Xenopus laevis*). The most prominent results are compiled in Table 3.

The main focus of the latter studies was on musk xylene (MX), musk ketone (MK) and the three polycyclic musks AHTN (7-acetyl-1,1,3,4,4,6-hexamethyltetraline), HHCB (1,3,4,6,7,8-hexahydro-4,6,6,7,8,8-hexamethyl-cyclopenta-(*g*)-2-benzopyran) and ADBI (4-acetyl-1,1-dimethyl-6-*tert*-butylindane). Additional data can be found for the three amino metabolites of MX and MK [29] as well as for musk moskene, tibetene, and ambrette [27]. Toxicity of either NMs or PCMs was observed at rather high concentrations of these respective compounds, i.e., in many cases at or exceeding the inherent water solubilities (Table 3).

The mechanism(s) involved in the acute toxicity of the NMs and PCMs is presently unknown. However, a generalized narcosis, as previously demonstrated for various other organic compounds in fish and amphibians [30], may be suggested in view of the high concentrations necessary to induce acute mortality [14, 26–28] and the erratic behavior noted with daphnia [26, 27]. The latter findings are contrasted by the report of Behechti et al. [25] who found acute toxicity of low concentrations of the amino metabolites of MX, especially of the 4-amino-MX in *D. magna* (EC_{50}=250 ng L^{-1}; 95% confidence interval 230–280 ng L^{-1}). A subsequent re-investigation of the findings of Behechti et al. [25] by Giddings et al. [26] demonstrated an $EC_{50\text{-}48h}$=490 µg L^{-1} (95% confidence interval 400–600 µg L^{-1}) and therefore that the previous findings by Behechti et al. overestimated the toxicity of 4-amino-MX by approximately a factor 2000, most likely as the result from highly toxic impurities present in their stock solutions. The acute toxic concentrations of 4-amino-MX for *Daphina magna* thus are also approximately four orders of magnitude greater than those concentrations found in surface waters

Table 3 Compilation of acute and subacute toxicity data obtained with nitro and polycyclic musks invarious species, modified from Dietrich and Chou [14]. All data in mg L^{-1}

Species	Endpoint	MX	2-NH_2-MX	4-NH_2-MX	2,4-di-NH_2-MX	MK	MM	AHTN	HHCB	ADBI	Ref.
Algae	$EC_{50\ growth}$	NE[a]				0.244	–	>0.797	>0.854	–	[4, 13, 23]
	$EC_{50\ biomass}$	NE[a]				0.118	–	0.468	0.723	–	
Daphnia magna	24 h EC_{50}	NE[a]	1.07	>0.93	23.3	NE[a]	NE[a]	–	–	–	[4, 13, 23, 25–27]
	48 h EC_{50}	>5,6		0.49							
	21 days LC_{50}	0.680				0.338–0.675	NE[a]	0.341	0.293	–	
Oncorhynchus mykiss	21 days $EC_{50\ repro.}$	–				0.169–0.338	NE[a]	0.244	0.282	–	[4, 23]
	96 h LC_{50}	>1000				–	–	–	–	–	
Lepomis macrochirus	21 days LC_{50}	–				>0.50	–	–	–	–	[4, 13, 23]
	96 h LC_{50}	1.20				–	–	–	–	–	
Danio rerio	21 days LC_{50}	–				–	–	0.314	0.452		[23, 28, 30, 48]
	14 days $LC_{50\text{-adult fish}}$	0.4				–	–	–	–	–	
	96 h $LC_{50\text{-embryo}}$	>0.4				>0.4	>0.4	>0.67	>0.67	>1.0	
	96 h $EC_{50\text{-embryo-hatching}}$	>0.4				>0.4	>0.4	>0.67	>0.67	>1.0	
	8 week $LOEC_{repro.}$					0.033					
Pimephales promelas	96 h $EC_{50\text{-embryo-teratogen}}$	>0.4				>0.4	>0.4	0.18	0.39	0.69	[13]
Xenopus laevis	96 h $EC_{50\text{-embryo-growth}}$	>0.4				>0.4	>0.4	>1.0	>1.0	>1.0	[28, 30]
	32 days $LC_{50\text{-embryo-adult}}$	–				–	–	0.100	>0.140	–	
	96 h $LC_{50\text{-embryo}}$	>0.4				>0.4	>0.4	>2.0	>2.0	>4.0	
	96 h $EC_{50\text{-embryo-teratogen}}$	>0.4				>0.4	>0.4	>4.0	>4.0	>4.0	
	96 h $EC_{50\text{-embryo-growth}}$	>0.4				>0.4	>0.4	>1.0	>2.0	>4.0	

[a] NE=no effect found at concentrations exceeding compound solubility in H_2O.

and treated sewage, demonstrating that nitro musks and their metabolites pose a negligible acute environmental hazard [26].

Indeed, the comparison of the NM and PCM concentrations found in environmental samples [2,3,10–12,19,31] with those concentrations inducing acute toxicity in various aquatic species, as discussed above, strongly suggests that NMs and PCMs do not pose an acute risk for the aquatic ecosystem. This conclusion is also supported by the instrumentalized risk assessment processes for NMs and PCMs using the EU-Technical Guidance Documents [4,23,24], which predict no effects of these musk fragrances in the aquatic environment.

3.1.1
Developmental Toxicity

In contrast to the more narcosis-like effects described above, more specific effects were reported when embryos of *X. laevis* and *D. rerio* are exposed to PCMs but not NMs [14]. Both *D. rerio* and *X. laevis* embryos presented with a significant increase in malformations [14]. Surprisingly, while all three PCMs (ADBI, AHTN, HHCB) induced malformations in zebra fish embryos, malformations were observed in ADBI treated *X. laevis* embryos only. While both species presented with ventro-dorsal curvature of the tail, the concentrations necessary to induce malformations in *D. rerio* were approximately one order of magnitude lower than those necessary to produce the same effects in the amphibian embryos. Of the PCMs tested, AHTN demonstrated the greatest degree of teratogenicity, with the steepest dose-response curve, while ADBI was teratogenic at high concentrations only. AHTN induced malformations appear to be specific for cyprinid embryos, as tail-loss was noted in *P. promelas* embryos exposed to 0.067 or 0.14 mg AHTN L^{-1}, while no malformations were observed in *X. laevis* embryos exposed to AHTN or HHCB [14] or in *P. promelas* exposed to HHCB [13]. Of the three PCMs tested in a semi-static embryotoxicity test with *X. laevis*, AHTN and HHCB demonstrated a significant and dose-dependent effect on growth at concentrations below those which were acutely toxic to the embryos. No effects on growth were observed in zebra fish embryos, as the doses necessary to induce a significant growth inhibition exceeded those inducing acute toxicity (Table 3). Similar effects were noted in *P. promelas* exposed to 0.140 mg HHCB l^{-1} but not for AHTN [13].

3.2
Subchronic/Chronic Toxicity

At present, only limited data are available for assessing the risk to the aquatic environment, i.e., the populations of aquatic species exposed subchronically or chronically to low concentrations of parent compounds and metabolites of NMs and PCMs. In general, there are three potential adverse interactions of xenobiotics with the health and sustainability of a population that are of primary importance: (i) an extremely high incidence of pathological changes, e.g., tumors [32] resulting from genotoxic or a tumor promoting activity; (ii) suppression of the immune system and thus higher susceptibility of the population to pathogens

[33]; and (iii) endocrine modulation affecting the reproductive success of the population.

Neither the parent compounds nor the metabolites of NMs and PCMs have been demonstrated to possess carcinogenic activity, with the exception of a species-specific promotion of liver tumors at high concentrations of MX observed in mice [34]. This process was shown to be not of genotoxic [35, 36], but rather of an epigenetic nature, i.e., driven by the induction of microsomal enzymes, particularly those of the CYP2B family [37], and the pattern of induction was consistent with that observed for phenobarbital, the classical CYP2B inducer and mouse liver carcinogen [38, 39].

No information is as yet available regarding the potential interaction of NMs and PCMs on immune parameters of aquatic species. However, the present expectation is that no immune-suppressive activity is to be expected in aquatic species as no evidence was found suggesting immune-suppressive activity of these compounds in mammalian species exposed subchronically or chronically to high concentrations of these compounds [34, 40, 41].

3.2.1
Endocrine Modulation

Although the present database on potential endocrine modulating activity of NMs and PCMs is still rather scant, the compilation of mammalian data and data from in vitro assays with cells and tissue homogenates from aquatic species suffices for a primary assessment, at least of the potential (anti)estrogenic activity of these compounds. Neither subchronic or chronic administration of NMs, PCMs, or mixtures of NMs and PCMs [34, 40, 41] suggests any form of (anti)estrogenic activity in rodent species. The basis for this assessment was organ weight and histopathological examination of the uterus, seminal vesicles, mammary gland, testes, ovaries, and vaginas. These findings are corroborated by a study of Seinen et al. [42] who exposed juvenile mice to high dietary levels of AHTN and HHCB and found no evidence for an increase in uterine weight. On the other hand, the same scientists reported a very weak estrogenic activity of both compounds using ERα- and ERβ-dependent gene transcription assays with human embryonal kidney 293 cells. The reported estrogenic activity was approximately six to eight orders of magnitude lower than the endogenous ligand estradiol (E_2). The latter findings demonstrated that only extremely high concentrations of AHTN and HHCB have measurable estrogenic potency and that the current levels of wildlife and human exposure to these compounds are too low to induce any estrogenic effects in the exposed species. The interaction of the PCMs with the hepatic estrogen receptor(s) of rainbow trout, carp, or the amphibian *X. laevis* was also shown in an in vitro competitive binding assay [14]. In comparison to the endogenous ligand E_2, approximately four orders of magnitude higher concentrations of AHTN were necessary to elicit the same degree of ligand competition (IC_{50}) in the *X. laevis* receptor binding assay. Very weak binding of AHTN and HHCB were found in the rainbow trout receptor binding [14], corroborating the findings by Seinen et al. [42]. Neither AHTN nor HHCB, but ADBI bound to the carp estrogen receptor [14], corroborating earlier findings by Smeets et al.

[43], who investigated AHTN and HHCB induced synthesis of vitellogenin in carp hepatocytes in vitro. Neither of the two compounds was capable of inducing vitellogenin in this system, suggesting that these compounds do not interact with the fish estrogen receptor(s) to the degree or with the high concentrations necessary for estrogen dependent gene transcription. Although metabolites of AHTN and HHCB, as found in environmental samples [3, 10, 12], were not analyzed for (anti)estrogenic activity, it can safely be assumed that these metabolites were also formed during incubation of the primary carp hepatocytes used as the screening method for estrogenic activity. If indeed these metabolites had any form of estrogenic activity the lack of vitellogenin induction in the carp hepatocyte system suggests that the metabolites were not formed in adequate concentrations to have an estrogenic effect. Overall it can be concluded that the current environmental PCM levels are too low to induce estrogenic effects in aquatic species.

In contrast to the PCMs neither of the two nitro musk parent compounds (MX and MK) had any competitive binding activity to either the rainbow trout or the Xenopus estrogen receptor(s). However, amino metabolites of MX and MK, formed during the sewage treatment process, were able to bind to the estrogen receptors of rainbow trout and *X. laevis*. The concentrations of the 2-NH_2-MX metabolite necessary to displace 50% of the endogenous ligand at the rainbow trout estrogen receptor(s) was approximately six orders of magnitude greater than that of the endogenous ligand (E_2) itself, again demonstrating that unrealistically high concentrations of these metabolites were needed to elicit any estrogenic activity in rainbow trout. Surprisingly the binding curves derived from the *X. laevis* estrogen receptor binding assay, demonstrated that all three known amino metabolites of MX and MK were able to compete with the endogenous ligand. The concentrations necessary for competition were only two to three orders of magnitude higher than those of E_2. Furthermore, the concentrations of 2-NH_2-MX necessary for E_2 competition at the *X. laevis* estrogen receptor(s) were nearly three orders of magnitude lower than those needed for competing at the rainbow trout estrogen receptor(s). The latter suggests that there are some species-specific susceptibilities with regard to potential estrogenic activities of nitro musk metabolites. Indeed, the findings in the *X. laevis* system are unique in that these in vitro findings were indicative for the endocrine modulating effects observed for bisphenol A (BA) in vivo [44]. Chronic exposure of *X. laevis* embryos to low concentrations of BA induced a feminization of male embryos [45]. Although the above in vitro systems may be indicative that some of the NM metabolites and PCMs may have the potential for endocrine modulation in aquatic species, the mere interaction of a xenobiotic with the estrogen receptor(s) of a given aquatic species does not imply that this interaction will also lead to all of the specific associated downstream events. The latter observations stand in stark contrast to the findings presented by Carlsson et al. [46] who reported reduced gonado-somatic and hepato-somatic indices as well as dose-dependent reduced fecundity and fertility in female zebra fish fed for eight weeks with feed containing 10 mg MK g^{-1} dry weight (dw). It is interesting to note that while exposed female fish fed with 10 mg MK g^{-1} dw for eight weeks were reported to contain 587 µg MK g^{-1} wet weight in fish muscle, no simultaneous analysis of the exposure water was conducted to demonstrate whether the high body concentration in the female fish actually resulted from the contaminated feed, or whether it was a mere reflection of the water concentrations of musk ketone resulting from decaying feed.

4 Conclusions

Although the present database for ecotoxicological effects of NMs and PCMs and of their respective metabolites is still too small for a concluding risk assessment, there is little evidence that would suggest that these compounds, despite their overt presence in environmental samples, generally would have an adverse impact on the aquatic ecosystem. The concentrations of musk fragrances in the aquatic environment are highly related to the distance to the STP [13]. Indeed, as indicated also via the comparison between the tissue levels of various ages of fish exposed to NMs and PCMs, no biomagnification within the same species (age classes) or various trophic levels appears to occur [16]. In consequence and contrary to the situation with PCBs, the potential for toxicological effects resulting from musk exposure stems largely from the actual concentrations the species are exposed to via the ambient water in situ [17] and this risk appears to be negligible when using the presently available database for risk estimation. However, as pointed out above, amphibians appear to be more susceptible to endocrine modulating compounds than most of the species investigated so far [45]. In light of this, the interaction of the MX and MK metabolites with the estrogen receptor of *X. laevis* [47] must be taken more seriously and should be encouragement to investigate the mechanisms of this interaction, the potential effects and risks associated with these amino metabolites for amphibians in more detail. Similarly, the fact that for the first time a laboratory study with zebra fish, although unconventional in its approach and lacking detail in many important aspects [46], has demonstrated a significant effect of high doses of musk ketone (or also its metabolite) on reproduction (fecundity and fertility), indicates that long-term effects should be investigated, especially including several generations of a given aquatic species, despite the overall environmentally unproblematic appearance of musk fragrances and their metabolites.

Acknowledgement We would like to acknowledge the Arthur und Aenne Feindt Foundation (Hamburg, Germany) for financial support of this project.

5 References

1. Gebauer H, Bouter T (1997) Euro Cosmetics 1:30
2. Rimkus GG, Hühnerfuss H, Gatermann R (1999) Toxicol Lett 111:5
3. Rimkus GG (1999) Toxicol Lett 111:28
4. Van de Plassche EJ, Balk F (1997) Environmental risk assessment of the polycyclic musks AHTN and HHCB according to the EU-TGD. 601503008. National Institute of Public Health and the Environment Bilthoven, The Netherlands
5. Heberer T, These A, Grosch U (2001) Occurrence and fate of synthetic musks in the aquatic system of urban areas: polycylic and nitro musks as environmental pollutants in surface

waters, sediments, and aquatic biota. In: Daughton DC, Jones-Lepp T (eds) Pharmaceuticals and personal care products in the environment: scientific and regulatory issues. American Chemical Society, Washington D.C., ACS Symposium Series 791:142
6. Eschke H-D, Traud J, Dibowski H-J (1994) Vom Wasser 83:373
7. Eschke H-D, Traud J, Dibowski H-J (1994) UWSF-Z Umweltchem Ökotox 6:183
8. Eschke H-D, Dibowski H-J, Traud J (1995) UWSF-Z Umweltchem Ökotox 7:131
9. Rimkus G, Wolf M (1995) Chemosphere 30:641
10. Biselli S, Gatermann R, Kallenborn R, Rimkus GG, Hühnerfuss H (2000) Poster. 10th Annual Meeting of SETAC Europe
11. Gatermann R, Hühnerfuss H, Rimkus G, Attar A, Kettrup A (1998) Chemosphere 36:2535
12. Gatermann R, Rimkus G, Hecker M, Biselli S, Hühnerfuss H (1999) Poster. 9th Annual Meeting of SETAC Europe
13. Balk F, Ford RA (1999) Toxicol Lett 111:57
14. Dietrich DR, Chou Y-J (2001) Ecotoxicology of musks. In: Daughton DC, Jones-Lepp T (eds) Pharmaceuticals and personal care products in the environment: scientific and regulatory issues. American Chemical Society, Washington D.C., ACS Symposium Series 791:156
15. Dietrich DR, Kehrer JP (1999) Toxicol Lett 111:1
16. Hajslova J, Gregor P, Chladkova V, Alterova K (1998) Organohalogen Compd 39:253
17. Rimkus G, Butte W, Geyer HJ (1997) Chemosphere 35:1497
18. Gatermann R, Biselli S, Hühnerfuss H, Rimkus GG, Hecker M, Karbe L (2002) Arch Environ Contam Toxicol 42:437
19. Gatermann R, Hellou J, Hühnerfuss H, Rimkus G, Zitko V (1999) Chemosphere 38:3431
20. Rimkus GG, Wolf M, Attar A, Gatermann R, Hühnerfuss H (1998) Proc 20th Int Symp Capillary Chromatography, Riva del Garda, Italy
21. van Dijk A (1996) Accumulation and elimination of [^{14}C]AHTN by bluegill sunfish in a dynamic flow-through system. 364825. RIFM, RCC Umweltchemie AG
22. van Dijk A (1996) Accumulation and elimination of [^{14}C]HHCB by bluegill sunfish in a dynamic flow-through system. 381418. RIFM, RCC Umweltchemie AG
23. Tas JW, Balk F, Ford RA, van de Plassche EJ (1997) Chemosphere 35:2973
24. Balk F, Ford RA (1999) Toxicol Lett 111:81
25. Behechti A, Schramm K-W, Attar A, Niederfellner J, Kettrup A (1998) Water Res 32:1704
26. Giddings JM, Salvito D, Putt AE (2000) Water Res 34:3686
27. Schramm K-W, Kaune A, Beck B, Thumm W, Behechti A, Kettrup A, Nickolova P (1996) Water Res 30:2247
28. Chou Y-J, Dietrich DR (1999) Toxicol Lett 111:17
29. Chou Y-J, Dietrich DR (1999) Toxicol Lett 111:27
30. McCarty LS, Mackay D, Smith AD, Ozburn GW, Dixon DG (1992) Environ Toxicol Chem 11:917
31. Bester K, Hühnerfuss H, Lange W, Rimkus GG, Theobald N (1998) Water Res 32:1857
32. Gardner GR, Yewich PP, Harshbarger JC, Malcolm AR (1991) Environ Health Perspectives 90:53
33. Prietz A, Fleischhauer V, Hitzfeld BC, Dietrich DR (2000) Poster. 10th Annual Meeting of SETAC Europe
34. Maekawa A, Matsushima Y, Onodera M, Shibutani M, Ogasawara H, Kodama Y, Kurokawa Y, Hayashi Y (1990) Fundam Chem Toxicol 28:581
35. Api AM, Ford RA, San RHC (1995) Fundam Chem Toxicol 33:1039
36. Api AM, Pfitzer EA, San RHC (1996) Fundam Chem Toxicol 34:633
37. Lehman-McKeeman LD, Caudill D, Vasallo JD, Pearce RE, Madan A, Parkinson A (1999) Toxicol Lett 111:105
38. Lehman-McKeeman LD, Caudill D, Young JA, Dierckman TA (1995) Biochem Biophys Res Commun 206:975
39. Lehman-McKeeman LD, Johnson DR, Caudill D (1997) Toxicol Appl Pharmacol 142:169
40. Api AM, Ford RA (1999) Toxicol Lett 111:143
41. Fukuyama MY, Easterday OD, Serafino PA, Renskers KJ, North-Root H, Schrankel KR (1999) Toxicol Lett 111:175

42. Seinen W, Lemmen JG, Pieters RHH, Verbruggen EMJ, van der Burg B (1999) Toxicol Lett 111:161
43. Smeets JM, Rankouhi TR, Nichols KM, Komen H, Kaminski NE, Giesy JP, van den Berg M (1999) Toxicol Appl Pharmacol 157:68
44. Lutz I, Kloas W (1999) Sci Total Environ 225:49
45. Kloas W, Lutz I, Einspanier R (1999) Sci Total Environ 225:59
46. Carlsson G, Örn S, Andersson PL, Söderström H, Norrgren L (2000) Mar Environ Res 50:237
47. Chou Y-J, Dietrich DR (1999) Toxicol Lett 111:27
48. Ministry of International Trade and Industry (MITI) (1992) Tokyo, Japan
49. Yamagishi T, Miyazaki T, Horii S, Kaneko S (1981) Bull Environ Contam Toxicol 26:656
50. Paradice A, Suprenant D (1984) Accumulation and elimination of ^{14}C-residues by bluegill (*Lepomis macrochirus*) exposed to P1618.01R (musk xylene). Unpublished report to RIFM
51. Boleas S, Fernandez C, Tarazona JV (1996) Bull Environ Contam Toxicol 57:217
52. Fromme H, Otto T, Pilz K (2001) Water Res 35:121

The Handbook of Environmental Chemistry Vol. 3, Part X (2004): 245–257
DOI 10.1007/b14123

Musk Fragrances and Environmental Fate Models – HHCB as an Example for Model Refinements

Stefan Schwartz · Volker Berding · Michael Matthies

University of Osnabrück, Institute of Environmental Systems Research, 49069 Osnabrück, Germany
E-mail: matthies@usf.uni-osnabrueck.de

Abstract The theory of environmental fate and distribution models is briefly introduced. Afterwards, by means of the models laid down in the European Union System for the Evaluation of Substances (EUSES) and the Geography-referenced Regional Exposure Assessment Tool for European Rivers (GREAT-ER) environmental concentrations of the polycyclic musk fragrance HHCB (1,3,4,6,7,8-hexahydro-4,6,6,7,8,8-hexamethyl-cyclopenta-[*g*]-2-benzopyrane) were calculated for the aquatic environment. Starting with a generic standard region a spatial refinement was carried out for the German river Ruhr region. The refinement was realised in different scenarios by successively replacing EUSES default parameters with realistic regional values and then applying the selected region to GREAT-ER. The results were compared to monitoring data from the region of the German Federal State of North Rhine-Westphalia (river Ruhr). It was shown that both EUSES and GREAT-ER estimate the median of the measured values very well in every scenario. Spatial refinement leads to lower concentrations. Even underestimations are possible if realistic regional parameters are inserted and a ready biodegradability is assumed. Furthermore, assuming the same region, the predicted concentrations of EUSES and GREAT-ER do not differ by more than a factor of 5. In addition, GREAT-ER delivers realistic regional information with visualised concentration profiles and maps.

Keywords Fate assessment · Polycyclic musk fragrances · HHCB · TGD · EUSES · GREAT-ER

1
Introduction

Since the early 1990s polycyclic musk fragrances have been detected in rivers and the sea, fish, human adipose tissue and human milk [1–6]. Some of the polycyclic compounds are the prevailing used musk fragrances in Western Europe with concentrations in aquatic biota that exceed those of the nitro musks by up to three orders of magnitude [7]. Seven single compounds are involved, of which HHCB (1,3,4,6,7,8-hexahydro-4,6,6,7,8,8-hexamethyl-cyclopenta-[*g*]-2-benzopyrane; CAS-No. 1222-05-5; trade name: e.g. Galaxolide) occurs with the highest concentrations in the environment.

Despite its widespread usage, a consumption rate of up to 2400 tons $year^{-1}$ in Europe, and environmental concentrations of up to 1.2 $\mu g\ L^{-1}$ in rivers, 63 $mg\ kg^{-1}$ dry weight in sewage sludge and 63.6 $mg\ kg^{-1}$ lipid weight in fish [8], these chemicals have not yet been investigated sufficiently. This situation hampers the comprehensive assessment of ecological risks posed by these fragrances. Further investigations are required to obtain a comprehensive evaluation. In addition to data concerning the effects of these substances, information on their environmental fate and distribution is also necessary.

In 1996 the European Commission published the Technical Guidance Documents (TGD) [9], which enable the assessment of the risk posed by new notified and existing chemicals to humans and the environment. The models and procedures laid down in the TGD have been made available in software form by the European Union System for the Evaluation of Substances (EUSES) [10]. EUSES makes default values for most of the parameters available; just four physicochemical substance properties and the tonnage are required as input parameters. The default values claim to represent a hypothetical "European average region", the so-called standard region [9]. Within the scope of risk assessments, environmental concentrations for HHCB were assessed by applying the TGD models [8, 11]. Schwartz et al. carried out HHCB fate assessments for different scenarios and recommended further assessments and the application of alternative models [12, 13].

Using the example of HHCB, our purpose is to calculate HHCB concentrations with different state of the art environmental fate models and to determine how accurately the models, parameters and procedures can be applied to an environmental risk assessment for this substance. The deviations of predicted environmental concentrations from measured values obtained by monitoring studies are of particular interest. A further objective is to determine, by consideration of various scenarios, the impact of different model assumptions on the results and to point out possibilities to reduce over- and underestimations. The German Federal State of North Rhine-Westphalia (NRW) was chosen for the comparison of measured with predicted concentrations. This chapter confirms the strength of the fate models and contributes to an extensive evaluation of HHCB.

2 Database

2.1 Emission Rates and Regional Parameters

Only a few production and consumption rates of HHCB have been published to date: Plassche and Balk quote a consumption of 2400 tons (1992) and 1482 tons (1995) in Europe [8], based on data from the producers. The substance is emitted into waste water during use, minus a negligible loss due to volatilisation. Thereafter, it reaches the surface waters and the aquatic food chain via municipal sewage treatment plants, or agricultural fields via sludge application.

Parameters characterising the region, e.g. area, number of inhabitants, precipitation rate etc. can be found in [15] for North Rhine-Westphalia and in [16] for the river Ruhr catchment.

2.2 Substance Specific Parameters

Besides emission rates and parameters characterizing the region, physico-chemical parameters are necessary to carry out environmental fate assessments. Particularly, partition coefficients are of major importance to the estimation of environmental substance distributions. As later described in the context of level models, distribution in accordance with partition coefficients requires thermodynamic equilibrium. In such a case, the quotient of the concentrations of a substance in two considered compartments adopts a substance-specific value. This value applies to the partition coefficient $K_{ij}=C_i/C_j$. Many partition coefficients can be determined by measurement, whilst others must be estimated by physico-chemical parameters or other partition coefficients using regression equations. Most environmental fate models require the input of the octanol-water partition coefficient (K_{OW}) which characterises the lipophilicity of a substance. Furthermore, data on water solubility, vapour pressure, melting point and molecular weight must be entered. By default, these values are used for the estimation of the requisite partition coefficients. Estimated partition coefficients may be replaced by more reliable (e.g. measured) values on demand. The physico-chemical data

Table 1 Physico-chemical data of HHCB [8]

Parameter	Value	Unit
Molar mass	258.4	$g\ mol^{-1}$
Vapour pressure	0.0727	Pa, 25 °C
Henry's law constant[a]	11.3	$Pa\ m^3\ mol^{-1}$
log K_{OW}	5.9	–
log K_{OC}[a]	4.86	$L\ kg^{-1}$
Water solubility	1.75	$mg\ L^{-1}$, 25 °C

[a] These parameters may be estimated by EUSES.

applied in this work are experimentally determined and listed in Table 1. Reasonable data on aquatic degradation have not yet been published; thus, for the purpose of a worst case estimation, no degradation is assumed. However, Plassche and Balk allude to a possible aquatic biodegradation [8]. In order to allow this eventuality, HHCB is classified as readily biodegradable in one scenario, and a half-life of 15 days in bulk surface water and 0.7 h in sewage treatment plants, respectively, was assumed.

2.3 Monitoring Data

Measured concentrations from the river Ruhr by Eschke et al. [2, 3] are used to compare the model's results with monitoring data. The river Ruhr, located in North Rhine-Westphalia (Germany), is a relatively large river with a length of 217 km, a catchment area of 4488 km^2 and an average annual mouth outflow of more than two billion m^3 [16]. This river system is of major importance to the drinking water supply for the surrounding conurbation. The measurement of 30 samples at the beginning of 1994 along a stretch of 160 km was carried out downstream from three sewage treatment plant discharges as well as at locations without the direct influence of emitters. Sampling was carried out daily over the period of a week. In the same period the contamination of nine fish samples from the river Ruhr was investigated. The given concentrations refer to the wet weight.

The HHCB content in the Ruhr is comparatively high, with a mean of 0.37 $\mu g\ L^{-1}$. In the river Elbe, the river Rhine and the river Lippe average concentrations of 0.12 $\mu g\ L^{-1}$ [6], 0.07 $\mu g\ L^{-1}$ [8] and 0.10 $\mu g\ L^{-1}$ [21] respectively, were measured.

3 Models

3.1 Level Models

In the risk assessment of substances, the prediction of regional background concentrations is of utmost importance. Average concentrations on a larger spatial scale are estimated without local distinctions. Local concentrations can be calculated with the use of possible refinement steps and special models. Multimedia models can be used to estimate background concentrations on larger spatial scales. Each environmental compartment, e.g. water, air or soil, is assumed to be a well-mixed, homogenous box that is connected to other boxes (environmental media). Differences within a single box are disregarded. For multimedia models, the term "level models" has been developed, which is explained briefly in the following section. A helpful reference for more comprehensive descriptions is [17]. Level models can be categorised into four parts that differ from each other in terms of equilibrium and their consideration of reactions. The first three parts are investigated in more detail.

Level 1: in a Level 1 model it is assumed that the whole mass of the investigated substance is located in the system, distributed according to its partition coefficients. There is no input or output, and no degradation. Moreover, there is no resistance between compartments. Consequently, the concentrations in the separate compartments can easily be calculated using these formulae:

$$C_1 = m/(V_1 + C_1K_{21}V_2 + \ldots + C_nK_{n1}V_n)$$

$$C_i = K_{i1}C_1, i = 2, \ldots, n$$

where m is the mass in the system, V_i is the volume of compartment i, C_i is the concentration in compartment i and K_{ij} is the partition coefficient between compartment i and j (see above). In thermodynamic equilibrium $K_{ij}=C_i/C_j$.

Level 2: as with the Level 1 model, thermodynamic equilibrium is assumed here, too. Thus, there is no resistance between compartments. The difference to the Level 1 model is the presumption of a continuous input into the system. Steady state is assumed, where the source and sinks are in equilibrium. The sinks represent the outflow out of the system and degradation in the compartments. In steady-state, therefore, dm/dt=0, and the sum of the inputs corresponds to the sum of all elimination processes.

$$\Sigma I_i = \Sigma (V_iC_i\lambda_i)$$

In a state of equilibrium the following applies:

$$C_1 = I/(\lambda_1V_1 + \lambda_2K_{21}V_2 + \ldots + \lambda_nK_{n1}V_n)$$

and

$$C_i = C_1K_{i1}, i = 2, \ldots, n$$

where λ_i is the total degradation rate in compartment i. First-order degradation is assumed.

Level 3: in contrast to the Level 2 model, a Level 3 model does not assume equilibrium partitioning, i.e. there are resistances between compartments. However, steady state with dm/dt=0 is still assumed. Advective and diffusive flows are possible between compartments. Furthermore, advective flows into the system and out of the system also take place. With N_{ij} as the sum of flows between compartments i and j, the following mass balance results for compartment i:

$$V_i\, dC_i/dt = I_i + \Sigma_j N_{ji}C_j - V_i\lambda_iC_i$$

A linear n×n-equation system must be solved in order to calculate the concentrations. This can be done, for example, using Gaussian elimination or with the inverse of the matrix. Since in a Level 3 model no thermodynamic equilibrium is assumed, the inputs into the various compartments are of importance.

Level 4: the Level 4 model corresponds to the Level 3 model, but no steady state is assumed, thus dm/dt≠0. In the case of a constant or a pulse input the solution can be carried out analytically, otherwise numerical integration routine, e.g. Runge-Kutta or Euler, are required.

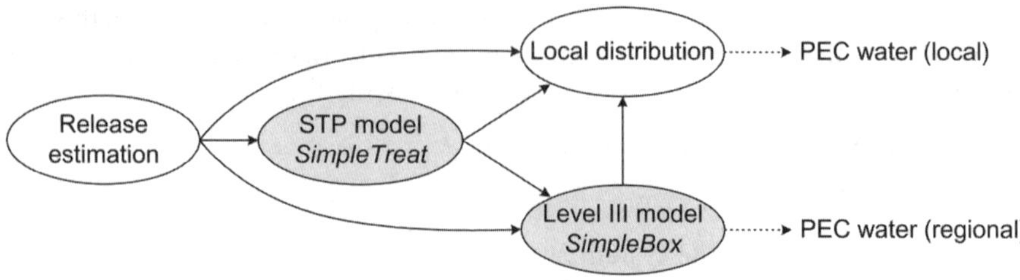

Fig. 1 Model structure of EUSES (STP – sewage treatment plant, PEC – predicted environmental concentration)

3.2 EUSES

The European Union System for the Evaluation of Substances (EUSES) was developed for the first and second tiers of risk assessment for chemicals. It consists of several modules and models that cover the various steps of risk assessment for a substance (Fig. 1). EUSES contains a separate module for the input of the main data to identify the substance and for its relevant physico-chemical parameters. The primary data entered are used for the estimation of secondary data, such as partition coefficients and degradation rates.

3.2.1 *Release Estimation*

Two different spatial scales are assumed: a local scale in the vicinity of a point source and a regional or continental scale, respectively, containing both point sources and diffuse sources. Since the amount of the substance released into the environment is of central importance for environmental concentrations, EUSES includes a release estimation module. Besides some physico-chemical data, it requires data about the main and industrial category, as well as production, import and export volumes. Releases are estimated by means of emission tables. Estimations are carried out for each important life cycle step and each use of the specific substance. Emissions take place into the atmosphere, industrial/urban soil, surface water and waste water. While the particular emissions are added to one overall emission on the regional scale, partitioning is retained on the local scale. Here, also the duration of each emission (episode) in which the total emission of the respective life cycle step occurs is estimated.

3.2.2 *SimpleTreat*

Another module contained in EUSES is the sewage treatment plant model SimpleTreat, which estimates the fate of substances in a sewage treatment plant (STP). According to the fraction of inhabitants connected to a sewer system, part of the estimated wastewater emissions is directed to the STP. Using partition coefficients and the degradation rate for STPs, the model calculates which fractions are emitted to the atmosphere and surface water and which parts remain in the sewage sludge. Thus, differences occur between the local and the regional/continental scale:

- Locally, for the duration of the episode estimated beforehand (and for each life cycle step) an indirect emission into the air takes place. For this period the substance's concentration in the STP effluent and in the sewage sludge is determined, and a predicted environmental concentration (PEC) for microorganisms in the STP is calculated.
- On the regional and continental scale, fractions of the substance are directed to the effluent of the STP (surface water), into the air and onto agricultural soil. Continuous emission is assumed. Life cycle steps are not distinguished. These emissions are the so-called indirect emissions to air, surface water and agricultural soil.

3.2.3
Local Environmental Distribution

Furthermore, a separate module exists for the calculation of local environmental concentrations. These are calculated for each relevant use and for each life cycle step, according to the previously estimated local emissions. The model comprises different submodels that calculate concentrations in air, surface water, sediment, soils and ground water.

3.2.4
Regional Distribution and Environmental Concentrations

In order to calculate environmental distributions of a substance, the SimpleBox model is included in EUSES. It estimates regional and continental background concentrations in air, water, sediment and three soils and is a Level 3 model. It consists of two nested spatial scales: the continental and the regional scale. Each scale contains the six compartments air, surface water, sediment, agricultural soil, industrial/urban soil and natural soil. Continuous emissions take place into air, water, agricultural soil and industrial/urban soil. First-order degradation is assumed in each compartment. A schematic illustration of the flows can be seen in Fig. 2.

3.3
GREAT-ER

The Geography-referenced Regional Exposure Assessment Tool for European Rivers (GREAT-ER) is a support tool for environmental risk assessment and river basin management which focuses on a more refined stage of risk assessment (tier 2) [18]. It uses a Geographic Information System (GIS) to produce a visualisation of PECs along a river [19]. Additionally, it is possible to produce pro-

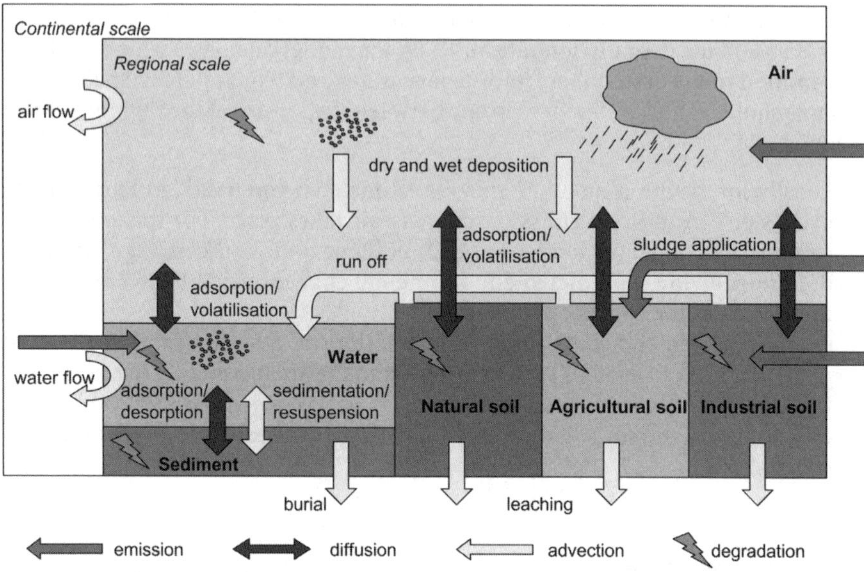

Fig. 2 Structure of the model SimpleBox

files of PECs throughout the studied catchment. Such profiles illustrate chemical fate from a river's headwaters down to its mouth, and can be used to compare model predictions with monitoring data. Furthermore, geo-referenced model results can be aggregated to obtain a spatially averaged PEC, which is representative of the river basin under investigation [20]. A comparison with level models can only be applied to concentrations in surface water. Furthermore, an investigation with the GREAT-ER system requires much more data than for the use of the regional distribution model.

4 Results and Discussion

First, a scenario analysis is performed with EUSES to discover how spatial refinement can ameliorate the predicted concentrations. The *Standard* scenario uses the default values of EUSES, based on the generic region. In the *NRW* realistic scenario regional data characterising North Rhine-Westphalia (NRW) and more realistic emission data and partition coefficients are used. A third scenario is added containing the river Ruhr catchment and which represents a further step of refinement (*Ruhr 1*). In a fourth scenario (*Ruhr 2*), supplementary ready biodegradation is assumed. In a subsequent step, the data of the Ruhr catchment are entered into GREAT-ER, and concentrations along the river are calculated. The aggregated results are compared with monitoring data and the results of the EUSES scenario calculations. Ready biodegradation was also assumed for this scenario.

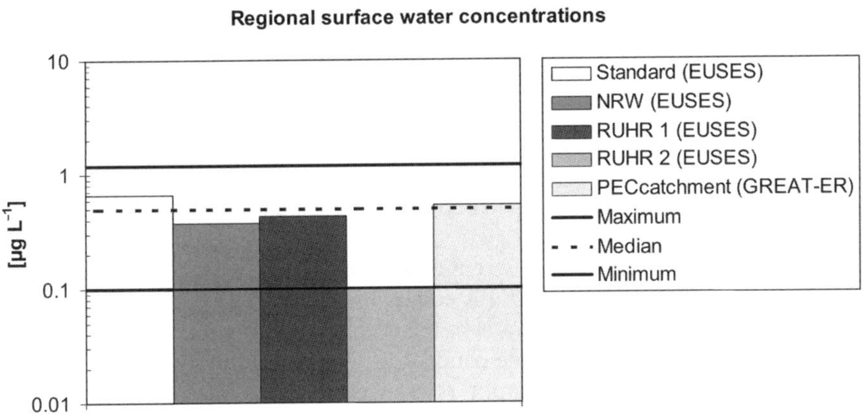

Fig. 3 Comparison of calculated surface water concentrations with monitoring data

The aquatic PECs estimated by EUSES represent regional background concentrations in surface water. The median of the measured values is chosen for a comparison of model results with monitoring data. The comparative GREAT-ER value is the $PEC_{catchment}$ which is the weighted spatial and temporal mean of all (polluted) stretches in a catchment and represents an average concentration of the Ruhr catchment.

Comparing calculated with measured HHCB surface water concentrations one must bear in mind that the range of the measurements is quite narrow (just between about 0.1 and 1 µg L^{-1}). Since the measurements were taken early in the year they perhaps represent only a part of possible concentrations. However, calculated concentrations in surface water show relatively small deviations to measured concentrations (Fig. 3).

On a regional scale, the first three scenarios are close to the measured median, over- and underestimating it slightly. The median of the concentrations measured in the Ruhr represents a mixture of local and regional concentrations, and can therefore be viewed as an increased regional contamination. This is also confirmed by the fact that lower contamination was measured in rivers such as the Elbe and the Rhine. On this basis EUSES delivers rather realistic estimations on a regional scale.

In the sense of the TGD, the measured values are rather to be classified as regional, since local concentrations are always defined by the direct vicinity of an emitter. The variation in the regional scale results of the individual scenarios does not exceed one order of magnitude. It is remarkable that the concentration of the *Ruhr 1* scenario nearly equals that of the *NRW* scenario. This shows that scaling down the environment from the North Rhine-Westphalian region to the Ruhr catchment only influences the resulting regional concentrations negligibly. The predicted concentration of the *Ruhr 1* scenario corresponds to the median of the measured values very well, but assuming a ready biodegradation (*Ruhr 2* scenario) leads to distinct underestimations. There can be several reasons for this: First, the supposed degradation rate in water could be too high, which could be reasoned by the fact that the *Ruhr 1* scenario delivers much better estimations. Second, the emissions to surface water are perhaps underestimated. Third, the underlying measured values for this investigation may perhaps not be representative and too high since they are all taken in the same month (February) along the river Ruhr (for a comparison with temporal variability see [13]). And another reason could be that EUSES generally underestimates regional concentrations and delivers only conservative predictions if the used estimations and defaults are conservative.

The latter presumption is strengthened by the GREAT-ER results: the $PEC_{catchment}$ predicted by GREAT-ER equals the measured median, although it uses the same high degradation rates as the underestimating *Ruhr 2* scenario. The reason for this is that the average concentration carried out by GREAT-ER is a weighted accumulation of geography-referenced simulation results while EUSES assumes one homogeneous box. Besides, the water volume assumed in EUSES needs not to be the same like in GREAT-ER. In the *Ruhr* scenarios and in the *NRW* scenario the same area fraction and depth for water are considered while the GREAT-ER calculation base on realistic river data. Another important functionality in GREAT-ER is to produce colour coded maps to visualise spatial distribution of estimated concentrations (Fig. 4).

A slight improvement of the EUSES model results can be gained by replacing the standard region by the NRW specific regional data and the use of measured partition coefficients. More precise emission rates lead to a considerable improvement of the model results. However, the values used are still afflicted by uncertainty: The consumption quantities are based on production figures provided by the fragrance industry. These figures do not contain information on import and export trade outside the EU, nor do they include all manufacturers. Furthermore, it is not known whether the reduction in consumption represents a general trend or fluctuations over a number of years [8]. This could be a significant source of uncertainty regarding statements on the comparison of measured vs predicted data. However, the figures can be taken as a best estimate, and it is expected that qualitative statements (i.e. over- or underestimations) do not change, because in our scenarios the emission rate shows a linear impact on the predictions and it is expected that variations in emissions do not exceed one order of magnitude.

The classification of the substance as readily degradable leads, of course, to noticeably lower environmental concentrations. Suspended matter can influence the distribution behaviour of substances considerably. Winkler et al. ascertained a dissolved fraction of 92% for HHCB in the Elbe with a medium partition coefficient between water and suspended matter of 4500 L kg^{-1} [6]. The values generated by EUSES are 88% and 1900 L kg^{-1}, respectively. According to this, adsorption to suspended matter has a negligible influence on the result.

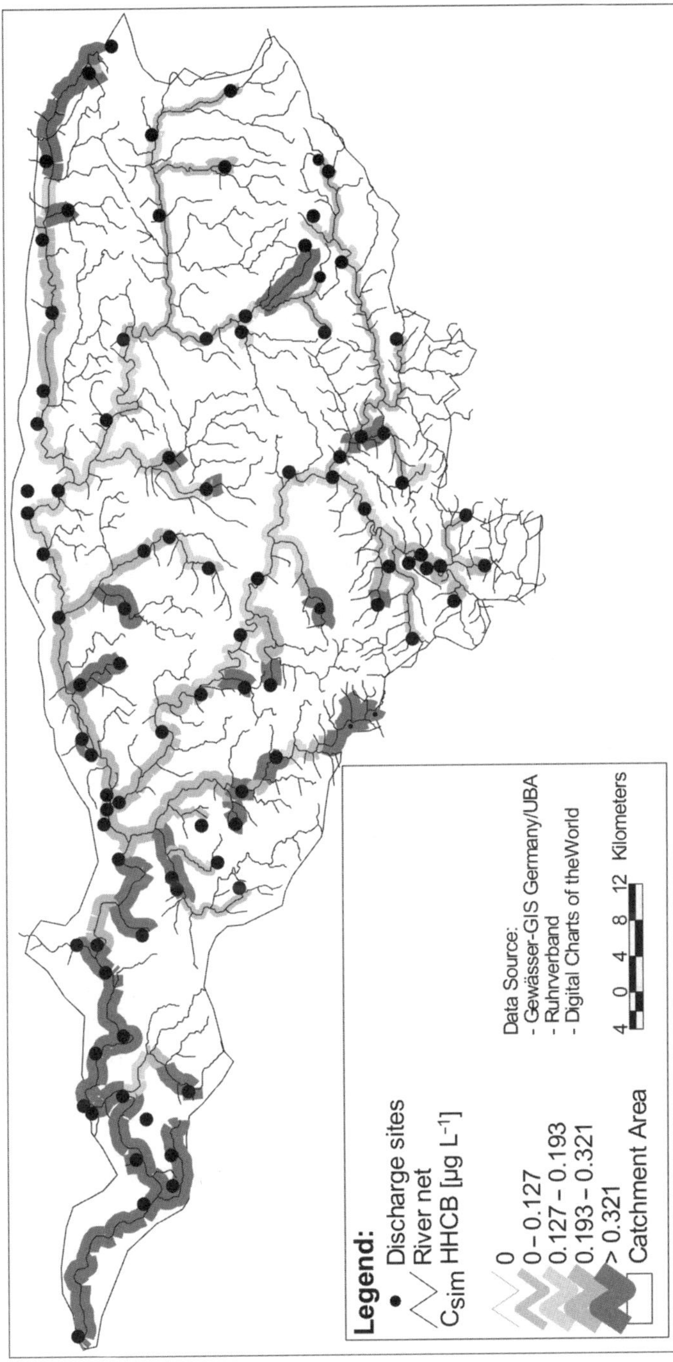

Fig. 4 Colour coded map of HHCB concentrations in the Ruhr catchment. C_{sim} is the average concentration in the appropriate river module

5 Conclusions

Both EUSES and GREAT-ER can be applied to evaluate the fate of the fragrance HHCB in the aquatic environment. In the EUSES standard scenario the predicted concentration is slightly higher than the measured median. Hence, the estimations are conservative with the default settings. However, EUSES does not necessarily deliver conservative estimations since the other scenarios underestimate the measured median. Using a high biodegradability leads to the highest underestimation but the value is still within the range of the measured values. However, the deviations are low for regional concentrations in the surface water. The variations of water concentrations between the individual scenarios are smaller than the difference between the minimum and maximum of the measured values. From the comparison of the concentrations of the investigated regions it can be concluded that an exact selection of representative environmental segments, including an adjustment of the emission rates to the selected region, leads to more realistic modelling of the contamination of the rivers. This requires closer scrutiny of the monitoring data and their spatial classification.

It is shown that the use of a geography-referenced model leads to good estimations of the median of the measured concentrations while on the same basis (*Ruhr 2* scenario) the compartment model underestimates the median. Additionally, only a geography-referenced model can show hot spots and can produce maps with visualised information. Nevertheless, the results of EUSES and GREAT-ER do not differ more than a factor of 5 if the same database is used. In view of this fact one must weigh if the gain of information is high enough to justify the large amount of additionally required data. In view of these results, the application of EUSES to other polycyclic musk fragrances seems to be possible, since AHTN (e.g. Tonalide) or ADBI (e.g. Celostolide) show similar environmental behaviour. Furthermore, correlations between these substances are reported [6], due to their use in mixtures. Individual statements for the other substances will either be more or less appropriate, depending on the respective sorption or accumulation behaviour.

In this chapter we dealt with potential uncertainties in an aquatic fate assessment using different scenarios and different types of model. The uncertainties of substance parameters were not analysed and sewage treatment process of HHCB was not considered in detail; the necessary database for such a study is not yet available.

Acknowledgement We would like to thank H.D. Eschke from the Ruhrverband, Essen for providing the monitoring data.

6 References

1. Bester K, Hühnerfuss H, Lange W, Rimkus G, Theobald N (1998) Water Res 32:1857
2. Eschke HD, Traud J, Dibowski HJ (1994) Z Umweltchem Ökotox 6:183
3. Eschke HD, Traud J, Dibowski HJ (1995) Z Umweltchem Ökotox 7:131
4. Eschke HD, Traud J, Dibowski HJ (1995) Dtsch Lebensm-Rundsch 12:375

5. Rimkus G, Wolf M (1996) Chemosphere 33:2033
6. Winkler M, Kopf G, Hauptvogel C, Neu T (1998) Chemosphere 37:1139
7. Gatermann R, Hellou J, Hühnerfuss H, Rimkus G, Zitko V (1999) Chemosphere 38:3431
8. Plassche EJ, van de Balk F (1997) Environmental risk assessment of the polycyclic musks AHTN and HHCB according to the EU-TGD. National Institute of Public Health and the Environment (RIVM), Report 601503008, Bilthoven
9. EC (1996) Technical Guidance Document in Support of the Commission Directive 93/67/EEC on Risk Assessment for New Notified Substances and the Commission Regulation (EC) 1488/94 on Risk Assessment for Existing Substances, Parts I–IV. Office for Official Publications of the European Communities, Luxembourg
10. EC (1996) EUSES – the European Union System for the Evaluation of Substances. Institute of Public Health and the Environment (RIVM), The Netherlands. Available from European Chemicals Bureau, Ispra
11. Balk F, Ford RA (1999) Toxicol Lett 111:57
12. Schwartz S, Berding V, Matthies M (1999) Umweltmed Forsch Prax 4:7
13. Schwartz S, Berding V, Matthies M (2000) Chemosphere 41:671
14. Ohloff G (1990) Riechstoffe und Geruchssinn: Die molekulare Welt der Düfte. Springer, Berlin Heidelberg New York
15. Berding V, Schwartz S, Matthies M (2000) Environ Sci Pollut Res 7:147
16. AWWR (1997) Ruhrwassergütebericht. The study group of the waterworks at the Ruhr (AWWR) and Ruhrverband, Essen
17. Trapp S, Matthies M (1998) Chemodynamics and environmental modeling – an introduction. Springer, Berlin Heidelberg New York
18. ECETOC (1999) GREAT-ER User Manual, Special Report No 16. ECETOC, Brussels
19. Matthies M, Wagner JO, Koormann F (1997) In: Alef K et al. (eds) Proceedings of ECO-INFORMA '97, vol 12, Eco-Informa Press, Bayreuth, p 523
20. Boeije G, Wagner JO, Koormann F, Vanrolleghem P, Schowanek D, Feijtel T (2000) Chemosphere 40:255
21. Dsikowitzky L, Schwarzbauer J, Littke R (2002) Org Geochem 33:1747

The Handbook of Environmental Chemistry Vol. 3, Part X (2004): 259–280
DOI 10.1007/b14121

Toxicology of Synthetic Musk Compounds in Man and Animals

Hubertus Brunn[1] · Nikola Bitsch[1] · Judith Amberg-Müller[2]

[1] Government Health Service Institute of Foodstuff and Veterinary Inspection, Marburger Strasse 54, 35396 Gießen, Germany
E-mail: Hubertus.E.Brunn@ernaehrung.uni-giessen.de
[2] Swiss Federal Office of Public Health, Food Toxicology Section, Stauffacherstrasse 101, 8004 Zürich, Switzerland

Abstract Synthetic musk compounds (musk fragrances) are chemically heterogeneous compounds. The available toxicological data are incomplete. The NOAEL (No Observed Adverse Effect Level) values, obtained in experimental toxicological studies, are of preliminary nature and can only serve as rough estimates. In the light of the available results and ongoing studies, it therefore seems prudent, in the sense of precautionary principle and consumer protection, to set temporary limits for persistent musk compounds in foodstuffs and cosmetics.

Keywords Musk fragrances · Metabolites · Toxicity · Endocrine activity

List of Abbreviations

ADBI 4-Acetyl-1,1-dimethyl-6-*tert*-butylindane (e.g. Celestolide)
AHDI 6-Acetyl-1,1,2,3,3,5-hexa-methylindane (e.g. Phantolide)
AHTN 6-Acetyl-1,1,2,4,4,7-hexamethyltetraline (e.g. Tonalide)
ATII 5-Acetyl-1,1,2,6-tetra-methyl-3-isopropylindane (e.g. Traseolide)
bw Body weight
CYP Cytochrome
DPMI 6,7-Dihydro-1,1,2,3,3-penta-methyl-4(5*H*)-indanone (e.g. Cashmeran)
EROD 7-Ethoxyresorufin-*o*-deethylase
HHCB 1,3,4,6,7,8-Hexahydro-4,6,6,7,8,8-hexamethylcyclopenta-gamma-2-benzopyran (e.g. Galaxolide)
ip Intraperitoneal
LOAEL Lowest observed adverse effect level
lw Lipid weight
MK Musk ketone
MROD Methoxyresorufin-*o*-demethylase
MX Musk xylene
NOAEL No observed adverse effect level
PB Phenobarbital
PROD 7-Pentoxyresorufin-*o*-dealkylase

1 Introduction

Like polychlorinated biphenyls (PCBs), musk compounds are used in vast amounts worldwide, and are both lipophilic and environmentally persistent. In addition, their potential health risk also became apparent only after the scientific community recognized the presence of large-scale environmental contamination and the substances had reached human beings either directly or through the food chain. Musk fragrances have also been detected in human breast milk. The available data on the toxicology of musk compounds is unsatisfactory and very heterogeneous. The aim of the present chapter is to summarize relevant data that will serve as a basis for human risk assessment and consequently focus on mammalian toxicity.

2 Metabolism and Pharmacokinetics

2.1 Percutaneous Absorption

2.1.1 *Nitro Musks*

The percutaneous absorption of musk xylene (MX) measured in vitro was high in hairless guinea pig skin and comparably low in human skin. Whereas total absorption (skin and receptor fluid) of MX in 24 h was 55% in guinea pig skin and 22% in human skin, the amount of compound diffused through the skin into the receptor fluid within 24 h was much higher in guinea pig skin (32% of the applied dose) compared to human skin (<5%). The MX contained in human skin at 24 h was mostly found in the stratum corneum. From day 1 to day 7, MX was slowly released into the receptor fluid, and by day 7 only 6% of the applied dose remained in the skin. Total recovery of MX was 70% for human skin and 80% for guinea pig skin [56]. In rats, after dermal application of 0.5 mg MX kg^{-1} bw maintained under occlusion for 6 h, 19% of the dose was absorbed during 48 h [57]. In human volunteers, percutaneous absorption of MX was 0.9% from a non-alcoholic formulation (body lotion) and 3.1% from an alcoholic solution (after shave) [58]. In another study in two human volunteers, after a single dermal dose of 0.02 mg cm^{-2} radiolabeled MX in an alcoholic solution, 90% of the dose was recovered from the site of application after 6 h, and 0.26% had been excreted in the urine and <0.1% in the faeces after 120 h. However, 10% of the applied dose was unaccounted [59].

For musk ketone (MK), 31% of a dermally applied dose of 0.5 mg kg^{-1} bw maintained under occlusion for 6 h was absorbed in rats during 48 h [57]. After a single dermal application of 0.02 mg cm^{-2} radiolabeled MK made to two human volunteers, 86% of the dose was recovered form the site of application after 6 h, and 0.42% and 0.05% of the dose had been excreted in the urine and faeces. However, 14% of the applied dose was unaccounted [60].

2.1.2
Polycyclic Musks

In a preliminary study on the in vitro absorption of AHTN and HHCB using rat skin, very low absorption for AHTN (0.28±0.14%, n=3) and for HHCB (0.07±0.03%, n=4) was observed after 24 h [74]; however, 50–60% of the applied dose was found within the skin at the end of the experiment.

Percutaneous absorption of ^{14}C-labelled AHTN and HHCB in rats after dermal application at doses of 4.5 mg kg^{-1} bw in ethanol to shaved skin for 6 h under occlusion was found to be approximately 19% for AHTN of the dose and 14% for HHCB after 120 h. The amount absorbed in 6 h was 9.3% of the applied dose for AHTN and 3.9% for HHCB, while substantial amounts were present in the treated skin [15]. Results from exposure experiments in human volunteers indicated that, based on radioactivity excreted in faeces and urine, less than 2% of AHTN is absorbed after a 6-h exposure at typical use levels of a 70% ethanol solution containing 4.0 mg HHCB ml^{-1} and 2.4 mg AHTN ml^{-1}. No detectable amounts of HHCB were absorbed under these conditions [15].

2.2
Biotransformation

2.2.1
Nitro Musks

The key step in the metabolism of MX in the rat after oral administration is the reduction of the 4- (i.e. *p*-) nitro group of MX to the amino group. Further metabolism of the *p*-amino (NH_2) metabolite, mainly hydroxylation of methyl groups, may proceed by decreased steric hindrance of functional groups. In the faeces, bile and urine, MX, 4-NH_2-MX, 4-acetylamino-MX, 4-NH_2-3-CH_2OH-MX and 4-NH_2-1-*tert*-BuOH-MX were found. 2-NH_2-MX and a non-identified metabolite were found in faeces and urine, 4-NH_2-3-CH_2OH-MX was found in urine and HO-MX in bile only [42]. Recent studies in rats showed that after dermal application of MX and MK, most of the absorbed dose was eliminated via bile mainly as polar conjugated metabolites. The major aglycones were hydroxylated analogues of MX and MK formed by oxidation of the methyl groups. They occurred in the bile mainly as glucuronide conjugates. For MK, at least two major less polar metabolites were present in the bile. The urine showed a more complex mixture of polar, though non-identified metabolites, although there was some evidence for the presence of MX glucuronides. Most of the metabolites of MK were not simple glucuronide and sulfate conjugates. The relatively complex pattern of urinary metabolites was proposed to be the result of enterohepatic circulation allowing further metabolism of simple aglycones [57].

Studies on the metabolism in humans are rare. It has been demonstrated that the 4-amino metabolite of MX is formed in humans after oral and dermal MX exposure and directly eliminated in the urine. Recoveries in urine were 0.12–0.53% of oral dose and 0.02–0.16% of dermal dose compared to the calculated total absorption of MX of 0.6–3.8% of the oral dose and 0.03–0.06% of the dermal dose.

The main metabolites excreted in the urine were not identified. The acetyl derivative of the 4-amino metabolite, a metabolite in the rat urine, was not found in human urine [55].

2.2.2
Polycyclic Musks

There are no data available on the metabolism of polycyclic musk fragrances in humans.

2.3
Distribution and Elimination

2.3.1
Nitro Musks

Dermal exposure of rats to 0.5 mg kg^{-1} bw MX and MK resulted in comparable tissue concentrations. Maximum levels were reached after 6–8 h. Highest concentrations were measured in the liver and the fat tissue, with peak concentrations of 0.2 μg nitro musk equivalents g^{-1} and 0.005 μg nitro musk equivalents g^{-1} after 120 h [57]. The major route of elimination for MX in rats was the faeces via bile. After oral application, 75.5% of the dose was excreted in the faeces and 10.3% in the urine [42]. In the rat, after single dermal application of MX and MK (0.5 mg kg bw^{-1} for 6 h under occlusion), most of the absorbed material was excreted within five days with only 1–2% of the applied dose remaining in the animal at this time. At 120 h total recoveries in the urine and faeces were 3.9% and 14.8% for MX and 8.5% and 20.5% for MK. Repeated daily doses for 14 days resulted in little bioaccumulation for MX and a threefold accumulation for MK [57]. Compared to humans, excretion in rodents is relatively fast with a half-life of a few days.

In human volunteers, dermal and oral application of 0.3 mg MX kg^{-1} bw resulted in peak plasma concentrations of 40–270 ng mL^{-1} and 1.6–5.5 ng mL^{-1} after 6 h. Pharmacokinetics of MX in plasma followed a two-compartment kinetic model. An initial rapid decrease due to distribution from the blood into a second compartment, likely the adipose tissue, was followed by a slow terminal elimination with an average half-life of 70 days [55]. Elimination of MX in humans occurs mainly in the urine. However, studies on the quantification of major metabolites have not been published. Maximal concentrations of the 4-amino metabolite in the urine were reached 18–24 h after administration of MX. After a short time of invasion, excretion followed a first-order kinetics with an average elimination half-life of 12 h [55]. From literature data on the body burden and the resulting half-life the average daily uptake was calculated to be around 11 μg per person or 160 ng kg^{-1} bw [58]. MX and MK were detected in human adipose tissue, the content in 15 samples ranging from 6.7–288 μg MX kg^{-1} lw and <1–173 μg MK kg^{-1} lw [73].

MX and MK have been detected in human breast milk. While earlier studies reported higher concentrations of about 100 μg MX kg^{-1} lw and about

40 µg MK kg^{-1} lw [61, 62], in more recent studies, lower levels of 8–20 µg MX kg^{-1} lw and about 8–10 µg MK kg^{-1} lw were found [63, 64].

2.3.2
Polycyclic Musks

Data on the distribution and elimination kinetics of polycyclic musk fragrances in humans and mammals are scarce.

After dermal application of radiolabeled AHTN and HHCB to rats, the principle route of excretion was in the faeces: For AHTN, 2.6% of the radioactivity was excreted in the urine and 14.5% in the faeces within 120 h. For HHCB, 1.3% was excreted in the urine and 11.6% in the faeces [15].

Compared to the nitro musks, relatively high concentrations of polycyclic musk fragrances have been reported in human adipose tissue and breast milk. Mueller et al. [73] detected HHCB, AHTN and ADBI in human adipose tissue, the content in 15 samples (10 females 3–100 years old and 5 males 49–77 years old) ranging from 12–171 µg HHCB kg^{-1} lw, 1.0–23 µg AHTN kg^{-1} lw and 0.12–3.5 µg ADBI kg^{-1} lw. All three substances were found in all samples. Eschke et al. [75] reported HHCB and AHTN concentrations in two samples of human fat (lipid content of 82%) of 145 and 149 µg kg^{-1} lw and 56 and 72 µg kg^{-1} lw, respectively. Concentrations of ADBI, AHDI, DPMI and ATII were close to the limit of determination (<1 to 9 µg kg^{-1} lw). Rimkus and Wolf [76] reported residual concentrations of polycyclic musk fragrances in 14 human adipose tissue samples (8 females 41–70 years old and 6 males 30–64 years old) of 28–189 µg HHCB kg^{-1} lw and 8–33 µg AHTN kg^{-1} lw. In some samples, ADBI, AHDI and ATII were also detected at low levels (1–10 µg kg^{-1} lw). There were no apparent trends with age, sex or time of obtaining the samples in any of these studies.

Eschke et al. [75] also identified polycyclic musk fragrances in human breast milk. Two samples (lipid content of 1.06% and 0.41%, respectively) were reported to contain 310 and 360 µg HHCB kg^{-1} lw, 290 and 250 µg AHTN kg^{-1} lw, 24 and 20 µg ADBI kg^{-1} lw, 13 and 15 µg AHDI kg^{-1} lw, 12 and 25 µg ATII kg^{-1} lw. DPMI could not be detected (<1 µg kg^{-1} lw). Comparably lower residual concentrations of polycyclic musk fragrances in human breast milk (n=5) were found by Rimkus and Wolf [76]. The concentrations reported ranged from 16–108 µg HHCB kg^{-1} lw, from 11–58 µg AHTN kg^{-1} lw and from 1–18 µg ADBI kg^{-1} lw. AHDI, ATII and ATTN were not detected (limit of determination 1–2 µg kg^{-1} lw).

In comparison, treatment of pregnant rats with repeated oral doses of 2 or 20 mg radiolabeled HHCB or AHTN kg^{-1} starting at the third week of pregnancy resulted in maximum levels in the milk at days 3 and 7 after parturition of 1350 µg HHCB L^{-1} and 940 µg AHTN L^{-1} at the lower dose and 23,590 µg HHCB L^{-1} and 12,140 µg AHTN L^{-1} at the higher dose [77].

3 Acute and Subacute Toxicity

3.1 Nitro Musks

Acute oral and dermal toxicity of the nitro musk compounds, MX, MK, musk tibetene and musk moskene is comparatively low in laboratory animals (LD_{50}> 5000 mg kg^{-1}) [1–4]. Repeated daily oral doses of 500 mg MK kg^{-1} bw for seven days were tolerated by mice [68], whereas similarly treated rats showed signs of severe intoxication and died within two days (cited in [23]). Various studies in mice and rats showed that acute exposure to high doses (200 to 500 mg kg^{-1} bw) of MX and MK for five to seven days induced significant effects on the liver, including increased liver weight, increased total microsomal protein and centrilobular hepatocellular hypertrophy due to enzyme induction (see also below) [18, 23, 68]. Musk ambrette has been shown to be acutely toxic in the rat at oral doses of 339 mg kg^{-1} (LD_{50}) [5]. Musk ambrette caused photoallergic symptoms in humans, MX, MK, musk moskene and musk tibetene, however, are not phototoxic or photosensitizing in humans [6, 7].

3.2 Nitro Musk Metabolites

There are no published toxicity studies in mammals. Four MX amino metabolites (*p*-amino-MX, *o*-amino-MX, *o,p*-diamino-MX, triamino-MX) were assayed for acute toxicity to *Daphnia magna*. A low EC_{50} of 0.00025 mg L^{-1}, i.e. a high acute toxicity was found for *p*-amino-MX. The other metabolites were less toxic. The sequence of the EC_{50} values for the metabolites tested was *p*-amino-MX<*o*-amino-MX<*o,p*-diamino-MX<triamino-MX [8]. These results could not be reproduced by Salvito et al. [9]. Under the same conditions, however using *p*-amino-MX of higher purity, toxicity was three orders of magnitude lower.

3.3 Polycyclic Musks

The Research Institute for Fragrance Materials (RIFM) reported an acute oral toxicity for HHCB in rats of >3.25 g kg^{-1} (LD_{10}, Moreno 1975a, cited in [10]). Another study reported an oral LD_{50} in rats of >3 g kg^{-1} (Avon Products 1977, cited in [10]). Moreno (1975a, cited in [10]) reported dermal toxicity in rabbits at >3.25 g kg^{-1}. Avon Products (cited in [10]) reported an LD_{50} value for the acute dermal toxicity in rats of >5 g kg^{-1}.

Moreno (1975b, cited in [10]) reported an LD_{50} value of 0.57 g kg^{-1} for acute oral toxicity of AHTN in rats. Avon Products (cited in [10]) listed an oral LD_{50} value for rats of 0.825 g kg^{-1}. Other studies reported an oral LD_{50} value for rats of 0.92 g kg^{-1} (Hercules 1985a, cited in [10]), or 1.4 g kg^{-1} (Givaudan 1982, cited in [10]). Acute dermal toxicity of AHTN in rats was reported with an LD_{50} value of >5 g kg^{-1} (Moreno 1975b, cited in [10]), or 7.94 g kg^{-1} (Avon Products 1977, cited in [10]).

From the reports cited above it can be generally concluded that the acute toxicity of polycyclic compounds is low. Nonetheless, a single large dose of AHTN (100 mg kg^{-1} ip) resulted in acute liver damage to rats. The morphological changes observed included localized liver-cell necroses, inflammation, swelling of the liver parenchyma cells, condensation of the cytoplasm, ultrastructural damage to the endoplasmic reticulum and mitochondria as well as an increase in the relative weight of the liver [11].

4 Subchronic Toxicity

4.1 Nitro Musks

In a 90-day study with oral treatment of rats with musk ambrette at 10–240 mg kg^{-1} bw day^{-1}, dosages of 80–240 mg kg^{-1} bw resulted in neurotoxic effects in males and females and in testicular atrophy in males [12].

In another 90-day study MK, MX, musk tibetene and musk moskene were applied dermally to rats (15 animals per sex and group) at daily doses of 240 mg kg^{-1} bw(MK and MX only) and 75, 24, and 7.5 mg kg^{-1} bw. Musk ambrette served as positive control, being clearly neurotoxic and causing testicular atrophy. The other nitro musk compounds failed to show such effects, but led to significant increases in relative liver weights in the high dose group of musk moskene (75 mg kg^{-1} bw, males only), MK (240 mg kg^{-1} bw, females only), and MX (240 mg kg^{-1} bw, males and females) as well as at the next lower dose (75 mg kg^{-1} bw, females only) of MX (no data given for musk tibetene). These effects were, however, not associated with histopathological changes [13]. From these experiments, the authors derived a dermal NOAEL value of 75 mg kg^{-1} bw day^{-1} (MK and musk tibetene) and 24 mg kg^{-1} bw day^{-1} (MX and moskene). They estimated the daily nitro musk exposure from average use of fragrances at 0.03–0.04 mg kg^{-1} bw indicating a safety factor of approximately 100 [13]. The significance of these NOAEL values has been clarified by studies on the effect of nitro musks on liver enzymes (see below), possibly the most sensitive parameter. Therefore the observed increases in relative liver weight may be the result of a liver enzyme induction by nitro musks [14]. Other studies in mice and rats indicated that continuous dietary exposure (10–17 weeks) to high doses of MX (70–80 mg kg^{-1} bw and 180 mg kg^{-1} bw, respectively) was not effective in inducing general hepatic changes (including increased liver weight), possibly due to adaptation processes [14, 24].

4.2 Polycyclic Musks

AHTN and HHCB were subjects of a number of studies on oral and dermal subchronic toxicity in the rat (cited by Ford in [10]). Most notable were the apparent dose dependent effects in the liver. Dermal exposure to female rats (15 per dose group) at daily doses of 0, 1, 10 or 100 mg kg^{-1} bw for 13 weeks led to an increase in liver weight at highest dosage tested for both AHTN and HHCB. The increase induced by HHCB was not accompanied by histopathological changes [10]. AHTN, in contrast, led to an increase in liver weight accompanied by hepatocytomegalic inclusions and discoloration as well as an increase in alkaline phosphatase activity in the plasma. There were no effects observed at the lower doses tested for both AHTN and HHCB.

Orally, HHCB at daily doses of 5, 15, 50 or 150 mg kg^{-1} bw for 13 weeks in the diet of rats (15 female and 15 male per group) led to a minimal and apparently reversible increase in absolute liver weight in males, independent of the dose. No effects on liver weight were observed in females at all doses tested. No histopathological changes were observed. Increased relative and absolute liver weights for male and female rats and an adverse effect on body weight was observed at oral doses of 341 mg kg^{-1} bw in a previous two week dose range finding study. Thus a NOAEL of 150 mg kg^{-1} bw was suggested for HHCB [10].

AHTN was tested at the same doses of 1, 5, 15, 50 and 150 mg kg^{-1} bw. At 15 mg kg^{-1} bw the liver weight increased in female animals, at 50 mg kg^{-1} bw in both sexes; however, no histopathological changes were observed to indicate potential disturbances of liver function. There were also minor changes in parameters of clinical chemistry [10]. It should also be noted, however, that in some animals of both genders, abnormal green discolorations of the liver and discoloured mesenteric lymph nodes were observed (50 mg kg^{-1} bw). Females were found to have green discolorations of the lachrymal glands, one animal at a dose of 5 mg kg^{-1} bw, four at a dose of 15 mg kg^{-1} bw and eight at a dose of 50 mg kg^{-1} bw. The discolorations in the liver did not display UV induced fluorescence indicating that there was no accumulation of porphyrins. From these results, the author [10] derived a NOAEL of 15 mg AHTN kg^{-1} bw.

While there is evidence that AHTN is hepatotoxic when administered at high doses, subchronic treatment of rats with AHTN at doses within the human exposure range (up to 300 µg kg^{-1} bw ip for 90 days) did not affect liver architecture or biochemical liver parameters [20].

Ford [10] related the NOAEL values of 150 mg HHCB kg^{-1} bw and 15 mg AHTN kg^{-1} bw obtained in these experiments to the calculated maximal daily systemic exposure of HHCB and AHTN by the use of cosmetic products of 0.11 mg HHCB kg^{-1} bw and 0.043 mg AHTN kg^{-1} bw, based on an estimated maximal dermal exposure of 0.76 mg HHCB kg^{-1} bw and 0.31 mg AHTN kg^{-1} bw and a conservative assumption of 14% percutaneous absorption and derived safety factors of 1400 for HHCB and 350 for AHTN.

Subchronic inhalation exposure of rats and hamsters to various fragrance mixtures including HHCB and MK at different doses did not result in any significant toxicological effects relating to diverse parameters including behaviour, body and organ weight, haematology, gross pathological or histopathological changes [17].

5 Induction of Xenobiotic Metabolising Enzymes

5.1 Nitro Musks

MX and MK induce cytochrome P450 (CYP) enzymes in the mouse as well as in the rat. The enzyme induction pattern in the mouse is comparable to that of phenobarbital (PB), the classic CYP2B inducer and liver carcinogen in the mouse [21, 65] (Table 1).

In the rat liver MX induces phase-I cytochrome P450 oxidases (CYP1A, CYP2B) as well as phase-II metabolic enzymes such as cytosolic DT-diaphorase, glutathione-*S*-transferase (GST Ya subunit) and UDP-glucuronyl transferase. The activities of the latter three enzymes were increased in a dose dependent manner after ip administration of MX to rats at 50–200 mg kg^{-1} bw for five consecutive days by a factor of 1.6–7, 2.6–3.1 and 1.6–2.0, respectively [19]. With respect to phase-I oxidases, MX strongly and preferentially induces CYP1A2 and to a lesser extent CYP1A1 and CYP1A3, with a ratio of CYP1A2/CYP1A1 protein content of 12 in rat hepatic microsomes after ip administration of 50 mg MX kg^{-1} bw for five days [18].

Table 1 Effects of musk xylene (MX) and musk ketone (MK) on microsomal liver enzymes in mice and rats (maximal effect during treatment with doses up to 200 mg kg^{-1} bw day^{-1} [up to 500 mg kg^{-1} bw day^{-1} with MK in mice] for seven days). Determination of enzyme activity (act) indirectly via activity of EROD (7-ethoxyresorufin-*o*-deethylase) for CYP1A1, MROD (7-methoxy-resorufin-*o*-demethylase) for CYP1A2 and PROD (7-pentoxyresorufin-*o*-dealkylase); for CYP2B resp., erythromycin *N*-demethylase for CYP3A. Determination of protein concentrations (prot) by western immunoblot analysis (according to [18, 21–23, 65, 67, 68])

	Mice			Rats		
	CYP1A	CYP2B	CYP3A	CYP1A	CYP2B	CYP3A
MX	act↑ 2.5–4× (A1), 2× (A2) prot↑ 3× (A2)	act no effect prot↑ 25× mRNA↑ 10×	act↑ 2× prot↑ 2×	act↑ 4× prot↑ 5× (A2)[a]	act↑ 3× prot↑ 50× mRNA↑ 10×	act↑ 2× prot↑ 5×
MK	act↑ 2× (A1), 4× (A2) prot↑ low mRNA↑ low	act↑ 28× prot↑ 6× mRNA↑ according to act and prot	act↑ 2× prot↑ low mRNA↑ low	act↑ 30×(A1), 20×(A2) prot↑10× (A1, A2)	act↑ 8× prot↑ 10× mRNA↑ 3×	act reduced by 50% prot no effect
PB	act↑ low comparable to MX	act↑ 25× prot↑ and mRNA↑ comparable to MX	act↑ low comparable to MX	act↑ 5× (A1), 1.5× (A2) prot↑ low	act↑ 50× mRNA↑ 15×	act↑ 2×

[a] CYP1A2/CYP1A1=12.

To elucidate the mechanism of microsomal enzyme induction by nitro musks various detailed studies have been carried out in mice and in rats. MX treatment of mice at oral doses of 200 mg kg^{-1} bw for seven days had a modest effect on CYP1A1 and 1A2 enzyme activities and increased both CYP1A1 and 1A2 protein levels; however, this induction was much lower than with the known CYP1A inducer 3-methylcholanthrene (3-MC). In contrast, whereas no increase in the CYP2B enzyme activity was observed, MX treatment induced a marked increase in CYP2B protein levels in the liver microsomes. At necropsy, an increase in the absolute and relative liver weight and hepatocellular hypertrophy was observed [21, 65]. To study the inhibition of CYP2B enzymes, PB stimulated male B6C3F1 mice, with a 25-fold increased PROD activity (as indicator of CYP2B activity), were treated with a single oral dose of 200 mg MX kg^{-1} bw. This led to a decrease of PROD activity by 90% after 18 h (but not after 2 h) as well as to covalent binding ^{14}C-labelled MX to microsomal proteins. If the gastrointestinal flora of the mice was reduced by antibiotic treatment prior to administration of MX to prevent the formation of the *p*-amino metabolite, inhibition of PB induced PROD activity did not occur [21,65]. Enzyme inhibition by *p*-amino-MX as well as the *o*-amino metabolite (*o*-amino-MX) was studied directly in vitro in PB induced mouse liver microsomes. Both amines inhibited PROD activity when preincubated in the absence of NADPH. If NADPH was present during preincubation, no further loss of PROD activity was observed with *o*-amino-MX, indicating that this metabolite does not inactivate CYP2B (i.e. CYP2B10, the major PB inducible enzyme in mice). In contrast, *p*-amino-MX induced a time-dependent decrease of PROD activity during preincubation with NADPH, with an inactivation rate of pseudo first-order displaying saturation kinetics. Further evidence for CYP2B10 inactivation by *p*-amino-MX was obtained in vivo by oral treatment of PB induced mice with *p*-amino-MX (180 mg kg^{-1} bw), which, in contrast to feeding equimolar amounts of MX, resulted in a complete loss of PROD activity after 2 h. Despite the loss of enzyme activity, there was no decrease of the CYP2B protein level [22].

The metabolites *o*- and *p*-amino-MX have also been evaluated for their potential to induce CYP2B10 and CYP1A2 mRNAs. After single doses of 180 mg kg^{-1} bw, both amines markedly induced CYP2B10 mRNA (about fivefold), whereas CYP1A2 mRNA, which is generally involved in the bioactivation of aromatic amines and frequently induced by aromatic amines, was only slightly induced (about 1.5-fold), a response similar to PB. With *o*-amino-MX, time course and levels of CYP2B10 induction were generally similar to that observed with both equimolar doses of MX and with PB, whereas, with *p*-amino MX, induction was of similar magnitude but much longer. Induction of CYP2B10 mRNA suggested that the amine metabolites may contribute to the enzyme induction seen with MX treatment. This hypothesis was supported by the results of a further study where MX administration (200 mg kg^{-1} bw) to antibiotic-treated mice did not result in microsomal enzyme induction, increases in liver weight, microsomal protein content, total P450 levels, and cytochrome c reductase activity, as seen with MX alone. No induction of CYP1A activity was observed when MX was administered to antibiotic treated mice [66].

Since MK lacks the *p*-nitro substitution, the *p*-amino metabolite cannot be formed. Thus, the enzyme inactivating and inducing effects observed with MX, which are related to the p-amino metabolite do not occur with MK. Oral treatment of mice with 200 mg kg^{-1} bw for seven days resulted in an increase in liver weight by 20%, while MX at similar doses caused an increase of 50%. MK induced a smaller increase in CYP2B protein levels in liver microsomes than MX. However, in contrast to MX, MK induced an increase in the CYP2B enzyme activity [67]. Treatment of mice with oral doses of 5–500 mg MK kg^{-1} bw for seven days resulted in increased liver weights (by 50% at the highest dose), increased total microsomal protein levels (twofold at highest dose) and centrilobular hepatocellular hypertrophy. At the highest dose tested, MK caused a 28-fold increase in CYP2B enzyme activity and a small (approximately twofold) increase in both CYP1A and CYP3A enzyme activities. Protein and mRNA analysis confirmed the relative levels of induction for CYP2B, CYP1A and CYP3A. For the CYP2B induction by MK a NOEL of 20 mg kg^{-1} bw was derived. Furthermore, it was shown that a single oral dose of MK (198 mg kg^{-1} bw) inhibited PB induced CYP2B enzyme activity by only 20%, whereas an equimolar dose of MX caused a 90% inhibition [68]. Apart from the effects on phase-I cytochrome P450 enzymes, MK induced an increased activity of the detoxifying enzyme glutathione *S*-transferase (phase-II enzyme) in the liver and small intestinal mucosa in mice [69].

In a more recent study, F344 rats were treated orally with MX (10, 50 or 200 mg kg^{-1} bw) or MK (20, 100 or 200 mg kg^{-1} bw) for seven days [23]. The effects of MX and MK on cytochrome P450 enzymes were similar to those previously observed in mice. MX treatment caused a two- to fourfold increase in the activity of CYP1A, 2B- and 3A enzymes, which was significant in all groups. For CYP1A and 3A, these changes were consistent with small (fivefold) increases in protein levels. However, for CYP2B, despite an only threefold increase in enzyme activity, protein levels were increased nearly 50-fold. This discontinuity between protein induction and enzyme activity is consistent with the inhibition of CYP2B observed previously for mice. In contrast to MX, MK treatment resulted in a similar increase in both CYP2B protein (tenfold) and enzyme activity (eightfold). The effect of MX and MK on CYP2B enzymes was also determined at the mRNA level. Results indicated that both MX and MK induced CYP2B1 mRNA and that transcriptional activation of CYP2B1 by MX was much greater than that observed with MK. MK treatment also induced a nearly 30-fold increase in the CYP1A enzyme activity, with significant effects at all doses. In this respect, the induction profile of MK was markedly different from MX, and very different from its effects in mice [68]. From these findings the authors concluded that in rats, MX is an inducer of CYP2B enzymes, which are, however, not functionally active. In contrast, MK also induces CYP2B enzymes, but with no concurrent inactivation due to the lack of a *p*-amino metabolite [23].

Overall, the enzyme induction profile of MK with large induction of both CYP1A and CYP2B enzymes in rats is unusual for nitro aromatic compounds or other aromatic amines [70]. Pronounced CYP1A1 induction is typically associated with an interaction between the compound and the Ah-receptor. Chemicals that induce CYP1A enzymes through a mechanism involving interactions with this cytosolic receptor, do not induce CYP2B enzymes. It is therefore very unusual for a compound to induce enzymes from both the CYP1A and the CYP2B family to this extent. The molecular mechanism underlying the induction profile of MK is presently unknown [23].

Developmentally specific effects of MX on cytochrom P450 enzymes have been studied in rats. Male and female Long Evans rats were fed MX with the diet (1–1000 mg kg^{-1} feed, i.e. 0.07–0.08 mg kg^{-1} bw up to 70–80 mg kg^{-1} bw) for at least ten weeks before mating. Exposure was continued through pregnancy, birth and lactation. The induction profile observed in the offspring was different compared to adult animals. MX caused a dose dependent induction of CYP1A1 and CYP1A2 enzymes in adult and developing rats at postnatal day 1 (PN1) and 14 (PN14). In adult males and females at 7–8 mg kg^{-1} bw, CYP1A1 and CYP1A2 activity increased 2.5 and fourfold, respectively. CYP2B activity increased about twofold in males and females. Protein levels of CYP1A were significantly induced in a dose dependent manner from 7–8 mg kg^{-1} bw to 70–80 mg kg^{-1} bw. No change occurred in CYP3A protein levels, whereas a marked induction of CYP2B protein was seen at the highest dose.

In the offspring, the lowest effective maternal dose for significantly increased CYP1A activity (1.7-fold for CYP1A1 and 1.9-fold for CYP1A2) in PN14 off-springs was 2–3 mg kg^{-1} bw. Maternal doses of 7–8 mg kg^{-1} bw resulted in a 3.0- and 3.7-fold increase in CY1A1 and CYP1A2 enzyme activities, respectively, at PN1 and PN14. Thus, the magnitude of CYP1A induction in the offspring was similar to that observed in adult animals. As in adult rats, induction of CYP1A protein occurred dose dependently in the maternal dose range of 7–8 mg kg^{-1} bw to 70–80 mg kg^{-1} bw and CYP2B was induced only at the highest dose of 70–80 mg MX kg^{-1} bw. However, in contrast to the adult stage, an increase in CYP3A protein level was observed in P14 offspring of rats exposed to the highest dose. High dose application (70–80 mg kg^{-1} bw) of MX in adult F1 males and F1 females resulted in an induction of CYP2B, but not of CYP3A, indicating a different sensitivity to MX during development [24]. From these data, a NOAEL of 0.7–0.8 mg kg^{-1} bw and a LOAEL of 2–3 mg kg^{-1} bw could be derived for P450 enzyme induction during development by maternal MX exposure.

5.2 Polycyclic Musks

In contrast to the nitro musks, induction of xenobiotic metabolising enzyme systems by the polycyclic compounds has not been tested extensively. AHTN is apparently not capable of acting as a phenobarbital- or 3-methylcholanthrene-like inducer in the rat liver (and also failed to show any peroxisome proliferating properties) [11]. AHTN and ADBI were shown to be substrates for CYP3A4 and were metabolised to cytotoxic compounds in transfected hamster lung fibroblasts (V79 cells), which stably express human CYP3A4 (CYPh3A4) [25].

6
Mutagenicity and Genotoxicity

6.1
Nitro Musks

MX and musk ambrette have been examined for mutagenic activity in the Ames test. While MX did not lead to the appearance of increased revertants in TA 98 and TA 100 strains of *Salmonella typhimurium*, even in the presence of S 9, musk ambrette proved to have concentration-dependent mutagenic properties after S 9 activation. [26]. A series of further studies on mutagenesis also included other nitro musk compounds [27, 28]. Musk ambrette proved to be mutagenic in *Salmonella typhimurium* TA 100, albeit only after metabolic activation by the S 9 fraction from rat liver.

MK, musk moskene and musk tibetene, however, were negative in this system (TA 98 and TA 100), even after metabolic activation [29]. In the mouse lymphoma assay, MK was also found to be non-mutagenic [30].

For musk ambrette a strong mutagenic activity in the Ames test in *Salmonella typhimurium* strain TA 100 with metabolic activation has also been reported by others [31], while MX, MK, musk moskene and musk tibetene, did not show such effects with *Salmonella typhimurium* strain TA 97, TA 98, TA 100 and TA 102 even with the addition of S 9. The authors concluded that the DNA damaging effect of musk ambrette is not the result of the activity of bacterial nitroreductases, as it is the case in other nitro aromatic compounds reacting positively in the Ames test without the addition of S 9, but that musk ambrette is more likely biotransformed by mammalian enzymes (in the S 9 fraction), possibly to an aryl hydroxyl amine. This kind of mechanism would suggest a mutagenic potential of musk ambrette for mammals [31].

MX, MK, musk ambrette, musk moskene and musk tibetene did not display genotoxicity in the SOS chromo test with *E. coli* PQ 37 [29]. Negative results have also been obtained with these substances in the sister chromatid exchange (SCE) assay [35]. No indication of a genotoxic potential was found for MX and MK in a chromosome aberration assay in Chinese hamster ovary (CHO) cells and in an unscheduled DNA synthesis (UDS) assay in primary rat hepatocytes [30, 36]. Musk ambrette, MX, MK, musk moskene and musk tibetene did not show a genotoxic activity in the micronucleus tests with human lymphocytes and the human hepatoma cell line Hep G2 [37].

6.2
Polycyclic Musks

HHCB, AHTN, ADBI, AHDI, DPMI and ATII have been examined in the Ames test with *S. typhimurium* strains TA97, TA98, TA100 and TA102, both with and without metabolic activation. None of the tested substances proved to be mutagenic in this system [32]. AHTN and ADBI were not mutagenic on the HPRT locus of V79 cells [33]. In an optimised Ames test with *S. typhimurium/E. coli*, AHTN and HHCB gave no indication of a mutagenic potential [34].

The polycyclic musk fragrances HHCB, AHTN, ADBI, AHDI, DPMI and ATII were tested for genotoxicity in the SOS chromo test with and without the addition of S 9. All of the substances tested exhibited no SOS inducing potency and were classified as non-genotoxic by the authors [38].

HHCB and AHTN were tested for genotoxicity in vitro in a chromosome aberration assay in CHO cells with and without the addition of S 9 as well as in a UDS assay in primary rat hepatocytes [34]. HHCB was negative in all the test systems used. For AHTN, it was only possible to show an increase in the number of aberrations in the group with the highest-tested dosage using CHO cells with S 9 activation. In the light of the test conditions, the authors considered this result as toxicologically nonsignificant. All other results for AHTN were negative [34]. Negative results have also been observed in a mouse in vivo micronucleus test for both AHTN and HHCB. It must be noted, however, that for both substances a moderate, not statistically significant reduction in the ratio of polychromatic erythrocytes to the total erythrocyte number of bone marrow cells was observed, providing evidence for the bioavailability of HHCB and AHTN to the bone marrow of the test animals [34]. No indication of genotoxic properties of AHTN was found in the sister chromatid exchange (SCE) test [11]. Negative results in the SCE test were also reported for HHCB, ADBI, AHDI, DPMI and ATII [72].

6.3 Cogenotoxicity

MX and MK induced xenobiotic metabolising enzymes (see above). The potential for synergistic effects between MX and MK and the pregenotoxic substances benzo[*a*]pyrene and 2-amino anthracene resulting in damage to DNA was studied. Benzo[*a*]pyrene and 2-amino anthracene were toxified in vitro by S 9 from rats pretreated with the two fragrances and then tested for positive results in the SOS chromo test (cogenotoxicity). While the S 9 fractions of untreated rats were hardly able to toxify benzo[*a*]pyren and 2-amino anthracene [31, 39], pretreatment with MK led to toxification of both benzo[*a*]pyrene as well as 2-amino anthracene. MX could only be shown to toxify 2-amino anthracene [31, 39].

MK also enhanced benzo[*a*]pyrene induced mutagenicity in human derived Hep G2 cells. While simultaneous treatment with MK (5–5000 ng mL^{-1}) and 0.2 µg benzo[*a*]pyrene mL^{-1} did not lead to synergistic effects, in cells pretreated with MK for 28 h and subsequently exposed to 0.2 µg benzo[*a*]pyrene mL^{-1} a pronounced comutagenic effect was observed. The LOAEL for MK was 0.05 µg mL^{-1}. Higher doses of 0.5–5 µg L^{-1} resulted in a significant increase of benzo[*a*]pyrene induced micronuclei frequencies, with induction rates of 50–88% [71].

As MK amplifies the genotoxic effects of benzo[*a*]pyrene in human derived cells, exposure of humans to MK might increase their susceptibility to the health hazards of benzo[*a*]pyrene and other polycyclic aromatic hydrocarbons.

The biochemical and molecular effects of the amine metabolites of MX are markedly different from those of other aromatic amines, but very similar to those of PB. Therefore, it appears that MX is a nongenotoxic chemical that may cause mouse liver tumours in a manner analogous to that of PB [66].

7 Long-term Toxicity and Carcinogenicity

7.1 Nitro Musks

Musk ambrette fed to male rats for 50 weeks and to female rats for 20 weeks at concentrations up to 4000 mg kg^{-1} in corn oil, equivalent to a dose of approximately 200 mg kg^{-1} day^{-1}, induced weight loss and an increasing weakness in the hind legs in both sexes. Additionally, in the females, a reduction in the number of erythrocytes, increased plasma levels of bilirubin and a decreased clotting time were observed. The increased bilirubin levels occurred at all dosages, whereas the other changes in the blood as well as indications of toxicity were only found at dosages of 1500 mg kg^{-1} and higher. The weakness in the hind legs resulted in a lack of movement and after 16–40 weeks in a complete loss of the ability to move the hind quarters. Microscopic examination revealed symptoms of muscle atrophy such as differences in size of the muscle fibres, but also degenerative alterations of the fibres, loss of cross striations and vacuolisation of the sarcoplasm [5].

MX was tested for its carcinogenic potential in an 80-week study in 50 6-week-old B6C3F1 mice of both sexes [14]. The mice received daily doses of 750 or 1500 mg kg^{-1} in the feed, equivalent to 90 and 170 mg kg^{-1} bw in the males and 100 and 190 mg kg^{-1} bw in the females, respectively. The total tumour incidence of the treated animals of both sexes was significantly increased at both doses over the rate in untreated controls. There was a particular increase in the incidence of liver tumours (hepatocellular adenomas and carcinomas). Hepatocellular carcinomas appeared to occur more numerously in male animals than in females. Tumour occurrence was not strictly organotropic since tumours were found in other organs, for instance Harderian's gland and the lung, however the incidence was not always significantly greater than in the controls. The authors evaluated their results according to tumour incidence, total number of tumours and individual types of tumours, as well as individually for benign and malignant tumours. By this means they determined that the incidence and total number of malignant tumours was significantly increased in the treated animals of both sexes over the untreated controls. The effect was, however, not dose dependent. The authors drew the conclusion that MX displays carcinogenic effects in both male and female B6C3F1 mice [14].

The work of Maekawa et al. [14] still remains the only published study on the long-term carcinogenic effects of musk fragrances in mammals. According to the results of this study, from a phenomenological standpoint, MX would be classified as a liver tumour initiator. This appears unlikely, however, since MX is neither mutagenic nor genotoxic. In addition, it behaves like the liver tumour promoter and classical CYP2B inducer phenobarbital if fed to mice (see above) [21]. Furthermore, the mouse characteristically displays a higher rate of spontaneous tumour formation than other species [40]. Therefore, it seems likely that the effects observed by Maekawa et al. [14] reflect a promotion of spontaneous lesions by MX and thus an epigenetic mechanism. The comparatively high dosage could, however, have led to formation of tumours by other means. Ames [41] postulated that treatment with large dosages of xenobiotics may result in the formation of tumours by chronic irritation, e.g. via inflammatory processes. It was stated by Ames [41] that when xenobiotics "are applied at dosages approaching those with acute toxic effects, carcinogenic effects can be expected for practically any chemical, regardless of whether synthetic or natural, genotoxic or not". MX therefore does not appear to be a (liver) tumour initiator. It must be noted, however, that after treatment of rats with ^{3}H-labelled MX, 2-amino-MX and other metabolites were identified in bile and urine and 4-amino-MX in faeces, bile and urine [42]. It therefore seems possible that MX metabolism follows nitroreduction to 2-nitroso-MX and further transformation to the corresponding hydroxylamine which, by way of a nitrenium ion in turn could form DNA adducts or be reduced to 2-amino-MX [43] or possibly via the 4-nitroso-MX to 4-amino-MX [44, 55]. There are no data on the formation of DNA adducts, however such effects are not being expected according to the results on mutagenesis and genotoxicity (see above). In contrast, haemoglobin adducts have been isolated from the haemoglobin of MX exposed as well as unexposed humans from which 4-amino-MX could be released by alkaline hydrolysis [44]. Consequently, there is considerable evidence for possible metabolic activation of MX to reactive intermediates, either by xenobiotic metabolism and/or by bacterial metabolism in the gastrointestinal tract [22] or on the skin. For risk assessment it must be noted that the biological half-life of MX in humans of about 80 days is much longer than in rats (a few days) [45].

7.2 Polycyclic Musks

The fact that AHTN does not seem to be effective as a phenobarbital- or 3-methyl cholanthrene (MC)-like inducer in the rat liver and is also apparently not a peroxisome proliferator [11] (see above) and the lack of mutagenic and genotoxic properties (see above) makes it unlikely that AHTN has hepatocarcinogenic properties. On the other hand, a single dose of 100 mg AHTN kg^{-1} bw resulted in an increase in the relative liver weight in the rat [11]. To detect potential liver tumour promoting properties of AHTN, rats were treated with diethyl nitrosamine as an initiating agent, followed by a 90-day treatment with AHTN. Rats treated with diethyl nitrosamine and phenobarbital served as positive control. While this control group developed a large number of glutathione-*S*-transferase P-positive liver foci, no such liver lesions were found in the AHTN treated animals. These results indicate that AHTN does not act as a liver tumour promoter at concentrations that reflect human conditions.

Even if there is no evidence of a genotoxic potential for both HHCB and AHTN, there still remains some question as to whether or not these substance are converted to potentially toxic metabolites and whether their metabolism in humans is similar to their metabolism in rodents. Metabolites of both HHCB and AHTN might be potentially (chronically) toxic; however, there have been no reports of studies on this question. Ongoing experiments should serve to clarify this point.

8
Reproduction Toxicity

8.1
Teratogenicity and Embryotoxicity

MX given to pregnant female rats is readily available to the embryo, apparently passing the placental barrier with little difficulty. It is also readily resorbed by the newborn offspring through milk from the mother [47, 49]. This may also be the case for other lipophilic musk fragrances.

When female Sprague-Dawley rats were orally exposed to HHCB (50–500 mg kg^{-1} bw day^{-1}), AHTN (5–50 mg kg^{-1} bw day^{-1}), MK (15–150 mg kg^{-1} bw day^{-1}) or MX (20–200 mg kg^{-1} bw day) from day 7 to day 17 of gestation no embryotoxic effects occurred in the 20-day-old embryos, removed by caesarean section, below maternotoxic concentrations. For AHTN and MX, the authors [48] did not find any adverse effects on embryo vitality, growth or morphology. Axial skeletal deformations were observed in the high dose HHCB group. In the high dose MK group there was an increase in post-implantation loss as well as reduced foetal birth weight [48].

From these experiments the authors derived NOAEL values for maternotoxicity of 50, 5, 15 and 20 mg kg^{-1} bw day^{-1} for HHCB, AHTN, MK and MX, respectively. The NOAEL values for developmental toxicity were 150, 50, 45 and 200 mg kg^{-1} bw day^{-1}, respectively. The oral rat maternal NOAEL values were well above the estimated maximal daily dermal human exposure of musk fragrances of 0.31 mg HHCB kg^{-1}, 0.76 mg AHTN kg^{-1} and 0.18 mg kg^{-1} for MK and MX [6, 10], and the authors [48] concluded that exposure to any of these fragrances during pregnancy would not pose a risk to human conceptus under conditions of normal usage [48].

8.2
Endocrine Effects

Endocrine activity of environmental contaminants has recently received considerable scientific interest [49]. Only few relevant studies have been carried out on the ubiquitous musk fragrances. Potential endocrine effects after the exposure of fish and amphibians to musk fragrances are presented by Dietrich et al. in this volume.

8.2.1
Nitro Musks

The estrogenic potential of various musk fragrances was examined in the E-Screen assay established by Soto et al. [51] and validated by Körner et al. [52] using a human breast cancer cell line (MCF-7) [50]. MX and MK proved to be partial agonists of 17β-estradiole at the oestrogen receptor. Their relative estrogenic potencies (17β-estradiole=1) were low and indicated as 3.3×10^{-5} (MX) and 7×10^{-5} (MK) [50].

8.2.2 *Nitro Musk Metabolites*

In the E-screen assay, 4-amino-MX was shown to be a weak agonist at the oestrogen receptor. Compared to the primary substance, MX, the same effect (increase of the cell proliferation rate) was achieved with only half the concentration. The relative estrogenic potency of 4-amino-MX compared to 17β-estradiole (=1) was 5×10^{-5}. The nitro musk metabolites, 2-amino-MX and 2-amino-MK were not active in the E-Screen [50].

8.2.3 *Polycyclic Musks*

HHCB was negative in the E-Screen assay [50, 53]. AHTN was shown to be a partial agonist in the E-Screen with a weak affinity to the oestrogen receptor. Its potency compared to 17β-estradiole (=1) was 1×10^{-4} [50].

AHTN and HHCB were tested for estrogenic activity in vitro in an ERα- and ERβ-dependent transcription assay using Human Embryonal Kidney 293 (HEK293) cells. Both AHTN and HHCB induced a weak dose dependent stimulation of transcription activity in ERα-transfected HEK293 cells. This weak estrogenic effect was not found in the ERβ-transfected cells, which, however, were also less sensitive to 17β-estradiol than ERα-transfected cells. In vivo, however, no uterotrophic activity was noted for AHTN and HHCB in the mouse uterine assay at dietary exposure of up to 6.5 mg kg^{-1} bw and 40 mg kg^{-1} bw, respectively [54].

A 90-day feeding study was undertaken to determine the subchronic oral toxicity of HHCB (see above). In supplementary histopathological examinations of the prostate, sperms, milk glands, vagina, uterus and testes no effects indicating a hormonal activity of HHCB were recorded [16].

9 Conclusions

Reviewing the data on the toxicology of synthetic musk fragrances showed that exposure to these compounds might present a potential hazard to human health. For consumer protection it is therefore imperative that, based on the available data, preliminary maximum limits or at least guidelines for the use of synthetic musk compounds in consumer products are established. Recently maximum authorized concentrations in cosmetic products were established by the European Commission for MX and MK (MX: 1.0% in fine fragrances, 0.4% in eau de toilets and 0.03% in other products, MK: 1.4% in fine fragrances, 0.56% in eau de toilets and 0.042% in other products) [78]. Also the Swiss authorities set comparable maximum limits for MX and MK in cosmetic products. Furthermore, the nitro musks MA, MM and MT have earlier been banned from the use in cosmetic products in the EU member states [79, 80]. In addition, the Scientific Committee on Cosmetic Products and Non-Food Products intended for Consumers (SCCNFP) of the EU Commission gave opinions on the safe use of the polycyclic musks AHTN and HHCB in cosmetic products [81, 82].

10 References

1. Opdyke DLJ (1975) Food Cosmet Toxicol 13(suppl):877
2. Opdyke DLJ (1975) Food Cosmet Toxicol 13(suppl):879
3. Opdyke DLJ (1975) Food Cosmet Toxicol 13(suppl):881
4. Opdyke DLJ (1975) Food Cosmet Toxicol 13(suppl):885
5. Davis DA, Taylor JM, Brouwer JB (1967) Toxicol Appl Pharmacol 10:405
6. Ford RA (1998) Dtsch Lebensm-Rundsch 94:192
7. Brunn H, Rimkus G (1997) Ernährungs-Umschau 44:4
8. Behechti A, Schramm KW, Attar A, Niederfellner J, Kettrup A (1998) Water Res 32:1704
9. Salvito D (2000) Water Res 34:2625
10. Ford RA (1998) Dtsch Lebensm-Rundsch 94:268
11. Steinberg P, Fischer T, Arand M, Park E, Elmadfa I, Rimkus G, Brunn H, Dienes HP (1999) Toxicol Lett 111:151
12. Spencer PS, Bischoff-Fenton MC, Moreno OM, Opdyke DL, Ford RA (1984) Toxicol Appl Pharmacol 75:571
13. Ford RA, Api AM, Newberne PM (1990) Food Chem Toxicol 28:55
14. Maekawa A, Matsushima Y, Onodera H, Shibutani M, Ogasawara H, Kodama Y, Kurokawa Y, Hayashi Y (1990) Food Chem Toxicol 28:581
15. Ford RA, Hawkins DR, Schwarzenbach R, Api AM (1999) Toxicol Lett 111:133
16. Api AM, Ford RA (1999) Toxicol Lett 111:143
17. Fukayama MY, Easterday OD, Serafino PA, Renskers KJ, North-Root H, Schrankel KR (1999) Toxicol Lett 111:175
18. Iwata N, Minegishi KI, Suzuki K, Ohno Y, Kawanishi T, Takahashi A (1992) Biochem Biophys Res Commun 184:149
19. Iwata N, Minegishi KI, Suzuki K, Ohno Y, Igarashi T, Satoh T, Takahashi A (1993) Biochem Pharmacol 45:1659
20. Steinberg P, Zschaler I, Thom E, Kuna M, Wust G, Schafer-Schwebel A, Muller R, Kramer PJ, Weisse G (2001) Arch Toxicol 75:562
21. Lehman-McKeeman LD, Caudill D, Young J, Dierckman T (1995) Biochem Biophys Res Commun 206:975
22. Lehman-McKeeman LD, Johnson DR, Caudill D, Stuard S (1997) Drug Metab Dispos 25:384
23. Lehman-McKeeman LD, Caudill D, Vassallo JD, Pearce RE, Madan A, Parkinson A (1999) Toxicol Lett 111:105
24. Suter-Eichenberger R, Boesterli UA, Conscience-Egli M, Lichtensteiger W, Schlumpf M (2000) Toxicol Lett 115:73; (1999) Toxicol Lett 111:117
25. Janzowski C, Vetter A, Burkart M, Thielen C, Eisenbrand G (1998) Lebensmittelchemie 52:71
26. Nair J, Ohshima H, Malaveille C, Friesen M, O'Neill IK, Hautefeuille A, Bartsch H (1986) Food Chem Toxicol 24:27
27. Rimkus G, Wolf M (1993) Dtsch Lebensm-Rundsch 89:103
28. Rimkus G, Wolf M (1993) Dtsch Lebensm-Rundsch 89:171
29. Emig M, Reinhardt A, Mersch-Sundermann V (1996) Toxicol Lett 85:151
30. Api AM, Ford RA, San RHC (1996) Food Chem Toxicol 34:633
31. Mersch-Sundermann V, Reinhardt A, Emig M (1996) Zentralbl Hyg Umweltmed 198:429
32. Mersch-Sundermann V, Kevekordes S, Jenter C (1998) Toxicol In Vitro 12:389
33. Janzowski C, Burkart M, Vetter A, Eisenbrand G (2000) In: Eisenbrand G, Dayan AD, Elias PS, Grunow W, Schlatter J (eds) Carcinogenic and anticarcinogenic factors in food. Wiley-VCH, Weinheim, p 455
34. Api AM, San RHC (1999) Mutat Res 446:67
35. Kevekordes S, Grahl K, Zaulig A, Dunkelberg H (1996) Environ Sci Pollut Res 3:189
36. Api AM, Ford RA, San RHC (1995) Food Chem Toxicol 33:1039
37. Kevekordes S, Zaulig A, Dunkelberg H (1997) Toxicol Lett 91:13

38. Mersch-Sundermann V, Kevekordes S, Jenter C (1998) Toxicol Lett 95:147
39. Mersch-Sundermann V, Emig M, Reinhardt A (1996) Mutat Res 356:237
40. Haseman JK, Hailey JR, Morris RW (1998) Toxicol Pathol 26:428
41. Ames BN, Gold LS (1990) Proc Natl Acad Sci USA 87:7772
42. Minegishi K, Nambaru S, Fukuoka M, Tanaka A, Nishimaki-Mogami T (1991) Arch Toxicol 65:273
43. Käfferlein HU, Göen T, Angerer J (1998) Crit Rev Toxicol 28:431
44. Riedel J, Birner G, van Dorp C, Neumann HG, Dekant W (1999) Xenobiotica 29:573
45. Kokot-Helbling K, Schmid P, Schlatter C (1995) Mitt Gebiete Lebensm Hyg 86:1
46. Rimkus G, Brunn H (1996) Ernährungs-Umschau 43:442
47. Suter-Eichenberger R, Altorfer H, Lichtensteiger W, Schlumpf M (1998) Chemosphere 36:2747
48. Christian MS, Parker RM, Hoberman AM, Diener RM, Api AM (1999) Toxicol Lett, 111:169
49. Kavlock RJ, Daston GP, De Rosa C, Fenner-Crisp P, Gray LE Jr, Kaattari S, Lucier G, Luster M, Mac MJ, Maczka C, Miller R, Moore J, Rolland R, Scott G, Sheehan DM, Sinks T, Tilson HA (1996) Environ Health Perspect 104 (Suppl 4):715
50. Bitsch N, Dudas C, Körner W, Failing K, Biselli S, Rimkus G, Brunn H (2002) Arch Environ Contam Toxicol 43:257
51. Soto AM, Sonnenschein C, Chung KL, Fernandez MF, Olea N, Olea-Serrano MF (1995) Environ Health Perspect 103:113
52. Körner W, Hanf V, Schuller W, Bartsch H, Zwirner M, Hagenmaier H (1998) Chemosphere 37:2395
53. Schuller W (1998) PhD thesis, University of Tübingen, Germany, Faculty of Biology
54. Seinen W, Lemmen JG, Pieters RHH, Verbruggen EMJ, van der Burg B (1999) Toxicol Lett 111:161
55. Riedel J, Dekant W (1999) Toxicol Appl Pharmacol 157:145
56. Hood HL, Wickett RR, Bronaugh RL (1996) Food Chem Toxicol 34:483
57. Hawkins DR, Ford RA (1999) Toxicol Lett 111:95
58. Kokot-Helbling KS (1995) PhD Thesis No 11152, ETH Zürich
59. Hawkins DR et al. (1984) Unpublished results cited in SCCNFP Opinion concerning musk xylene, 8 December 1999
60. Hawkins DR et al. (1984) Unpublished results cited in SCCNFP Opinion concerning musk ketone, 8 December 1999
61. Liebl B, Ehrenstorfer S (1993) Chemosphere 27:2253
62. Rimkus G, Wolf M (1994) Chemosphere 28:421
63. Waizenegger W, Kypke K, Fetteroll B, Hahn J, Sohnius E, Feucht M (1998) Dtsch Lebensm-Rundsch 94:120
64. Sönnichsen C, Mayer R, Ommer S, Koletsko B (1999) Unpublished data in SCCNFP Opinions concerning musk xylene and musk ketone, 8 December 1999
65. Lehman-McKeeman LD, Johnson DR, Caudill D (1997) Toxicol Appl Pharmacol 142:169
66. Lehman-McKeeman LD, Stuard SB, Caudill D, Johnson DR (1997) Mol Carcinog 20:308
67. Stuard SB, Johnson DR, Caudill D, Lehman-McKeeman LD (1996) 35th Annual Meeting of the Society of Toxicology (SOT), Abstract 1406
68. Stuard SB, Caudill D, Lehman-McKeeman LD (1997) Fundam Appl Toxicol 40:264
69. Zheng GQ, Kenney PM, Lam LKT (1992) J Pharm Sci 81:950
70. Degawa M, Nakayama M, Yoshinari K, Hashimoto Y (1995) Cancer Lett 96:95
71. Mersch-Sundermann V, Schneider H, Freywald C, Jenter C, Parzefall W, Knasmüller S (2001) Mutat Res 495:89
72. Kevekordes S, Mersch-Sundermann V, Diez M, Bolten C, Dunkelberg H (1998) Anticancer Res 18:449
73. Müller S, Schmid P, Schlatter C (1996) Chemosphere 33:17
74. Ashcroft J-A, Hotchkiss SAM (1996) Fund Appl Toxicol 30 (No 1, Part 2 Suppl):169
75. Eschke HD, Dibowski HJ, Traud J (1995) Dtsch Lebensm-Rundsch 91:375
76. Rimkus GG, Wolf M (1996) Chemosphere 33:2033

77. Hawkins DR, Ford RA (1996) ISSX International Meeting 10:154
78. Council Directive 2002/34/EC (2002) Off J Eur Comm L 102:19
79. Council Directive 95/34/EC (1995) Off J Eur Comm L 167:19
80. Council Directive 98/62/EC (1998) Off J Europ Comm L 253:20
81. Opinion of the Scientific Committee on cosmetic products and non-food products intended for consumers (SCCNFP) concerning AHTN, SCCNFP/0609/02
82. Opinion of the Scientific Committee on cosmetic products and non-food products intended for consumers (SCCNFP) concerning HHCB, SCCNFP/0610/02

The Handbook of Environmental Chemistry Vol. 3, Part X (2004): 281–310
DOI 10.1007/b14124

Risk Evaluation of Dietary and Dermal Exposure to Musk Fragrances

Premysl Slanina

Toxicology Division, Department of Research and Development,
National Food Administration, 756 21 Uppsala, Sweden
E-mail: prsl@slv.se

Abstract Synthetic musk compounds comprising nitro musks, polycyclic musks and macrocyclic musks are used in fragrances which are added to a variety of cosmetic as well non-cosmetic household products. The main source of human exposure to musks is fresh water fish, as a result of water contamination from the sewage plants, and a dermal application of cosmetics. The musks accumulate in human adipose tissue and significant levels are present in breast milk. The most frequently used polycyclic compounds AHTN and HHCB as well as nitro musks musk xylene (MX) and musk ketone (MK) show a low acute and subchronic toxicity and no indication of teratogenicity or reproduction toxicity. The substances do not show a mutagenic potential but MX was found to be tumorigenic in animal experiments probably due to non-genotoxic mechanism based on a significant hepatic cytochrome P450 enzyme induction. The hepatic microsomal enzyme induction activity under the development seen in the animal experiments appears to be the most sensitive toxicological endpoint. The calculation of a tolerable intake based on the data available indicates a good safety margin for the consumers of musk contaminated fish. In comparison, the exposure of infants to MX via breast milk and the dermal exposure to MX and AHTN from the cosmetics may reach the range of estimated safe intake. More toxicological studies as well as data on the occurrence and exposure are needed to enable a reliable risk assessment and management of the synthetic musk compounds after dietary and dermal exposure.

Keywords Nitro musks · Polycyclic musks · Toxicology · Risk evaluation

List of Abbreviations

ad lib Ad libitum
ADBI 4-Acetyl-1,1-dimethyl-6-*tert*-butyl-dihydroindene
AHDI 6-Acetyl-1,1,2,3,3,5-hexamethyl-dihydroindene

AHTN	7-Acetyl-1,1,3,4,4,6-hexamethyl-tetrahydronaphtalene
ATII	5-Acetyl-1,1,2,6-tetramethyl-3-isopropyl-dihydroindene
bw	Body weight
CHO	Chinese Hamster Ovaria
EROD	7-Ethoxyresorufin-*O*-deethylase
HHCB	1,3,4,6,7,8-Hexahydro-4,6,6,7,8,8-hexamethylcyclopenta[*g*]-2-benzopyrane
ip	Intraperitoneally
LD_{50}	Lethal Dose (50%)
LOEL	Lowest Effect Level
lw	Lipid weight
MA	Musk ambrette
MK	Musk ketone (4-acetyl-1-*tert*-butyl-3,5-dimethyl-2,6-dinitro-benzene)
MM	Musk moskene
MRL	Maximum Residue Limit
MROD	7-Methoxyresorufin-*O*-deethylase
MT	Musk tibetene
MX	Musk xylene (1-*tert*-butyl-3,5-dimethyl-2,4,6-trinitrobenzene)
NOAEL	No Adverse Effect Level
NOEL	No Effect Level
po	Per os (oral)
PTDI	Provisional Tolerable Daily Intake
RIFM	Research Institute for Fragrance Materials
SCE	Sister Chromatide Exchange
SPF	Specific Pathogen Free
TDI	Tolerable Daily Intake
UDS	Unscheduled DNA Synthesis
ww	Wet weight

1 Introduction

Synthetic musk compounds have a widespread use as a substitute for natural musks in fragrances and can be found in a number of consumer products such as decorative cosmetics, perfumes, shampoos and toilet soaps. They are also added to various non-cosmetic household products, e.g. household cleaners, laundry detergents and fabric softeners [1].

The main classes of synthetic substances mimicking the natural musk scent are nitro musks such as musk ketone (MK) and musk xylene (MX) and polycyclic musks HHCB (e.g. Galaxolide) and AHTN (e.g. Tonalide). These compounds are estimated to represent approximately 95% of the currently used synthetic musks [2]. More recently a new group of macrocyclic musks including, e.g. cyclopentadecanone (e.g. Exaltone) and ethylene brassylate with a structural resemblance to the natural musk compounds, has been gradually introduced on the market.

The major source of environmental contamination with musk compounds is the water from the sewage plants. A high lipophilicity and chemical stability of nitro and polycyclic musks render them persistent in various compartments of the aquatic ecosystem. Because of a relatively high volume of use and their physical-chemical properties they have been found in sea- and fresh water, sediment, sludge as well as in fish and other marine organisms [3]. More recent studies have also confirmed significant concentrations in the human breast milk and fat [4].

Despite a long and extensive use of synthetic musks there have until recently been relatively few studies on their fate and effects in the environment or their potential for harmful effects on human health. With exception of MX (for review see, e.g. [4]) the studies available were mostly concerned with the health impact from the use of musk compounds in the cosmetics [5, 6].

In the last few years a number of new analytical studies on the concentration of nitro and polycyclic musk compounds in foodstuffs and in human tissues have been presented. Although an overall information on the global occurrence is still limited the data available now permit a general assessment of the exposure as well as body burden of these compounds in various populations. This makes it compelling to provide a comprehensive up-to-date toxicological evaluation of possible adverse effects of synthetic musk compounds on human health.

This chapter attempts to compile and critically evaluate the data available on the toxicity of musks in the diet and, where possible, to suggest a tolerable daily intake (TDI). Such an information is necessary as a basis for the establishment of maximum residue limits (MRLs) for individual musk compounds in relevant foodstuffs. Since a significant part of the exposure may be expected from the use in cosmetics a dermal application of the musks is also taken into account in the evaluation.

2 Risk Analysis of Synthetic Musks

2.1 Risk Assessment

2.1.1 *Hazard Identification*

2.1.1.1 Nitro Musks

Nitro musks are highly substituted benzenes with at least two nitro groups in the molecule. The most often used compounds, which are also frequently detected in biotic samples as well as in human tissues, are musk xylene (MX) (1-*tert*-butyl-3,5-dimethyl-2,4,6-trinitrobenzene) and musk ketone (MK) (4-*tert*-butyl-3,5-dinitro-2,6-dimethylacetophenone). Other substances of commercial interest from this group are musk tibetene (MT), musk moskene (MM) and musk ambrette (MA). The MA has been withdrawn from the market because of severe toxic effects including neurotoxicity, testicular atrophy and phototoxicity observed after the application to the laboratory animals (cf. [5]). Only limited data is available on the occurrence and biological effects of MT and MM and therefore this review will primarily deal with MX and MK.

2.1.1.1.1
Occurrence in Food

The only foodstuff known to contain appreciable amounts of MX, MK and other nitro musks is fish. In earlier studies musks were also detected in shrimps and mussels but the levels were relatively low and the marine organisms probably do not represent a significant exposure source of the compounds [3]. The musks were not found in fat or other foods of animal origin (cf. [4]). For newborn children the most important source of exposure to synthetic musk compounds is breast milk.

2.1.1.1.2
Use in Cosmetics and Other Non-Food Products

The highest concentrations of MK and MX as well as other nitro musks (MM, MT) are used in alcoholic fragrances (7% corresponding to 0.6% in the final product) [5]. Taking into account the application quantity, frequency of use and dermal retention the most important sources of exposure for these compounds in the cosmetic products are eau de toilettes, face creams and body lotions [5].

2.1.1.2
Amino Metabolites of MK and MX

In addition to parent compounds the German research groups (see, e.g. [1]) have identified a microbial reduction of *o* and *p* nitro groups to corresponding 2-amino and 4-amino-MX as well as 2-amino-MK as a significant metabolic pathway of nitro musks in the environment. The metabolism takes place under the anaerobic conditions in the sewage plants and significant levels of the metabolites could be found in the sludge and water samples.

The main amino-metabolite invariably detected in the fish samples was 4-amino-MX. The concentrations exceeded that of the parent compound. In comparison lower levels of 2-amino-MX and 2-amino-MK as compared to MX and MK were present in the fish [1].

2.1.1.3
Polycyclic Musks

The most important materials from this group are a tetraline derivative AHTN (6-acetyl-1,1,2,4,4-hexamethyltetraline) (e.g. Tonalide, Fixolide; CAS No 1506-02-01) and an isochroman-type HHCB (1,3,4,6,7,8-hexahydro-4,6,6,7,8,8-hexamethyl-cyclo-penta-γ-2-benzo-pyran) (e.g. Galaxolide, Abbalide, Pearlide; CAS No 122-05-5). The latter compound consists of two diastereoisomers (the main isomers) and two minor isomers. In addition there are other compounds such as

ADBI (e.g. Celestolide), AHDI (e.g. Phantolide) and ATII (e.g. Traseolide) of a commercial interest with a low use volume.

2.1.1.3.1
Occurrence in Food

The main source of dietary exposure to AHTN and HHCB is freshwater fish. Concentrations of polycyclic musks in the fish are often higher than those of nitro musks probably due to higher environmental degradation of the latter compounds [3].

2.1.1.3.2
Use in Cosmetics and Other Non-Food Products

Hydroalcoholic cosmetics such as eau de toilettes contain highest level of fragrances with a high percentage of the polycyclic musks. Face creams and body lotions are also an important source of the exposure due to their retention in the skin. The estimated maximum concentration of AHTN and HHCB in the final products has been estimated to 0.96% and 2.4% respectively [6].

2.1.1.4
Macrocyclic Musks

No data is available in the public domain on the chemical properties and environmental distribution of macrocyclic musks. Examples of the compounds used are ethylenedodecandioate ($C_{14}H_{24}O_4$) (e.g. Musk MC-4) and 15-pentadecanolide ($C_{15}H_{28}O_2$) (e.g. Exaltolide), 3-methylcyclopentadecanone (e.g. Muscone) and ethylene brassylate.

2.1.1.5
Bioconcentration and Bioaccumulation of Musks in Fish

According to investigations by Rimkus et al. [3] and others (see, e.g. [4]) the lipophilicity of polycyclic musks such as AHTN and HHCB expressed by the log K_{ow} was comparable to that of organochlorine pesticides, e.g. *p,p'*-DDT, dieldrin, HCB as well as some PCBs. In comparison MX and MK showed a somewhat lower log K_{ow} [3]. However in difference from the other persistent contaminants the musk compounds only show a bioconcentration (i.e. a direct uptake from the polluted water) in the fish. The bioaccumulation of any class of musk compounds in the food chain has not been observed [3, 4].

2.1.2
Hazard Characterization

2.1.2.1
Toxicokinetics

2.1.2.1.1
Nitro Musks

Animal Experiments

Experimental studies on absorption, distribution, metabolism and excretion were conducted in laboratory animals after both the oral and dermal application.

Oral application
Seven days after a single po application of ^{3}H-MX (70 mg kg^{-1} bw) to male Wistar rats the highest concentration of radioactivity was detected in the adipose tissue and the liver. The total residue of radioactivity in tissues was less than 2% of the administered dose. The major excretion route for MX was via bile into faeces. Urinary and faecal excretion accounted for 10 and 75% of the dose, respectively. The elimination half-time in rat was less than a few days [7].

The reduction of nitro groups to corresponding amines appears to be the principal metabolic pathway of nitro musks. The reduction of 2-nitro group (in para position to *tert*-butyl group) to corresponding 2-NH_2-MX was identified as a key step in the metabolism [7].

In a study of an accumulation of MX under the development the male and female rats (Long Evans) were fed with MX containing feed (0.001, 0.01, 0.033 and 0.1 g kg^{-1} diet) for ten weeks before mating. Treatment continued during pregnancy and lactation. The highest levels of the substance in adult animals were found in the fat. Female fat levels were about four to seven times higher than in males. Measurable concentrations of MX were detected in the liver of one-day-old animals (highest dose group). In the young animals (14 days old) there was a dose dependent accumulation of MX in the adipose tissue corresponding to approximately 50–75% of the mothers' levels. No sex difference was observed. Maternal milk levels were comparable to adult female adipose tissue. The results indicate a placental transfer to foetus and an exposure to musks via mother's milk in neonates [8].

Topical Application
After a dermal application of ^{14}C-MX or ^{14}C-MK to rat 20% of the applied dose of MX and 28% of MK was absorbed after 6 h. Virtually all the absorbed material was excreted in 96 h, primarily via bile into faeces [5].

Whole-body autoradiography distribution studies after a dermal application of ^{14}C-MX or ^{14}C-MK following the 14 repeated daily doses to male CD rats showed an incomplete absorption with large amount of radioactivity remaining at the site of application. There was no indication of a time-dependent tissue accumulation of the labelled compounds [5].

Human Studies

Oral Application

Toxicokinetic studies were conducted in human volunteers. Six persons (three males and three females) were given a single dose of ^{15}N-MX in gelatine capsule (0.3 mg kg^{-1} bw) and MX and its metabolites were measured in the blood and urine. Maximal concentration of ^{15}N-MX in the plasma was seen 6 h after the application and ranged between 36–262 ng MX mL^{-1} corresponding to 0.6–3.8% of the oral dose. The toxicokinetics of the ^{15}N-MX was described by a two-compartment kinetic model with an initial rapid decrease (distribution to the tissues) followed by a slower terminal elimination phase with an average half-life of 70 days [9]. This is in an agreement with an estimated a half-time body elimination of 100 days based on plasma kinetics in the earlier study with ^{15}N-MX in man [10].

^{15}N-4-amino-MX was identified in the urine after the application of the ^{15}N-MX to humans. Maximum concentration of the metabolite was observed 18–24 h after the exposure representing 0.1–0.5% of the applied dose with an elimination mean half-life calculated to 10.7±1.1 h [9]. Analysis of the metabolite 4-amino-MX in the blood of ten volunteers not knowingly exposed to the musk revealed a binding of the metabolites to haemoglobin. A possible use of haemoglobin adducts as a biomarker for MX exposure in man was suggested [11].

Topical Application

A single topical application of ^{14}C-MX (1 mg kg^{-1} bw, duration 120 h) studied in two human volunteers resulted in a poor absorption. At the site of application, 90% of the applied dose was present after 6 h. After 120 h a mean of 0.26% and <0.1% of the dose respectively had been excreted in the urine and faeces [5]. Similar results were obtained after the dermal application of ^{14}C-MK to human volunteers [5].

2.1.2.1.2
Polycyclic Musks

There appears to be even fewer studies on the toxicokinetics polycyclic musks as with the nitro musk compounds.

^{14}C-AHTN and ^{14}C-HHCB respectively were applied on the skin in rats (4.5 mg kg^{-1} bw) and the radioactivity in the tissues was measured during the five days after the treatment. The total amount of AHTN absorbed was estimated to 19% and for HHCB to 14%. The radioactivity could be detected in the plasma after 30 min with a continuing diffusion from the skin depot. The highest concentration of the label was seen in the intestine content and faeces indicating a biliary excretion. High concentration of the radioactivity was also measured in the adipose tissue [12].

In the same study ^{14}C-AHTN and ^{14}C-HHCB were applied in normal perfume concentrations on the skin of three human volunteers. The concentration in the blood were non detectable during the whole period (five days) of the study. Most of the absorbed fraction was excreted in urine. The total dose absorbed was estimated to 1% for AHTN and 0.1% for HHCB. More than 20% of the dose given evaporated and a substantial part was removed through the desquamation of the skin [12].

2.1.2.1.3
Nitro and Polycyclic Musks in Human Adipose Tissue

Rimkus et al. [13] were the first to detect nitro musks in samples of human adipose tissue. The samples of fat were collected from 32 individuals in Northern Germany. MX and MK were present in all the samples tested. The concentrations varied between 20–220 μg MX kg^{-1} lw and 10–220 μg MK kg^{-1} lw. In some samples MA, MM but not MT could also be detected in low levels. In another study on human adipose tissue samples (n=14) collected at the same area the polycyclic musk were measured. All the samples contained AHTN (8–58 μg kg^{-1} lw) and HHCB (16–189 μg kg^{-1} lw) and in some samples low levels of ADBI, AHDI and ATII were also observed [14]. Similar findings were also reported from studies in Switzerland [15]. Both the nitro and polycyclic musk compounds were also invariably found in the breast milk (see "Exposure Assessment" in this chapter, p. 299).

2.1.2.1.4
Enzyme Induction

Nitro Musks

Several studies have demonstrated that musk compounds may act as an inducer of liver microsomal enzymes in laboratory rats and mice [4, 16].

In series of experiments Iwata et al. [17, 18] have shown that an acute treatment (ip) with high doses of MX (50–200 mg kg^{-1} bw) induces CYP450 mixed function oxidase activities, particularly CYP1A2 microsomal enzyme in the liver of Wistar rats. The induction of NAPDH-cytochrome reductase and several phase II enzymes (DT-diaphorase, glutathione-S-transferase and UDP-glucoronyl-transferase) was also seen in these studies [18]. In more recent investigations [19, 20] MX given orally to B6C3F1 mice in doses up to 200 mg kg^{-1} bw caused a hepatocellular hypertrophy and markedly increased CYP2B protein but at the same time inhibiting the activity of the enzyme. The subsequent experiments with an inhibition of intestinal microflora by antibiotics given concomitantly to the animals, indicated that the CYP2B activity inhibition is related to the formation of MX-amino metabolites [21].

It has been speculated [4, 20–22] that in laboratory animals MX induces microsomal P-450 enzyme system with accompanying cell proliferation thus inducing a phenobarbital-like liver tumour formation. The enzyme induction was seen with relatively high doses and a possible significance for the man is not clear [16].

In more recent investigations [23] male and female rats (Long Evans) were given MX in feed for up to 18 weeks. In another experiment the females received MX in the diet before mating, during pregnancy and up to 14 days post partum in the concentrations corresponding to 0.07–0.08, 0.7–0.8, 2–3, 7–8 and 70–80 mg MX kg^{-1} bw. The results confirmed the capacity of MX to induce CPY1A1 enzyme activity (measured by EROD activity) and CPY1A2 (measured by MROD activity) in both the adult and developing animals. In addition an increase of CYP2B and CYP3A proteins were also detected after the treatment. The authors concluded that the P-450 microsomal enzyme was affected at fat tissue levels exceeding human fat levels by a factor of about 85 (males) and 400 (females).

Both prenatal and postnatal exposure seemed to be effective in inducing CYPA1 and CYPA2 activities in developing rats (postnatal days 1 and 14). There were however differences in sensitivity to the enzyme induction between the adult animals and the pups studied. The NOAEL 0.7–0.8 mg MX kg^{-1} bw was derived from the developmental study [23].

The different patterns of microsomal enzyme induction were seen with MK. When given orally (gavage) to mice MK induced primarily CYP2B enzyme activity and to a lower degree also CYP1A and CYP3A enzymes [24]. In another study in rats (Long Evans) an exposure to MK or MX resulted in the induction of CYP2B and CYP1A. MK was more potent in inducing CYP1A and in difference to MX did not inhibit the enzyme activity of CYPB2. The results indicate a different mechanism of enzyme inducing activity between MX and MK as well as differences in effects of MK in mice and rats [16].

The induction microsomal enzyme metabolic activity with MX and MK was also shown to result in an increased genotoxic activity of benzo[*a*]pyrene, 2-amino-anthracene and aflatoxin B_1. Thus the synergistic effect of nitro musks and pregenotoxicants seems possible [25].

Polycyclic Musks

Experiments in rats showed no induction by AHTN of CYP2B or CYP1A [26].

2.1.2.1.5
Summary of the Toxicokinetics

The results of studies available on the bioavailability and fate of nitro and polycyclic musk compounds indicate a generally low absorption from gastrointestinal tract in experimental animals and man. The compounds can be detected in the blood after an oral application with the highest accumulation seen in the adipose tissues followed by liver and tissues with a high content of fat in laboratory animals. In man significant levels were measured in the adipose tissue. The musks were shown to cross the placental barrier and accumulate in the breast milk. The dermal absorption of both the nitro- and polycyclic compounds is limited but appreciable amounts of the musks are found in the tissues after a topical application. In experimental models using laboratory animals the absorption via skin was estimated to 20% and 28% for MX and MK respectively and 14–19% for HHCB and AHTN. In humans the dermal uptake of MX, MK, AHTN and HHCB did not seem to exceed 1% of the applied dose.

Nitroreduction resulting in the formation of amino metabolites was suggested to be a principal metabolic pathway. In humans 4-amino-MX was identified as a main metabolite in the urine. Elimination half-life for MX was estimated to several days in laboratory animals while in man it was calculated to 70–100 days.

Results of the limited studies in man and laboratory animals indicate both biliary and urinary excretion but the relative importance of these pathways in different animal species and humans is at present unclear.

MX and MK and its amino metabolites were shown both to induce microsomal cytochrome P-450 mixed oxidases enzymes as well as to inhibit the enzyme activities in mice and rats. The effects of individual nitro musks on various subpopulations of cytochrome P-450 enzyme system appear to be different and very complex. At the present the exact mechanism of the enzyme inducing activity is still unclear. The induction of mixed oxidase activities was however also observed in young laboratory animals after a prenatal and postnatal treatment with MX. The NOAEL 0.7–0.8 mg kg^{-1} bw was determined in the experimental studies of the enzyme induction in developing organism.

2.1.2.2
Acute Toxicity

2.1.2.2.1
Nitro Musks

Acute toxicity of nitro musks is low. Oral LD_{50} for MX in mice and rats was found to be 4 g kg^{-1} bw [26] and 10 g kg^{-1} bw respectively [28]. A similar range of oral LD_{50} values was reported for MK and MT [28, 29]. MA showed a somewhat lower oral LD_{50} (approximately 0.3 g kg^{-1} bw) in rats [5].

An estimated dermal LD_{50} in rabbits ranged from >5 g kg^{-1} bw for MX to 10 g kg^{-1} for MK and >15 g kg^{-1} bw observed with MT and MM [5, 28].

2.1.2.2.2
Polycyclic Musks

According to references (see [6, 27, 28]) to the older studies conducted in the mid 1970s by Research Institute for Fragrance Materials (RIFM) acute toxicity of the most important polycyclic musks in laboratory rodents were also low. The reported LD_{50} varied between 0.57 g kg^{-1} bw to 5 g kg^{-1} bw for AHTN (rat-oral, rabbit-dermal application) as well as ADBI (rat-oral application, rabbit-dermal application) [30].

For HHCB LD_{50} was given in the range of 3.25 g kg^{-1} bw to 5 g kg^{-1} bw after oral or dermal application to rats and rabbits (see, e.g. [6, 28, 30]). Other musk compounds such as ADBI showed oral and dermal LD_{50} of a similar magnitude [30].

2.1.2.3
Repeated Dose Toxicity

2.1.2.3.1
Nitro Musks

There is very little data available on subacute and subchronic toxicity of nitro musks after oral application to the laboratory animals. In dose finding experiments groups of 10 male and female mice ($B6C3F_1$) were fed MX in the diet for 17 weeks [26]. The musk concentrations mixed in the diet were 0.0375, 0.075, 0.15, 0.3 and 0.6%. In the highest two dosings all of the animals with exception of two males in the 0.3% group died during the study. In the lower dose groups (≤0.15%) corresponding to approx. 0.2 g MX kg^{-1} bw the only effects observed under the experimental conditions used were slight histopathological changes in the liver. No toxic effects other that a decrease in body weight gains and a lower organ weight were seen in a 90-day dermal study in rats after a dosing with MX, MK or MM and MT (7.5, 75, 24 mg kg^{-1} bw and 240 mg kg^{-1} bw, MX and MK only). NOAELs derived from the dermal study were 24 mg kg^{-1} bw (females) (75 mg kg^{-1} bw, males) for MX, 75 mg kg^{-1} bw for MK as well as MT and 24 mg kg^{-1} bw (75 mg kg^{-1} bw, females) for MM [31].

2.1.2.3.2
Polycyclic Musks

Oral Application

Subchronic (13 weeks) toxicity studies on AHTN and HHCB were conducted in rats. AHTN was administered in the diet at daily doses of 1.5, 5, 15 or 50 mg kg^{-1} bw to groups of rats consisting of 15 male and 15 female animals. After the last treatment three males and three females from the highest dose groups were maintained on musk free diet four weeks before the sacrifice. The results were summarized in a review article [6] with no details given concerning, e.g. statistical significance of the findings. A compliance of the experimental design with GLP could not be determined from the information available.

There was a decreased body weight gain in the highest dose group. The liver weights of females at 15 mg kg^{-1} and in both sexes at the highest dose were also increased but there was no histopathological findings indicating that this may represent an increased demand for liver function. A dose-independent green discoloration was seen in tissues such as liver, lymph nodes and lacrimal glands in some experimental animals. The discoloration and the slight changes in some haematological parameters and blood chemistry observed in all the dose groups were not considered toxicologically significant. NOAEL of 15 mg AHTN kg^{-1} bw was suggested based on the findings from this study [6].

An identical experimental protocol was applied in the testing of HHCB but the dose regimen used was 5, 15, 50 or 150 mg kg^{-1} bw. Apart from some changes in blood chemistry and a slightly increased mean absolute liver weight in males in all dose groups (dose-independent) there were no toxicologically effects recorded in the study. The proposed NOAEL from this study was 150 mg HHCB kg^{-1} bw [6].

Dermal Application

Both AHTN and HHCB were also subjects to several subchronic (13 weeks and 26 weeks) studies in rats using dermal application of the compounds [6, 32]. The estimated highest doses were 100 mg kg^{-1} bw and 36 mg kg^{-1} bw in the 13 weeks and 26 weeks study respectively. Only limited histopathology was performed. A decrease in body weight gains, hypertrophy of the liver as well slight changes in haematology were among the changes observed in the studies. The design of the studies included also neuropathology and behavioural tests. No evidence of neurotoxicological potential was observed. The limited nature of these studies do not allow for the conclusion of NOAEL.

2.1.2.4
Reproduction Toxicity and Teratology

2.1.2.4.1
Nitro Musks

Only two embryotoxicity, teratogenicity and foetotoxicity studies with MX and MK given orally to rats were cited in the review articles or toxicology assessments [5, 33]. The results were however presented in a brief format only with no primary data shown which makes it difficult to evaluate in detail the significance of the effects observed.

Based on the results of a dose-finding study in one teratotology investigation MX was administered in corn oil by gavage daily to groups (n=25) of female Sprague Dawley rats on days 7–17 of gestation. The doses used were 0, 20, 60 and 200 mg kg^{-1} bw day^{-1}. A significant reduction of a food consumption accom panied by a decrease in weight gains in the pregnant animals was seen in the highest two dose groups. No changes in litter parameters or foetal variations and malformations related to the treatment were observed in the foetuses. The proposed NOAEL for maternal toxicity is 20 mg MX kg^{-1} bw and 60 mg MX kg^{-1} bw for developmental toxicity [33].

A corresponding study on teratogenicity and foetal toxicity was conducted in rats with MK. The overall experimental protocol was same as with MX but the doses used (based on pilot experiments) were 0, 15, 45 and 150 mg kg^{-1} bw. The dams in the two highest dose groups showed a reduced or significantly reduced body weight gains and clinical signs of toxicity (e.g. salivation, dehydration, tremors) in the highest group. The NOAEL was proposed to be 15 mg MK kg^{-1} bw for maternal toxicity and 45 mg MK kg^{-1} bw for developmental toxicity [33].

No multiple generation reproduction and developmental studies on nitro musks were found in the literature.

2.1.2.4.2
Polycyclic Musks

The studies on embryo-, teratogenicity and foetal toxicity of polycyclic musk available were part of the previously cited investigation which also included nitro musks [6, 33].

No reproduction (perinatal-, postnatal) investigations of the polycyclic musks could be at present time found in the literature.

Developmental studies with AHTN and HHCB were conducted in rats. The pregnant animals were dosed daily by gavage on days 7 through 17 of gestation. AHTN was dosed at 5, 15 or 50 mg kg^{-1} day^{-1}. The treatment caused clear maternal toxicity at the highest dose and a body weight reduction in the middle dose. There were no adverse effects on embryo or on foetal development observable at any dose [6, 33]. HHCB dosing was 50, 150 or 500 mg kg^{-1} day^{-1}. The highest dose group showed maternal toxicity. There were also a lower foetal body weight and an increase in skeletal malformations (vertebra/rib) at the highest dose. The effects were assumed to be related to the maternal toxicity. No direct adverse developmental effects were reported at any dose [33]. The proposed NOAELs based on maternal toxicity were 50 mg and 5 mg kg^{-1} bw day^{-1} for HHCB and AHTN respectively.

2.1.2.5
Genotoxicity

2.1.2.5.1
Nitro Musks

MX, MK, MA, MM and MT were screened in a number of studies for genotoxic potential using a battery of in vitro and in vivo tests. The assays used followed the International Guidelines on mutagenicity testing of chemicals [34]. A full account of all the studies published is beyond the scope of this monograph. Examples of representative investigations are given below.

MX, MK, MM and MT were all negative in assays for gene mutations both using the prokaryotic (Ames test) and mammalian cells (e.g. mouse lymphoma test; cultured human lymphocytes). The same results were seen with or without a metabolic activation (see, e.g. [35–38]). The lack of genotoxic potential was also shown in in vitro and in vivo assays for chromosomal aberrations (bone marrow micronucleus test) as well as when testing repair of primary DNA-damage (UDS; SOS chromosome test) [37, 38].

MA was positive in the Ames test in the presence of metabolic activation [35]. The results of assays for chromosomal mutations, e.g. SCE (sister chromatide exchange) and micronucleus tests using human lymphocytes were negative [39, 40] indicating the absence of damaging potential to hereditary material of the substance.

2.1.2.5.2
Polycyclic Musks

Both AHTN and HHCB have been evaluated for potential genotoxicity in assays for gene mutations as well as for chromosomal damage. Studies of chromosomal aberrations have also been published with ADBI, AHDI and ATII.

No indications of gene mutagenicity were found for AHTN or HHCB in the battery of bacterial tests (*Salmonella typhimurium*, *Escherichia coli*) with or without metabolic activation [6]. Similarly, neither compound showed any significant increases in structural or numerical chromosome aberrations in CHO with or without activation [41, 42].

ADBI, AHDMI and ATII were studied in the in vitro micronucleus test using cultured human lymphocytes and hepatocytes. None of the substances showed any genotoxic potential in the assays [43]. Results of more recent studies with polycyclic musks have also confirmed a lack of chromosome breaks or loss (SCE) in cultured human lymphocytes for AHTN, HHCB, ADBI, AHDI and ATII [44].

2.1.2.5.3
Macrocyclic Musks

No studies have until now been available in the literature on the genotoxicity of macrocyclic musk compounds.

However, recently the potential of three compounds namely cyclopentadecanone, ethylenedodecandioate and ethylene brassylate to induce gene mutations or chromosomal aberrations was assessed using the in vitro Ames test and in vivo mouse micronucleus bone marrow assay. The results of the studies showed the absence of genotoxic activity in the Ames test with or without metabolic activation. No induction of chromosomal aberrations in micronucleus assay was seen after ip injection of up to 2 g kg^{-1} bw of the macrocyclic musks studied into mice. Overall the data indicate an absence of a significant mutagenic potential for these compounds [45].

In conclusion both the nitro musks (MX, MK, MA, MM, MT) and polycyclic musk compounds AHTN, HHCB, AHDI, ATII) used in the fragrances have been studied in the battery of relevant in vitro and in vivo genotoxic tests (gene mutations, chromosomal aberrations) in accordance with current toxicological requirements. The results of the studies provided a convincing evidence of the lack of potential of the musk compounds tested to damage the hereditary material and induce mutations. The negative results in mutagenicity assays (Ames test, mouse micronucleus bone marrow assay) were also recently obtained in studies with macrocyclic musk compounds.

2.1.2.6
Carcinogenicity

There is only one investigation available on a tumorigenic potential of musk compounds. In the carcinogenicity study MX was fed (0.075% or 0.15%) ad lib to groups (n=50) of male and female SPF B_6C_3 mice for 80 weeks [26]. Dietary exposure corresponded on the average to 91 and 170 mg MX kg^{-1} bw for males, and to 101 and 192 mg kg^{-1} bw for females. All the surviving animals were killed on week 90 and a complete histopathological examination of the animals was performed.

The overall tumour incidence in all treated groups of both sexes were significantly higher as compared to controls. Malignant and benign liver cell tumours (adenomas, carcinomas) were significantly increased in both sexes. In males the incidence of Harder's gland tumours was also significantly higher in both treated groups than in controls. In addition, the range of tumour types was wider in the treated groups as compared to controls. The observed tumorigenicity was dose independent although in females there was an indication of dose-dependent increase in a number of animals with liver adenomas and carcinomas.

The exact mechanism of MX carcinogenicity observed in mice is not clear. However in view of the absence of genotoxicity it may be assumed that other mechanism, presumably a phenobarbital-like liver cell proliferation is involved and consequently a threshold for this effect can be determined. However, from the present study no NOEL could be established. The lower dose used corresponding to approximately 0.1 g MX kg^{-1} bw may be considered LOEL for the tumour inducing effects.

2.1.2.7 Photoallergenicity

Due to earlier findings on the photoallergenic potential of MA other nitro musk compounds such as MX, MK, MM, MT were tested in animal and human studies. Overall the results showed no indication of a photosensitising activity of any of the musk compounds studied [5]. Likewise neither AHTN or HHCB produced photoallergenicity in humans (cf. [6]).

2.1.2.8 Summary of Experimental Toxicological Studies

The results of experimental toxicological studies show that musk compounds have generally a low acute toxicity. Oral LD_{50} ranged between 4 and 10 g kg^{-1} bw for nitro musks and between 0.5 and 5 g kg^{-1} bw for polycyclic musks with dermal LD_{50} showing even higher values. A lower acute oral toxicity was recorded for MA (0.3 g kg^{-1} bw) which is no longer in use.

There is a very limited amount of experimental data on subchronic oral toxicity of nitro musks available in the open literature. In the 90-days dermal treatment study with MX, MK, MM and MT in rat only slight pathological changes were seen such as reduced body weight gain, lower organ weights and indication of cell proliferation in the liver. Similar observations were reported from more extensive but inadequately reported subchronic toxicity studies (13 weeks) after oral application of AHTN and HHCB to rats. Apart from the discoloration of some organs, minor changes in blood chemistry and haematological parameters as well as slight changes in body and organ weights there were no toxic effects recorded in this study. The lowest NOAELs derived from these experiments were 24 mg kg^{-1} bw for MX, 75 mg kg^{-1} bw for MK (90 days dermal study), 15 mg kg^{-1} bw for AHTN and 150 mg kg^{-1} bw for HHCB (oral application).

The results of the poorly presented teratogenicity (embryotoxicity, foetotoxicity) studies with nitro musks (MX and MK) in rats indicated a low maternal toxicity and no signs of a direct adverse effects on the developing organism.

Teratological studies using gavage treatment with AHTN and HHCB were each conducted in rats. Similar was the case with nitro musks where a maternal toxicity was recorded with the highest doses used. Skeletal malformations observed in the foetuses of highest HHCB dose groups, but not with AHTN, were considered to be due to maternal toxicity.

Similarly, in one generation developmental study in rats treated with MX and MK only reduced body weight gains (MX) and/or a delayed sexual maturation (MK) were seen in F_1 pups from the treated groups. No effects on the behaviour or reproductive capacity could be seen with any of the two compounds. The lowest NOAELs, from the developmental studies, all based on the maternal toxicity, were 20, 15, 5 and 50 mg kg^{-1} bw for MX, MK, AHTN and HHCB, respectively.

Nitro musk compounds such as MX, MK, MM and MT tested negative in a battery of in vitro and in vivo assays for gene mutations and chromosomal aberrations. MA (currently withdrawn from the market) was positive in Ames test (with metabolic activation) but did not show any mutagenic activity in any of the in vivo testing for gene and chromosomal mutations. The same negative results were also seen in all the assays on mutagenicity using the polycyclic musk compounds AHTN HHCB, ADBI, AHDI and ATII. Overall the results of the mutagenicity studies indicate that the nitro or polycyclic musks can be regarded as having no significant potential to cause a damage to hereditary material. The lack of mutagenicity was also shown in a recent study with macrocyclic compounds.

The results of a carcinogenicity study with MX given in feed showed a significant increase of malign and benign tumours in the liver (both sexes) and the Harder's glands (males). The observed tumorigenicity was dose-independent. It has been suggested that in the absence of genotoxicity (see above) the increased occurrence of tumours is due to phenobarbital-like induction of CYP2B liver enzymes in mice by MX and is not relevant for man. The carcinogenicity study with MX using a second animal species was not done. Similarly, carcinogenicity testing with other nitro, polycyclic or macrocyclic musk substances has not been reported.

None of the major nitro or polycyclic musks was a skin irritant or showed a potential for sensitisation in animal and human studies.

2.1.2.9 Calculations of PTDIs

A summary of NOEL/NOAELs derived from various toxicological experiments with individual musk compounds is shown in Table 1.

It should be noted that the NOEL/NOAELs as proposed from the developmental investigations other than the enzyme-induction experiments, were based on the results which were not presented in detail. It is therefore difficult to assess the quality (statistical evaluation) of the studies.

2.1.2.9.1 Nitro Musks

The lowest NOAEL for nitro musks was 0.7–0.8 mg MX kg^{-1} bw derived from the well designed and conducted developmental studies on the induction of monooxidase enzyme system in the liver of weaning rats. There are no corresponding developmental studies with MK or other nitro musks.

Based on the lowest NOAEL 0.7–0.8 mg MX kg^{-1} bw and using safety factor 100 the PTDI for MX can be calculated to be 0.007 mg kg^{-1} bw corresponding to 0.42 mg $adult^{-1}$ day^{-1}. Taking into account a structural similarity of nitro musk

Table 1 NOEL/NOAELs for nitro and polycyclic musks from oral and dermal repeated dose and developmental studies in rats

Substance	Administration	Type of study	Toxicological endpoint	NOEL/ NOAEL mg kg^{-1} bw	Reference
MX	Oral (feed)	Developmental	Enzyme – induction	0.7–0.8	[23]
	Oral (gavage)	Teratogenicity	Maternal toxicity Foetotoxicity	20 60	[33]
	Dermal	Repeated dose (rat females) (rat males)	 Liver weight Liver weight	 24 75	[31]
MK	Oral (gavage)	Developmental	Maternal toxicity Foetotoxicity	15 45	[33]
	Dermal	Repeated dose	Body weight	75	[31]
MA	Oral (feed)	Repeated dose	Neurotoxicity; testicular atrophy	25	[46]
HHCB	Oral (gavage, feed)	Repeated dose	Hematology Liver weight	150	[47]
		Developmental	Maternal toxicity	50	[33]
AHTN	Oral (feed, gavage)	Repeated dose	Organ changes (discolouring)	15	[6]
		Developmental	Maternal toxicity	5	[33]

the PTDI 0.42 mg $person^{-1}$ day^{-1} may be considered for all the nitro musk compounds.

In comparison, the apparent lack of mutagenic mechanism of the MX induced tumorigenicity makes it feasible to presume a threshold for the effect and to estimate a tentative PTDI from the carcinogenic study. Based on LOEL 100 mg MX kg^{-1} bw and taking into account that only LOEL is available as well the severity of the effects a safety (uncertainty) factor 2000 might be appropriate. The PTDI from the carcinogenicity study would therefore be roughly 50 µg MX kg^{-1} bw corresponding to 3 mg MX $person^{-1}$ (adult).

No sufficient toxicological studies are available to determine PTDIs for amino metabolites of MX and MK.

2.1.2.9.2
Polycyclic Musks

The extent of toxicological testing of the polycyclic musks is limited as compared to the nitro musks. There are virtually no studies of the enzyme induction capacity corresponding to those with the nitro musks. As a result NOEL/NOAELs proposed for the polycyclic musks were much higher based on the limited subchronic and/or developmental studies. The lowest NOAEL determined in these studies was 5 mg kg^{-1} bw and 50 mg kg^{-1} bw for AHTN and HHCB, respectively. Based on these NOAELs and using a safety factor of 100 the estimated PTDI for AHTN is 50 µg kg^{-1} bw corresponding to 3 mg $adult^{-1}$ day^{-1} and for HHCB 500 µg kg^{-1} bw corresponding to 30 mg $adult^{-1}$ day^{-1}, respectively. The toxicological database for other less used polycyclic musks is too limited to permit PTDI estimation.

2.1.2.9.3
Macrocyclic Musks

No data are at present available to permit an establishment of PTDI for this group of musk compounds.

2.1.3
Exposure Assessment

The main exposure routes for synthetic musk compounds are dietary intake via contaminated foodstuffs and dermal absorption from the musk containing cosmetics and/or household products. In addition a respiratory exposure by inhalation may occur but this pathway is considered less important [15].

2.1.3.1
Musk Levels in Fish

The occurrence and levels of individual musk compounds in various fish species is in detail discussed in other chapters of this monograph and will be only briefly summarized here. An excellent survey on the occurrence of polycyclic and other musk compounds in the fish and environment was recently presented by Rimkus [3].

Yamagushi et al. [48] were the first to report 1981 the presence of synthetic musks in fish (*Carassius auratus longsdorfii*) from Tama river in Japan. The concentrations of MX and MK detected in the study (0.20 mg and 0.05 mg kg^{-1} ww, respectively) were higher than those of dieldrin and *p,p'*-DDE measured in the same samples.

In the 1990s extensive studies of synthetic musks in wild fish from the European rivers, aquaculture produced fish as well as in sea fish were conducted in Germany (for references see, e.g. [1, 3, 49–51]. Later on investigations concerning musk levels in fish were also presented from Italy [52], Czech republic [53] and Sweden [54]. A comparison of the nitro and polycyclic musk levels in Canadian and European fish and marine foods were presented by Gatermann et al. [55].

There are great differences in the reported musk concentrations probably reflecting the inter-species differences as well as geographical variations in the environmental contamination.

Marine fish contains low levels of musk compounds as compared to the river and lake fish species.

Samples collected from the European aquatic systems generally show higher levels of the polycyclic musks HHCB and AHTN as compared to nitro musks MX and MK. This differs from the findings in the fish and marine samples from Canada with higher levels of nitro musks, probably reflecting the domination of this class of musk compounds on the North American market [55].

As can be expected the highest concentrations of the musk compounds and/or its metabolites were reported in fish from sewage ponds or the vicinity of the sewage plants [1, 3, 14, 54, 56]. In comparison relatively low levels of musks were found in the limited number of fish samples studied in Sweden, with the highest levels (outside sewage plants) measured for AHTN (1.0–7.4 μg kg^{-1} ww) and HHCB (1.9–11.7 μg kg^{-1} ww) in the perch liver and arctic char (Riddarfjärden, Stockholm, lake Vättern) [54, 57]. It is also of interest that low but clearly detectable levels of HHCB and AHTN (2.9 and 1.0 μg kg^{-1} ww respectively) were recently measured in perch from some Swedish lakes without a known industrial pollution [57].

2.1.3.2
Estimation of Musk Exposure from Fish

An estimation of the dietary exposure to musks involves both the levels of the compounds determined in the foodstuffs and consumption of the foodstuffs in a general population as well as in risk groups. In order to compensate for the limited database available on the occurrence of the musks in the foodstuffs and to provide sufficient safety margins, the worst case situation should be used when calculating the dietary exposure to musks.

Apart from the fish from the sewage ponds high concentrations of the musk compounds including AHTN (10–1408 μg kg^{-1} ww), HHCB (15–4130 μg kg^{-1} ww), MX (1–170 μg kg^{-1} ww)and MK (1–380 μg kg^{-1} ww) were detected in eel collect-

Table 2 Estimations of the daily intake of MX, MK, its amino metabolites 4-NH_2-MX, 2-NH_2-MX, 2-NH_2-MK as well as AHTN, HHCB, ADBI, AHDI and ATII in fish

Compound	Concentration (μg kg^{-1} wet weight)	Reference	Normal consumer[a] (ng kg^{-1} bw day^{-1})	High consumer[b] (ng kg^{-1} bw day^{-1})
MX	170	[56]	85	340
MK	380	[56]	190	760
HHCB	4131	[56]	2066	8262
AHTN	1408	[56]	704	2816
ADBI	8	[56]	4	16
AHDI	5	[52]	2.5	10
ATII	0.65	[57]	0.0003	0.001
4-NH_2-MX	78	[1]	39	156
2-NH_2-MX	4.2	[1]	2.1	8.4
2-NH_2-MK	6.2	[1]	3.1	12.4

[a] Normal consumer = 30 g fish (60 kg) $person^{-1}$ day^{-1}.
[b] High consumer = 120 g fish (60 kg) $person^{-1}$ day^{-1}.

ed from relatively highly polluted river and lake systems of Berlin area. These values may be used in the intake calculations. Less information is available concerning the concentrations of nitro musk amino metabolites and therefore the results from fish (carp liver) originating from sewage ponds may be used for the conservative estimate of the intake. The estimation includes both high (120 g day^{-1} $person^{-1}$) and low (30 g day^{-1} $person^{-1}$) consumers of fish based on the Swedish consumption data. The summarized calculations are given in Table 2.

Assuming the worst case situation the intake via fish (high consumer) may be estimated to approximately 340, 760, 8260 and 2816 ng kg^{-1} bw corresponding to 20 µg MX, 46 µg MK, 500 µg HHCB, and 170 µg AHTN $person^{-1}$ (60 kg) day^{-1}.

2.1.3.3
Musk Levels in Breast Milk

The results of investigations dealing with the musk compounds in breast milk have been subjected to several reviews (see, e.g. [4, 58]). German research groups have during the last decade conducted several studies which also indicate time trends of musk compounds in the breast milk [13, 58, 59]. Analytical data are also available from, e.g. Switzerland and Sweden [4, 54].

Generally, the polycyclic musks HHCB and AHTN were invariably found in high concentrations in the samples of breast milk studied. The concentrations ranged between 16 and 1316 µg kg^{-1} lw for HHCB and 16 and 148 µg kg^{-1} lw for AHTN. Other polycyclic musks such as ADBI, AHDI, ATII were present in much lower concentrations (Table 3). The earlier investigations reported also high levels of the nitro musks MX (up to 1220 µg kg^{-1} lw) and MK (up to 240 µg kg^{-1} lw) (Table 3). However the results of more recent studies indicate much lower concentrations, roughly by factor 10, reflecting probably both the methodological problems in the early analyses and gradual phasing out of the nitro musks from the market [58].

It should be noted that the average levels of musk compounds detected in the Swedish samples were of the same order of magnitude as in Germany and Czech

Table 3 Estimations of the daily intake of MX, MK, MM, HHCB, AHTN, ADBI, AHDI and ATII in breast milk

Musk compound	Concentration (µg kg^{-1} lw)	Reference	Exposure[a] µg kg^{-1} bw day^{-1}
MX	1220	[59]	6.32
MK	240	[59]	1.24
MM	3	[58]	0.02
HHCB	1316	[58]	6.81
AHTN	148	[58]	0.76
ADBI	18	[13]	0.09
AHDI	20	[58]	0.10
ATII	51	[58]	0.26

[a] The estimated daily consumption of 0.7 L milk with 3.7% lipid content in 5-kg child.

Republic [54]. Since the contamination of fish appears to be low in Sweden the use of cosmetics is probably the source of the breast milk contamination with the musk compounds.

2.1.3.4
Estimation of Musk Exposure from Breast Milk

Calculations of the estimated daily intake of musk compounds via breast milk in infants are summarized in Table 3. Adopting the worst case situation the highest levels reported in the studies for individual musk compounds were used. The calculation was done for 5 kg newborn with a daily intake of 0.7 l breast milk with approximately 3.7% fat content.

The calculation based on the worst-case situation (highest level in breast milk reported, the most susceptible age in term of milk intake) shows that breast fed infants are exposed to relatively high levels of MX and HHCB with the intake exceeding 6 $\mu g\ kg^{-1}$ bw day^{-1} for each substance. The intake of MK is also higher than 1 $\mu g\ kg^{-1}$ bw day^{-1}. In comparison the exposure of a high consumer of fish to, e.g. MX and MK was calculated (worst-case situation) to be in the region of 0.34 $\mu g\ kg^{-1}$ bw day^{-1} and 0.76 $\mu g\ kg^{-1}$ bw day^{-1}, respectively (see Table 2).

2.1.3.5
Musk Concentrations in Cosmetics and Household Products

According to surveys on the use of musks in fragrance products conducted by the industry, the most important source of exposure are skin creams because of their retention in the skin, followed by hydroalcoholic products (eaux de toilette) which have a high concentration of musk containing fragrances [5, 6]. On the other hand hygienic products such as soaps and bath foams, which are applied on large body area, will be washed off and are probably less important dermal source of musk compounds [6].

The musk compounds are used in fragrances (oils) at varying levels. The fragrances are then added in varying amounts to the consumer products resulting in different concentrations of musks in the final products. The fragrance industry has reported that the 97.5% use levels in formulated fragrance are around 7% for nitro musks (MX and MK). Taking into account the percentage of fragrance mixture added to the final product the concentration of, e.g. MX in the eau de toilette was estimated to approximately 0.6% [5]. By the same calculation approximately 0.96% AHTN and 2.4% HHCB was estimated to be present in the eau de toilette used by the consumers (see, e.g. [6]).

2.1.3.6
Estimation of Dermal Exposure to the Main Nitro and Polycyclic Musks from Cosmetics and Hygienic Products

In addition to the concentration in the toiletry products, the knowledge about the application quantity and frequency as well as the retention factor in the skin is needed for the overall calculation of dermal exposure to musks. Finally, the most

Table 4 Calculation of an exposure to MX, AHTN and HHCB from the most important cosmetic products[a] (adopted from [5, 6])

Compound	Product	Application quantity (g) × application frequency day^{-1}	Fragrance oil in product (%)	Compound in fragrance oil (%)	Compound in product (%)	Exposure $adult^{-1}$ [b] (mg)	Calculated exposure × absorption 10% $adult^{-1}$ (mg)
AHTN	Body lotion	8×0.71	0.4	12	0.048	2.73	0.27
	Eau de toilette	0.75×1	8	12	0.96	7.20	0.72
	Fragrance cream	5×0.29	4	12	0.48	6.96	0.70
HHCB	Body lotion	8×0.71	0.4	30	0.15	6.82	0.68
	Eau de toilette	0.75×1	8	30	2.40	18.0	1.80
	Fragrance cream	5×0.29	4	30	1.20	17.4	1.74
MX	Body lotion	8×0.71	0.4	7.1	0.03	1.61	0.16
	Eau de toilette	0.75×1	8	7.1	0.57	4.26	0.43
	Fragrance cream	5×0.29	4	7.1	0.28	4.12	0.41

[a] Retention in the skin assumed to be 100%.
[b] Adult = 60 kg bw.

difficult part of the exposure estimation is to determine the absorption (uptake) via the skin.

The results of in vivo studies in human volunteers or in the models using the human skin show that the absorption rate of individual musks are probably in the range 0.3–5% [5]. However because of the uncertainties such as up to 14% of the given dose unaccounted for in the experiments, a more conservative absorption rate may perhaps be justified. Apart from the difficulties in mimicking realistic conditions of use in the studies the absorption rate can also be affected by other materials present in the product [5, 6].

Examples of the estimated exposure to MX, AHTN and HHCB from the selected cosmetic products, representing the main exposure sources of the compounds, are shown in Table 4.

As can be seen from Table 4, the dermal exposure to musks from the body lotions, fragrance creams and eau de toilets taken together can be estimated roughly up to 10 mg MX, 17 mg AHTN and 42 mg HHCB person^{-1} day^{-1}, respectively. The total exposure from all the relevant cosmetic and household products was estimated to approximately 11 mg MX, 18 mg AHTN and 46 mg HHCB person^{-1} day^{-1} [5, 6]. Given the inherent uncertainties concerning the rate of penetration through the skin an absorption factor 10% was considered appropriate in the final calculation of the exposure to the individual musk compounds.

Under the assumptions given above, including 10% absorption in the skin, the systemic exposure (bioavailable fraction) after the dermal use of the musk containing cosmetics/hygienic products may be estimated to 1.1 mg MX, 1.8 mg AHTN and 4.6 mg HHCB person^{-1} day^{-1}. It should however be emphasized that the tentative estimation of the dermal exposure reflects a highly conservative approach in terms of use, skin retention and, particularly the expected absorption rate. In real life the exposure of an average consumer using cosmetic/hygienic products may be expected to be substantially lower.

2.1.4
Risk Characterization

2.1.4.1
Dietary Intake

The toxicological documentation required for a meaningful and reliable risk analysis of the synthetic musk compounds in the diet is at present inadequate. The most detailed experimental data were obtained from the toxicological studies with MX. Much less information is available on MK or polycyclic musk compounds AHTN and HHCB. Toxicological investigations of the other nitro and polycyclic musks are generally missing.

Even where available, many studies with both the nitro and polycyclic musk compounds are only presented in the review form which makes it difficult to evaluate the quality and significance of the experimental data. This applies to, e.g. reproduction studies which constitute a very important endpoint with respect to assessment of the safe intake in infants and children. Moreover, some of the studies particularly on toxicokinetics and short-term toxicity, were performed 30–40 years ago and do not fully comply with the current testing standards.

Therefore an additional extensive testing based on relevant toxicological endpoints, with exception of genotoxicity, is required before the NOAELs proposed in various studies for the individual musks can be fully accepted. Consequently, the establishment of reliable safe levels for single musk molecules is not possible. All the calculations of the estimated TDI/PTDIs must be at the present stage of knowledge regarded as a recommendation rather than an administrative norm.

MX in high doses was shown to induce tumours in mice. However the absence of positive response in an extensive genotoxicity testing indicates a threshold for the effect. A comparison of the tumour inducing dose in the animals and exposure levels in the diet shows a negligible risk for humans. There is however

Table 5 Estimation of the intake of selected musk compounds from different sources compared to PTDI

Chemical compound	PTDI (μg kg^{-1} bw)	Intake day^{-1} (μg kg^{-1} bw)		
		Fish	Breast milk	Cosmetics
MX	7.0	0.3	6.3	18
MK	7.0	0.8	1.2	?
AHTN	50	2.8	0.8	30
HHCB	500	8.2	6.8	77

remaining uncertainty as concerns the lack of more exact quantitative data on the tumour-inducing potential of MX as well as the precise mechanism of tumour formation. In this respect additional experimental studies of the mechanistic processes behind the enzyme induction which is considered to play the main role in the musk carcinogenicity are needed. The absence of the carcinogenicity data on compounds other than MX, particularly the chemically related MK and other nitro musks, is another uncertainty which needs to be resolved.

Individuals with a regular high consumption of fish from the contaminated areas represent a major risk group for the potential harmful effects of musks in the diet. However, even assuming the worst case situation with the highest levels of nitro musks detected and high consumers of fish (120 g day^{-1}) the estimated daily intake will be approximately 0.3 μg MX kg^{-1} bw and 0.8 μg MK kg^{-1} bw corresponding to 18 μg MX and 48 μg MK 60-kg-$person^{-1}$ day^{-1}, respectively. For a person with a normal fish consumption (30 g day^{-1}) the corresponding daily intakes will be 0.08 μg MX kg^{-1} bw and 0.19 μg MK kg^{-1} bw. The intakes are significantly lower than the estimated PTDI 7 μg kg^{-1} bw (corresponding to 420 μg $person^{-1}$) based on the lowest NOAEL determined from enzyme induction effect by MX (see Table 5).

A similar calculation can be made for polycyclic musks AHTN and HHCB, which are considered the most prevalent contaminants in fish from the European waters. The calculated highest daily intake of 8.2 μg HHCB kg^{-1} bw compares favourably, from the health risk point of view, with the estimated PTDI 500 μg kg^{-1} bw (30,000 μg $person^{-1}$). The intake of 2.8 μg AHTN kg^{-1} bw day^{-1} is also significantly lower than PTDI 50 μg kg^{-1} bw (3000 μg $person^{-1}$ day^{-1}) (Table 5).

Women of childbearing age and breast-fed infants represent the most important risk groups as concerns the dietary exposure due to mobilization of musk compounds from the body fat into breast milk. Relatively high levels of musks detected in breast milk may be related at least partly to the use of cosmetics since a significant contamination of mother's milk is seen even in the geographical area such as, e.g. Scandinavia, where the levels of musk compounds in the sea fish normally consumed are generally low.

The estimated exposure of infants to nitro musks via breast milk using the highest concentration of individual substances reported in the literature, shows an intake of approximately 6 μg MX and 1.2 μg MK kg^{-1} day^{-1} respectively. In case of MX it is roughly equal to the recommended TDI (7 μg kg^{-1} bw). Given the uncertainties concerning the toxicological effects as well as the limited knowledge about the levels of musks in breast milk this is certainly a cause for concern.

2.1.4.2
Dermal Exposure

The absorption of musks from fragrances through the skin, although limited, results in appreciable uptake in the body organs. The toxicological documentation is however based on the limited experimental studies after topical application to laboratory animals and the uncertainties about possible health effects are identical with those mentioned in relation to the dietary intake.

Under the assumption of the worst case situation in terms of use and particularly in respect to the absorption through the skin the expected systemic exposure from the cosmetics may be estimated to approximately 18 μg MX kg^{-1} bw, 30 μg AHTN kg^{-1} bw and 77 μg HHCB kg^{-1} bw, respectively. This is about 2.5 times higher than the proposed PTDI, for the MX (7 μg MX kg^{-1} bw), in the same range as PTDI for the AHTN (50 μg kg^{-1} bw) and about 6.5 times lower than the PTDI for HHCB (500 μg kg^{-1} bw) (Table 5).

It should be pointed out that there are considerable uncertainties in the assumptions employed in the intake calculations particularly as concerns the absorption of various musk compounds after the topical application. The real life exposure for the musks from fragrances in cosmetics may be expected to be lower. However, because of the incomplete toxicological data the potential risks from the direct use of fragrances in cosmetics should be viewed with caution.

In conclusion the toxicological data at the present state of knowledge do not indicate a risk for adverse health effects from the residues of the musk compounds in foodstuffs. However there is a concern about the intake of the musks in newborns via breast milk which probably is contaminated through the use of musk-containing cosmetics and household products by the mothers. The dermal exposure of the adults to individual musk compounds especially MX and AHTN from topical application of musk-containing cosmetics may also be expected to be higher or in the range of tentative PTDI for the substances. However there are substantial uncertainties both concerning the toxic effects and the exposure estimations. Until a more complete database on toxicology and systemic exposure is available the cosmetics as a source of exposure should be viewed with caution. Overall, due to high volume of use and a widespread occurrence in the environment and food chains a cautionary attitude to the risk analysis of the musk compounds is in place.

The tentative calculations of PTDIs based on the critical evaluation of the toxicological end-points presented in this survey can be used as a recommendation when discussing the administrative regulation of the musk compound exposure. The gaps in toxicological information and the limited knowledge about the levels of the musk compounds in various fish arts from different geographical locations does not allow at the present stage of knowledge to implement MRLs for the individual musk compounds in foodstuffs.

2.2 Risk Management

The main effort in the management of the musk exposure from the dietary sources should be aimed at the expansion of database on possible toxicological effects of the individual compounds present in the diet. There is an urgent need for more extensive toxicological studies of the polycyclic musk compounds using the internationally accepted toxicological end points and experimental protocols. Basic research aiming at the elucidation of mechanism of action for carcinogenic effect as observed with MX and the importance of enzyme induction is also needed. In addition, toxicological data on new substances from macrocyclic musk group should be made available as soon as possible.

The toxicological investigations should be accompanied by a continuous long-term monitoring and control of the occurrence of various classes of musk compounds in the foodstuff represented by the relevant fish species. A special effort should be made to follow the exposure of the mothers by monitoring the levels of the individual musk compounds in breast milk.

Taking into account the ubiquitous presence of the compounds in the aqueous environment and a low biodegradability of the substances the precautionary principle should be applied in the risk assessment and risk management of musk compounds. The importance of cosmetics as a major source of body exposure to synthetic musks has been recognized by the EU authorities. Recently maximum authorized concentrations in cosmetic products were established for MX and MK (MX: 1.0% in fine fragrances, 0.4% in eaux de toilette and 0.03% in other products, MK: 1.4% in fine fragrances, 0.56% in eaux de toilette and 0.042% in other products) [60]. Furthermore, the nitro musks MA, MM and MT have earlier been banned from the use in cosmetic products in the EU member states [61, 62]. In addition, the Scientific Committee on Cosmetic Products and Non-Food Products intended for Consumers (SCCNFP) of the EU Commission gave opinions on the safe use of the polycyclic musks AHTN and HHCB in cosmetic products [63, 64]. Given the toxicological uncertainties it would be prudent for industry to act towards gradual reduction of the synthetic musk based fragrances in the cosmetics and household products towards the lowest concentrations technologically achievable.

The information available on the toxicity and occurrence in the diet does not make it at present possible to propose reliable administrative MRL for musks in foodstuff. The PTDIs as proposed in this survey may be used as a benchmark against which the results of the monitory studies can be compared.

Last but not least, the efforts by the industry to introduce, where possible, toxicologically inert classes of musk substances should continue. The withdrawal of potentially toxic musk ambrette and use of the polycyclic musks instead of nitro musks in Europe shows that these goals are achievable.

2.3
Risk Communication

The experience with other natural or anthropogenic toxicants in foods shows that it is important to provide consumers with a balanced, objective and transparent information in order to retain the credibility for the proposed risk management. In case of previously not regulated substances there is a problem of striking a balance between the legitimate right of the public to relevant information on potential adverse effects and the risk of causing an unnecessary and scientifically unfounded concerns among the perceptive groups in the population.

In view of the pertaining uncertainties about the toxicological potential and the extent of food contamination with musks the information should be aimed primarily at the risk groups. The two groups identified in risk analysis are high fish consumers and pregnant women.

Based on the results of the environmental monitoring a dietary recommendation concerning the excessive consumption of fish from the polluted localities directed at high fish consumers (sport fishermen and others) may be considered.

The professional counselling at, e.g. maternity units should be available in order to explain the conclusions and process of risk analysis. This would hopefully help to calm down possible concerns among women of child bearing age about the health risks arising from the dietary intake of the musk compounds. Accompanying information about the sources of musk compounds as well as the advisory about how to reduce the exposure via, e.g. cosmetics would be a part of the risk communication with this group.

A continuous exchange of information with the producers as concerns the best measures to reduce possible health risk from the synthetic musks should be further encouraged.

Finally, an international harmonization of the scientifically based legislation would make it much easier to explain possible health risks to the consumers and contribute to maintaining the public trust.

3
References

1. Rimkus G, Gatermann R, Hühnerfuss H (1999) Toxicol Lett 111:5
2. OSPAR Convention for the Protection of the Marine Environment of the North East Atlantic. Working Group on Diffuse Sources (DIFF), 6–9 Oct, 1998 Helsinki, Finland
3. Rimkus GG (1999) Toxicol Letters 111:37
4. Käfferlein HU, Göen T, Angerer J (1998) Crit Rev Toxicol 28:431
5. Ford RA (1998) Dtsch Lebensm-Rundsch 94:192
6. Ford RA (1998) Dtsch Lebensm-Rundsch 94:268
7. Minegishi K-I, Nambaru S, Fukuoka M, Tanaka A, Nishimaki-Mogami T (1991) Arch Toxicol 65:273
8. Suter-Eichenberger R, Altorfer H, Lichtensteiger W, Schlumpf M (1998) Chemosphere 36:2747
9. Riedel J, Dekant W (1999) Toxicol Appl Pharmacol 157:145
10. Kokot-Helbling K, Schmid P, Schlatter C (1995) Mitt Gebiete Lebensm Hyg 86:1
11. Riedel J, Birner G, Van Dorp C, Neumann I-G, Dekant W (1999) Xenobiotica 29:573
12. Ford RA, Hawkins DR, Schwarzenback R, Api AM (1999) Toxicol Letters 111:133

13. Rimkus G, Rimkus B, Wolf M (1994) Chemosphere 28:421
14. Rimkus GG, Wolf M (1996) Chemosphere 33:2033
15. Müller S, Schmid P, Schlatter C (1996) Chemosphere 33:17
16. Lehman-McKeeman LD, Caudill D, Vassallo JD, Pearce RE, Madan A, Parkinson A (1999) Toxicol Letters 111:105
17. Iwata N, Minegishi K, Suzuki K, Ohno Y, Kawanishi T, Takahashi A (1992) Biochem Biophys Res Commun 184:149
18. Iwata N, Minegishi K, Suzuki K, Ohno Y, Ygarashi T, Satoh T, Takahashi A (1993) Biochem Pharmacol 45:1659
19. Lehman-McKeeman LD, Caudill D, Young JA, Dierckman TA (1995) Biochem Biophys Res Commun 206:975
20. Lehman-McKeeman LD, Johnson DR, Caudill D (1997) Toxicol Appl Pharmacol 142:169
21. Lehman-McKeeman LD, Johnson DR, Caudill D, Stuard SB (1997) Drug Metab Dispos 25:384
22. Lehman-McKeeman LD, Stuard SB, Caudill D, Johnson DR (1997) Molecul Carcinogenesis 20:308
23. Suter-Eichenberger R, Boelsterli UA, Conscience-Egli M, Lichtensteiger W, Schlumpf M (2000) Toxicol Letters 115:73
24. Stuard SB, Caudill D, Lehman-McKeeman LD (1997) Fund Appl Toxicol 40:264
25. Mersch-Sunderman V, Emig M, Reinhardt A (1996) Mutation Res 356:237
26. Maekawa A, Matsushima Y, Onodera H, Shibutani M, Ogasawara H, Kodama Y, Kurokawa Y, Hayashi Y (1990) Food Chem Toxicol 28:581
27. Steinberg P, Fischer T, Arand M, Park T, Elmadafa I, Rimkus G, Brunn, H Dienes H (1999) Toxicol Letters 111:151
28. Nylander A (2001) Master Thesis, Project Report No 77. Dept Environ Toxicol, Uppsala Univ, Uppsala, Sweden, pp 1–63 (in Swedish)
29. Opdyke DLJ (1975) Food Cosmet Toxic 13:877
30. Opdyke DLJ (1976) Food Cosmet Toxic 14:793
31. Ford RA, Api AM, Newberne PM (1990) Food Chem Toxicol 28:55
32. Gressel Y, Troy WR, Foster GV (1980) Toxicol Letters Special Issue 1:134
33. Christian MS, Parker RM, Hoberman AM, Diener RM, Api AM (1999) Toxicol Lett 111:169
34. OECD (1997) OECD Guidelines II. Health effects IV:471. OECD, Paris, France
35. Nair J, Ohshima C, Malaveille C, Friesen M, O'Neill IK, Hautefeuille A, Bartsch H (1986) Food Chem Toxicol 24:27
36. Emig M, Reinhardt A, Mersch-Sundermann V (1996) Toxicol Lett 85:151
37. Api AM, Ford RA, San RHC (1995) Food Chem Toxicol 33:1039
38. Api AM, Pfitzer EA, San RHC (1996) Food Chem Toxicol 34:633
39. Kevekordes S, Grahl K, Zauling A, Dunkelberg H (1996) ESPR-Environ Sci Pollut Res 3:189
40. Kevekordes S, Zauling A, Dunkelberg H (1997) Toxicol Lett 91:13–17
41. Api AM, Pfizer EA, San RHC (1996) Fund Appl Toxicol 30:232
42. Pfizer EA, Api AM, San RHC (1996) Fund Appl Toxicol 30:233
43. Kevekordes S, Mersch-Sundermann V, Diez M, Dunkelberg H (1997) Mut Res 395:145
44. Kevekordes S, Mersh-Sundermann V, Diez M, Bolten C, Dunkelberg H (1998) Anticancer Res 18:449
45. Abramsson-Zetterberg L, Slanina P (2002) Toxicol Lett 135:155
46. Davis DA, Taylor JM, Jones WI, Brouwer JB (1967) Toxic Appl Pharmacol 10:405
47. Api AM, Ford RA (1999) Toxicol Lett 111:143
48. Yamagishi T, Miyazaki T, Horii S, Kaneko S (1981) Bull Environ Contam Toxicol 26:656
49. Rimkus G, Wolf M (1993) Dtsch Lebensm-Rundsch 89:171
50. Eschke HD, Traud J, Dibowski HJ (1994) UWSF-Z Umweltchem Ökotoxikol 7:131
51. Rimkus G, Brunn H (1996) Ernährungs-Umschau 43:442
52. Draisci R, Marchiafava C, Ferretti E, Palleschi L, Catellani G, Anastasio A (1998) J Chromatogr A 814:187
53. Hajslova J, Gregor P, Chladkova V, Alterova K (1998) Organohalogen Compd 39:253

54. Eriksson S, Darnerud P-O, Aune M, Bjerselius R, Slanina P, Crattingius S, Glynn A (2003) Report 7. Swedish National Food Admin, Uppsala, Sweden (in Swedish)
55. Gatermann R, Hellou J, Hühnerfuss H, Rimkus G, Zitko V (1999) Chemosphere 38:3431
56. Fromme H, Otto T, Pilz K, Neugebauer F (1999) Chemosphere 39:1723
57. Broman D (1999) Swedish Environmental Resesearch Foundation for Strategic Environmental Research News, Annual Report, Stockholm (in Swedish)
58. Liebl B, Mayer R, Ommer S, Sönnichsen C, Koletzko B (2000) In: Koletzko B (ed) Short and long term effects of breast feeding on child health. Kluwer Academic Plenum Publishers, chap 26
59. Liebl B, Ehrenstorfer S (1993) Chemosphere 27:2253
60. Off J Europ Comm (2002) Council Directive 2002/34/EC L102:19
61. Off J Europ Comm (1995) Council Directive 95/34/EC L167:19
62. Off J Europ Comm (1998) Council Directive 98/62/EC L253:20
63. Opinion of the Scientific Committee on cosmetic products and non-food products intended for consumers (SCCNFP) concerning AHTN, SCCNFP/0609/02
64. Opinion of the Scientific Committee on cosmetic products and non-food products intended for consumers (SCCNFP) concerning HHCB, SCCNFP/0610/02

The Handbook of Environmental Chemistry Vol. 3, Part X (2004): 311–331
DOI 10.1007/b14119

Recent Studies Conducted by the Research Institute for Fragrance Materials (RIFM) in Support of the Environmental Risk Assessment Process

Froukje Balk[1] · Daniel Salvito[2] · Han Blok[1]

[1] Royal Haskoning, P.O. Box 151, 6500 AD Nijmegen, The Netherlands
E-mail: f.balk@royalhaskoning.com, han.blok@royalhaskoning.com
[2] Research Institute for Fragrance Materials, 50 Tice Boulevard, Woodcliff Lake, NJ 07677, USA
E-mail: dsalvito@rifm.org

Abstract Although the environmental risk assessments for the nitro musks, musk ketone and musk xylene, and the polycyclic musks, AHTN and HHCB, showed that the risks of these fragrance ingredients were low, questions regarding their environmental behaviour and fate stimulated further research. The removal during the sewage treatment process of AHTN, HHCB, musk ketone and musk xylene is 89, 92, 83 and 99%, respectively, which is higher than predicted on the basis of log K_{ow}. This may, at least partly, be explained by the biotransformation of the polycyclic musks into more polar metabolites and the transformation of the nitro musks into their amino metabolites. Because of higher polarity, the transformation products of AHTN and HHCB have a lower log K_{ow} and, therefore, it is likely that they will have a lower accumulation potential and a lower aquatic toxicity.

Contrary to earlier findings, RIFM studies showed that the toxicity of the 4-amino-metabolite of musk xylene for *Daphnia magna* was of the same order of magnitude as musk xylene, ranging from 0.37 to 0.51 mg L^{-1}. The phenomenon of missing tail fins observed in a fish-early-life stage test with AHTN was confirmed in another species. This effect occurred at the concentrations where growth and survival were affected as well. For the polycyclic musks with a lower market volume, ADBI, AHDI and ATII, a preliminary risk assessment was carried out on the basis of structural similarity and the proportionality of the environmental concentrations. The present findings support the earlier conclusion that the risks posed by these products are low.

Keywords Nitro musks · Polycyclic musks · Environmental fate · Aquatic toxicity · Risk assessment

1 Introduction

Musk fragrance ingredients have attracted attention during the past decade following the detection of various representatives in samples of surface water, fish and human adipose tissue and milk in Germany [1–6]. These findings prompted more environmental sampling in other countries and initiated discussions on the environmental safety of these substances. In the Netherlands, on behalf of the Ministry of the environment, environmental risk assessments were carried out by the National Institute of Public Health and the Environment (RIVM) for the nitro musks, musk ketone and musk xylene [7, 8], and for the polycyclic musks, AHTN and HHCB [9–11]. Within the scope of the OSPAR (Oslo and Paris Commissions for the Prevention of Marine Pollution) Action Plan of 1998, musks were included in the category of "diffuse sources and groups of substances" to be considered for action and in this connection, the risks for the aquatic environment were evaluated [12]. These risk assessments considered the ratio of the predicted environmental concentration and the predicted no-effect concentration (PEC/PNEC) and did not indicate a risk for the aquatic or terrestrial environment based on the available data. Human health effects of the nitro musks and the polycyclic musks were evaluated by the EU Scientific Committee on cosmetic products and non-food products intended for consumers (SCCNFP). In the meantime, musk xylene and musk ketone were also included on the EU's Third Priority List related to the EU Existing Chemicals Regulation (Council Regulation (EEC) No. 793/93), whereas AHTN and HHCB are on the Fourth Priority List. These risk assessments were carried out in close cooperation with the Research Institute for Fragrance Materials (RIFM), a non-profit research institute responsible for testing and evaluation of the human health and environmental safety of fragrance ingredients. During the process, more knowledge of the environmental fate and behaviour and of effects was generated, paving the way again to more questions and more studies. Recently RIFM initiated more research on (1) the polycyclic musks in view of (a) their biotransformation and removal during sewage treatment, (b) the effect on fish early life stages and (c) the risks of the "minor" musks, polycyclic musks with a lower market volume, and (2) the nitro musks to clarify (a) their removal during the sewage treatment process and (b) the environmental impact of the 4-amino-metabolite of musk xylene. The results will be reviewed here.

2
Polycyclic Musks

The risk assessment for AHTN and HHCB was carried out according to the procedures described in the EU Technical Guidance Documents [13]. In this procedure, the environmental concentrations in water, soil and fish was compared to a Predicted No Effect Concentration for the environment. For both AHTN and HHCB an extensive database was available with concentrations measured in Germany, the Netherlands and some other countries. During the analysis of the environmental exposure, it became clear that the concentrations predicted by EUSES (European Union System for the Evaluation of Substances, [14]) were systematically overestimated. In general, a large discrepancy was observed between measured and predicted concentrations in effluents, sludge and surface water. This may be due, on the one hand, to a series of "reasonable worst case assumptions" on the emission of the substances to a waste water treatment plant and, on the other hand, to assumptions concerning their persistence and poor removal during waste water treatment. Preliminary results in studies on biodegradation and removal were referred to in an earlier paper [10] and are reviewed here.

The risks assessment for AHTN and HHCB also drew attention to other polycyclic musks that are used in lower volumes, such as ADBI, AHDI and ATII. Therefore, the available data for these substances were collected, both on environmental concentrations and on aquatic toxicity in order to assess the risks of those substances.

For the risk assessment, the aquatic toxicity of AHTN and HHCB was studied for algae, *Daphnia* and fish. The toxicity test with AHTN on the early life stages of the Fathead minnow showed an unusual phenomenon of fish larvae lacking a tail fin occurring at concentrations near those inducing mortality. The results of a further study into this effect are reported here.

2.1
Environmental Concentrations

Due to their application in detergents, household cleaning products and bath and hair care products, most of the production volume of the synthetic musks is released to the sewer after use. As a consequence, surface water concentrations of AHTN and HHCB are highly variable, from extremely low in remote areas (<1 ng L^{-1}) to higher concentrations in rivers beyond the discharge points of sewage treatment plant effluents. The data of more than 200 surface water samples taken in Germany, The Netherlands and Switzerland (freshwater) were combined into a frequency distribution. The median concentrations were 0.07 μg L^{-1} for both AHTN and HHCB and the 90th-percentiles were 0.3 and 0.5 μg L^{-1} for AHTN and HHCB, respectively [10]. Extreme values are found where effluents are discharged to surface waters virtually without dilution. Concentrations in fish caught in the natural environment (i.e. excluding effluent ponds and aquaculture) were reported from Germany, The Netherlands and Italy [1, 2, 10, 15–18]. AHTN and HHCB were detected in all fish samples except in herring caught in the open ocean. When all data were combined, the overall median and 90th-percentile concentrations of 60 fish samples were 0.01 and 0.10 mg kg^{-1} wet weight (ww) for AHTN as well as for HHCB.

Occasionally, concentrations of the other polycyclic musks (i.e. ADBI, AHDI, ATII) were reported in surface water samples. Often they remained below the detection limits. When detected, they were consistently lower than these main products in the same samples by a factor of 10–25. This reflects the ratio in the use volume, where AHTN and HHCB constitute more than 95% of the use [19]. Concentrations of the other polycyclic musks in fish were below those of AHTN and HHCB by a factor of at least 10, and up to a factor of 60.

Domestic sewage treatment plants are considered to be the main port of entry of these substances into the environment. Concentrations in influents measured in Germany [3], The Netherlands [20] and in Ohio, USA [21] varied between 2 and 11 μg AHTN L^{-1} and 4 to 20 μg HHCB L^{-1}, thus reflecting the difference in use volumes for both substances. The data of the USA fall in the middle of the range of the observations in Europe. These three studies also reported concentrations in effluents, but there are additional data from Switzerland [22] and Sweden as well [23]. The median and 90th-percentile of the reported effluent concentrations were 1.2 and 2.9 μg AHTN L^{-1} and 1.5 and 4 μg HHCB L^{-1}. Concentrations of the other polycyclic musks in influent and effluent were occasionally above the detection limits. Reported values for the minor musks are below AHTN concentrations in the same sample by a factor of around 20.

Concentrations of AHTN and HHCB measured in digested sludge in Germany [24], The Netherlands [25, 20] and Switzerland [26] ranged between 1 and 30 mg/kg. As sludge is always collected during a longer time before it may find an application onto agricultural land, variations in the concentrations are evened out. Therefore the median concentration, 5 and 10 mg kg^{-1} for AHTN and HHCB, respectively, is more meaningful than the 90th-percentile of these samples (20 and 28 mg kg^{-1}, respectively). The concentrations of the other polycyclic musks were below the concentration of AHTN by a factor of 13.

The environmental concentrations were also estimated from the use volume and the modelling of release, removal and dilution processes according to the model concepts in the EU Technical Guidance Document and the model EUSES [13, 14]. When the measured concentrations for AHTN and HHCB are compared to the predicted concentrations, the predicted concentrations are overestimated by a factor of 5 to 15 for most compartments [9]. The discrepancy between predictions and measurements was studied in more detail by analysis of the environmental fate and distribution of these substances.

2.2
Fate and Behaviour

As reported by RIFM, use volumes in Europe for 1995 were 585 tonnes for AHTN and 1481 tonnes for HHCB and in 1998 these volumes were 385 and 1473 tonnes, respectively. The other polycyclic musks ADBI, ATII and AHDI are used at a lower level, with, in 1995, 26, 40 and 3 tonnes, reported respectively (1998 volumes were 18, 2, and 19 tonnes, respectively). For the purpose of the risk assessment, it was assumed that the whole use volume is ultimately disposed of through the sewer. Based on standard test results for mineralisation, this process is considered negligible during sewage treatment. The removal in the sewage treatment plant (STP) by adsorption processes is estimated on the basis of an equilibrium partition model and the partition coefficient log K_{oc} (in L kg^{-1}) being 4.80 for AHTN and 4.86 for HHCB. By this model 81% of AHTN and 83% of HHCB would remain on the sludge and 19 respectively 17% would be released via the effluent.

Several explanations for the discrepancy between predicted and measured concentrations can be postulated. First, the use volumes may show fluctuations in time, but these seem to be in the order of 10–20% at most. Secondly, assuming complete discharge to the sewer implies that volatilisation from skin, fabric or other surfaces is not taken in account. And finally, although mineralisation during sewage treatment is not expected, a primary degradation process may take place that leads to the formation of more polar metabolites.

2.2.1
Biotransformation

As a follow-up to earlier studies on the biodegradability of AHTN and HHCB summarised by Balk and Ford [10], the fate of these substances in activated sludge and river water is being studied using ^{14}C- (ring) radio-labelled material [27, 28]. To date, die-away studies in activated sludge have been completed with HHCB and AHTN, and a die-away study in effluent diluted into river water has been completed with HHCB. In addition, a continuous activated study (CAS) was performed with AHTN. These studies measured mineralisation of the compounds to CO_2, disappearance of the parent compounds, formation of metabolites and incorporation into microbial biomass. To maximize realism, tests were conducted at the lowest practical concentration given the specific activity of the test materials, which ranged from 5 to 50 μg L^{-1}.

In activated sludge, both AHTN and HHCB were biotransformed to more polar metabolites. In a similar manner, HHCB was biotransformed in effluent diluted into river water. RP-HPLC K_{ow} analysis with on-line radioactivity detection of the HHCB metabolites revealed that K_{ow} values for the metabolites were 2–4 orders of magnitude lower than that of the parent compound. The CAS study with AHTN indicated removal of parent was >85% with approximately half the removal the result of sorption to wasted solid with the balance resulting from biotransformation to polar metabolites. These results indicate AHTN and HHCB, although not biodegradable in standard tests, are not persistent. Both are biotransformed to more polar metabolites that would be predicted to be less toxic and bioaccumulative than the parent molecule.

2.2.2
Removal in the STP

A new analytical method was used to measure the levels of fragrance materials in influent and effluent of two STPs in the USA. Over three days, daily composite flow-proportional samples were taken from the influent, primary effluent and final effluent of an activated sludge and a trickling filter STP. The total concentration of fragrance materials in the influent showed a considerable diurnal variation around the mean (±60%), whereas the effluent concentration was very stable. The measurements in influents and effluents showed that 89% of AHTN is removed and 91.5% of HHCB after treatment in an activated sludge plant [21].

These data indicate a higher removal than predicted by EUSES on the basis of log K_{ow}, 81% for AHTN and 83% for HHCB. This means that the effluent contained 11% instead of the predicted 19% for AHTN (11/19 or 58% of the predicted fraction) and 8.5% instead of the predicted 17% (8.5/17 or 50% of the predicted fraction). This difference between predicted and measured level can be explained by and is consistent with the biotransformation of AHTN and HHCB to more polar metabolites.

2.2.3
Volatilisation

The large discrepancy between predicted and measured data was a strong argument to base the risk characterisation on measured data. A careful analysis of the data from surface water shows a relationship between the measured concentration and the proximity of discharge points for STP effluent. Volatilisation, adsorption, degradation and dilution are held responsible for the decrease in concentrations downstream. By application of the model EUSES to predict the regional concentration in surface water it can be concluded that volatilisation is responsible for a decrease from 1.2 $\mu g\ L^{-1}$ directly after discharge to 0.03 $\mu g\ L^{-1}$ further downstream [29].

2.3
Aquatic Toxicity

The toxicity data for AHTN and HHCB were described by Balk and Ford [11] and are summarized in Table 1.

2.3.1
Fish Early Life Stage Test

As part of the ongoing risk assessment, the toxicity of AHTN and HHCB was assessed in an early life stage test with fish according to OECD Test Guideline 210. The eggs of the test species, *Pimephales promelas*, were exposed under flow-

Table 1 Aquatic and terrestrial toxicity data for AHTN and HHCB [11]

Test species	AHTN	HHCB
Algae *Pseudokirchneriella subcapitata (=Selenastrum capricornutum)*	72-h-NOEC 0.374 mg L^{-1}	72-h-NOEC 0.201 mg L^{-1}
Crustaceae *Daphnia magna*	21-day-NOEC 0.196 mg L^{-1}	21-day-NOEC 0.111 mg L^{-1}
Fish *Lepomis macrochirus* (Bluegill sunfish)	21-day-NOEC 0.089 mg L^{-1}	21-day-NOEC 0.093 mg L^{-1}
Pimephales promelas (Fathead minnow, early life stage)[a]	36-day-NOEC 0.0625 mg L^{-1}	36-day-NOEC 0.068 mg L^{-1}
Brachydanio rerio (Zebrafish)[a]	34-dayd-NOEC 0.035 mg L^{-1}	–
Worms *Eisenia foetida* (Earthworm)	56-day-NOEC 105 mg kg^{-1}	56-day-NOEC 45 mg kg^{-1}
Insecta *Folsomia candida* (Springtail)	28-day-NOEC 45 mg kg^{-1}	28-day-NOEC 45 mg kg^{-1}

[a] Reference [30].

through conditions. Triethylene glycol was used as a solvent for the preparation of the test medium. Actual concentrations in the test medium were measured at regular intervals during the 36-day test period. Mean measured concentrations ranged from 0.01 and 0.14 mg L^{-1}. The results are summarized in Table 2.

For HHCB, effects were observed on the growth and swimming behaviour of the larvae and the NOEC was 0.068 mg L^{-1} (measured concentration). The NOEC for growth was 0.035 mg L^{-1} for AHTN, but in addition, a particular observation was the absence of tail fins in the majority of the surviving larvae at concentrations of 0.067 mg L^{-1} and higher. This effect was not observed at lower concentrations. For the fish that did have tail fins in concentrations ≥0.067 mg L^{-1}, the relative tail fin length was not significantly different from the control and it was concluded that AHTN caused the total absence of the tail-fin rather than a shortening. The abnormal development occurred at the concentration just below the onset of mortality [30].

The test was repeated for the zebrafish, *Brachydanio rerio*, to determine whether this unusual effect was induced also in other species. The test was carried out in accordance with OECD TG 210 in an intermittent-flow-through system over 34 days. Triethylene glycol was used as the solvent to prepare test concen-

Table 2 Toxicity of AHTN and HHCB on fish early life stages of *Pimephales promelas* and *Brachydanio rerio* [30]

Test	NOEC, μg L^{-1}	Remarks
HHCB *P. promelas*		Test range 9 to 140 μg L^{-1}, 36 days
Hatching		No effects at 140 μg L^{-1}
Development		No effects at 140 μg L^{-1}
Growth (dry weight)	68	54% reduction at 140 μg L^{-1} (LOEC)
Growth (standard length)	68	20% reduction at 140 μg L^{-1} (LOEC)
Larval survival	68	78% at 140 μg L^{-1} (LOEC) Less active, deviating behavior
AHTN *P. promelas*		Test range 8 to 140 μg L^{-1}, 36 days
Hatching		No effects at 140 μg L^{-1}
Development	35	84% of survivors at 67 μg L^{-1} (LOEC) and all survivors at 140 μg L^{-1} missed caudal fin
Growth (dry weight)	35	7% reduction at 67 μg L^{-1} (LOEC), 75% reduction at 140 μg L^{-1}
Growth (standard length)	35	7% reduction at 67 μg L^{-1} (LOEC), 38% reduction at 140 μg L^{-1}
Larval survival	67	18% at 140 μg L^{-1} (LOEC)
AHTN *Brachydanio rerio*		Test range 10 to 75 μg L^{-1} (with solvent); 65 μg L^{-1} without solvent, 34 days
Hatching		No effects at 75 μg L^{-1}
Development	35	41% missed tail fin, $^1/_3$ had curled bodies/curved spines at 50 μg L^{-1}, all survivors missed tail fin, $^2/_3$ had also curled bodies/curved spines at 75 μg L^{-1}
Growth (standard length)	50	13% reduction[a] at 65 μg L^{-1}, 15% reduction[a] at 75 μg L^{-1}
Growth (total length)	35	12% reduction at 50 μg L^{-1}
Growth (dry weight)		No statistically significant effects at 75 μg L^{-1}
Larval survival		No effects at 75 μg L^{-1}

[a] Difference is statistically significant.

trations ranging from 0.010 to 0.075 mg L^{-1}. The actual concentrations were between 104 and 110% of the nominal concentrations and therefore the results are expressed as nominal. To investigate the possible contribution of the solvent triethylene glycol to the effects, eggs/larvae were also exposed to an AHTN solution prepared using a generator column. The mean actual concentration was 0.065 mg L^{-1}. The results are included in Table 2. In the range finding test, malformations in the egg stage were induced at concentrations ≥0.200 mg L^{-1}. The eggs/larvae in the definitive test showed a normal development and in concentrations up to 0.035 mg L^{-1} the phenomenon of the missing tail fin was not observed. Higher concentrations induced the lack of the caudal fin and curling or curving of the body. These effects were concentration-dependent. The reduced length of the fish and the observed disturbed swimming behaviour at concentrations above 0.035 mg L^{-1} (NOEC) is explained by the absence of the tail fin [30]. These findings confirm the results of the study on fathead minnow.

2.3.2
Aquatic Toxicity Data for the Other Polycyclic Musks

The acute toxicity of ATII was tested on *Daphnia magna*, whereas for AHDI acute tests were carried out on the alga *Selenastrum capricornutum*, on *D. magna* and the fish *Brachydanio rerio*. The tests were carried out according to the OECD Test Guidelines and EEC Directive 92/69/EEC L383 A part C. Concentrations in the test media were determined by HPLC (ATII) or by GC (AHDI). For both substances, the test concentration decreased rapidly during the test period. The results, expressed as mean measured concentrations, are included in Table 3 [19].

The results with AHDI and ATII for *D. magna* compared well to the effects after three days in the prolonged toxicity tests with AHTN (EC_{50}=0.8 mg L^{-1}) and HHCB (EC_{50}>0.9 mg L^{-1}). The fish data for AHDI were also in line with the interpolated short term results from the chronic fish tests with AHTN and HHCB. The sensitivity of alga to AHDI seemed to be higher than to AHTN or HHCB.

The aquatic toxicity of the polycyclic musks may be predicted using Quantitative Structure Activity Relations (QSAR) based on log K_{ow}. Log K_{ow} for these substances range from 5.2 to 5.9. A summary of these predictions is presented in Table 3. Although the selected QSAR is supposed to present the "minimum toxicity" of the substance, generally, the experimental observations showed that the toxicity was less than the minimum toxicity by an order of magnitude for the acute tests, whereas the discrepancy in the prolonged study with *Daphnia magna* was a factor of 50. On the other hand, the longer term predicted fish toxicity is lower than observed by a factor of 3–10 only [31].

The oral LD_{50} for the rat was above 3000 mg kg^{-1} for HHCB and ADBI, and ranged from 570–1380 mg kg^{-1} for AHTN, from 820 to 2200 mg kg^{-1} for ATII, and from 300 to 800 mg kg^{-1} for AHDI. For the assessment of the risks for the predatory food chain, the results of a 90-day oral study for the rat were used. The NOAEL in the diet was 15 mg kg^{-1} for AHTN and 150 mg kg^{-1} for HHCB [32, 33].

It was already demonstrated that the structural similarity between AHTN and HHCB is reflected in a similar behaviour of the two substances. In view of the

Table 3 Aquatic and terrestrial toxicity of polycyclic musks; comparison of observation and prediction by QSARs (in italics) [19, 30]

	AHDI mg L^{-1}	ATII mg L^{-1}	ADBI mg L^{-1}	AHTN mg L^{-1}	HHCB mg L^{-1}
Algae					
Selenastrum capricornutum					
72-h-NOEC	0.044	–	–	0.374	0.201
72-h-EC_{50}	0.200	–	–	>0.835	>0.854
QSAR 72 h–96 h EC_{50}[a]	*0.094*	*0.031*	*0.057*	*0.03*	*0.02*
Crustaceans					
Daphnia magna					
48-h-EC_{50}	0.32	0.42	–	0.8[b]	>0.9[b]
QSAR 2-day-EC_{50}[a]	*0.14*	*0.05*	*0.09*	*0.05*	*0.03*
21-day-NOEC	–	–	–	0.196	0.111
QSAR 16-day-NOEC[a]	*0.013*	*0.004*	*0.008*	*0.004*	*0.002*
Fish					
4-day-LC_{50}	>1.13[f] 0.9 NOEC nom.[f]	–	–	1.41[b,d] nom.	1.36[b,d]
QSAR 4-day-LC_{50}[a]	*0.39*	*0.15*	*0.26*	*0.15*	*0.10*
QSAR 14-day-LC_{50}[c]	*0.225*	*0.525*	*0.359*	*0.208*	*0.139*
21-day-LC_{50}	–	–	–	0.314[d]	0.452[d]
36-day-LC_{50} (e.l.s.)	–	–	–	0.1[e]	>0.14[e]
21-day-NOEC growth	–	–	–	0.089[d]	0.093[d]
36-day-NOEC e.l.s.	–	–	–	0.035[e]	0.068[e]
34-day-NOEC e.l.s.				0.035[f]	–
Worms					
Eisenia foetida					
56-day-NOEC	–	–	–	105 mg kg^{-1}	45 mg kg^{-1}
Insecta					
Folsomia candida					
28-day-NOEC	–	–	–	45 mg kg^{-1}	45 mg kg^{-1}

nom.=nominal concentration.
[a] QSAR proposed in EU-Technical Guidance Document.
[b] Interpolated from chronic test.
[c] QSAR from EPA-ECOSAR (neutral organics).
[d] Test fish *Lepomis macrochirus*.
[e] Test fish *Pimephales promelas*.
[f] Test fish *Brachydanio rerio*.

close structural similarity between AHTN, ATII and AHDI and, although to a slightly lesser extent, between AHTN and ADBI, it is reasonable to assume that the experimental toxicity data for AHTN are the best prediction of the toxicity of the other polycyclic musks. The presently available data on aquatic toxicity and the acute oral toxicity to the rat confirm this assumption [19].

3 Nitro Musks

The risk assessment for musk ketone and musk xylene [7, 8] was also carried out in accordance with the EU-TGD [13]. The database of measured environmental concentrations was rather limited at the time of the risk assessment and therefore the risk assessment was based both on predicted and measured concentrations. The risk ratios indicated that the risks of both musk ketone and musk xylene for organisms in the aquatic compartment and for predators were low [7]. The risk ratio for soil organisms, based on the predicted concentration in sludge, was around 1; however, later studies showed that concentrations measured in sludge were considerably lower than predicted and as a consequence the risk for soil organisms was low as well [29].

As for the polycyclic musks, for musk ketone and musk xylene a large discrepancy was also observed between measured and predicted concentrations in effluents, sludge and surface water. Also here this is partly due to "worst case" assumptions concerning their persistence and poor removal during waste water treatment. Recently it was shown that during the sewage treatment process, amino derivatives of musk ketone and musk xylene may be formed and these substances were also detected in surface water and fish. An acute toxicity study reported an EC_{50} of one of these derivatives below 1 μg L^{-1} [34]. This low level was reason to repeat the test. The risks of the nitro musks are reviewed in view of the potential impact of amino metabolites.

3.1 Environmental Concentrations

Musk ketone and musk xylene, as well as the polycyclic musks, are mostly used in detergents, other cleaning products and cosmetics that finally end up in domestic wastewater. Surface water concentrations are found to vary largely, with higher concentrations clearly related to samples in close proximity to STP discharge points. As for the polycyclic musks, the data for freshwater samples taken in Germany, The Netherlands and Switzerland were combined into a frequency distribution. The median of more than 170 samples was <0.005 μg L^{-1} and the 90th-percentile was 0.04 μg L^{-1} [1, 35–37]. The concentrations of musk xylene were considerably lower. The median of the concentrations measured in the highly loaded German River Ruhr was 0.01 μg L^{-1} [1]. Likewise, the median and 90th-percentile of the concentrations measured in freshwater fish from Germany, The Netherlands, Switzerland and Italy were around 10 and 100 μg kg^{-1} ww for musk ketone [1, 38–42], whereas the concentrations of musk xylene were of the same order of magnitude.

Concentrations measured in influent were reported from the same countries as listed above. They varied between 0.07 and approximately 2 μg L^{-1} for both musk ketone and musk xylene [1, 20, 43]. In two STPs in the USA, influent concentrations were in the same range [21]. Concentrations in the effluent were around 0.1 μg L^{-1} for musk ketone and 0.03 μg L^{-1} for musk xylene. These are similar to the values as reported from the USA [21]. Concentrations measured in digested sludge ranged between the detection limit and 0.06 mg kg^{-1} for both musk ketone and musk xylene [20, 24–26].

3.2
Fate and Behaviour

The use volumes in Europe reported for 1995 were 61 tonnes for musk ketone and 110 tonnes for musk xylene. Comparison of environmental concentrations predicted on the basis of the use volume by the model EUSES to measured concentrations in the Rivers Ruhr, Rhine and Meuse shows an overestimation of the actual concentrations by a factor of 17 to more than 50 for musk ketone and at least a factor of 10 for musk xylene [8]. The discrepancy between predicted and measured concentrations is one aspect that is not yet satisfactorily solved. The difference may be explained in several ways. There may be a significant loss by evaporation from the skin or fabric and furthermore, the standard assumptions in the model for emission and dilution are intended to present an overestimation. In addition there is growing evidence for the formation of metabolites which are less hydrophobic. Thus the predicted removal in sewage treatment plants, exclusively based on adsorption of the parent substance, will not be correct.

Recently, accurate mass balance studies for the removal of musk ketone and musk xylene in sewage treatment plants showed significantly higher removal percentages [21] than predicted by the partition model. The model predicted 63% removal by adsorption to sludge for musk ketone and 87% for musk xylene. In the study of Simonich et al. [21], primary and secondary removal together amount to 82.6% for musk ketone and 98.7% for musk xylene. These differences would imply that the fraction going to effluent and thus to surface water is lower by a factor of 2 for musk ketone and by a factor of 10 for musk xylene.

The presence of amino metabolites of musk ketone and musk xylene was shown in various environmental samples; see Table 4. Concentrations of 4-amino musk xylene (4-AMX) found in the River Elbe (Germany) were significantly higher than the parent material. The isomer 2-AMX was found as well, but at the same level as musk xylene. Also for musk ketone, an amino derivative, 2-AMK, was detected in a higher concentration than the parent. In several sewage treatment plants the relative concentration of amino musks increased from influent to effluent, whereas the parent substance concentrations decreased [15, 44]. In a sampling programme of excess sludge from 12 Swiss sewage treatment plants, musk xylene was found in only one out of 12 samples, whereas its amino metabolite was measured in 6 out of 12 samples. Musk ketone was detected in 7 out of 12 samples, whereas the metabolite 2-AMK was detected only in one out of 12 samples. These results indicate that in particular musk xylene is readily metabolised to its amino derivative.

Table 4 Concentrations of amino metabolites of nitro musks in environmental samples (median and range)

Sample	Musk ketone	AMK	Musk xylene	AMX	Reference
Influent (ng L^{-1})	550	<0.5	150	<0.5	[44]
Effluent (ng L^{-1})	6	250	10	34	[44]
Sludge (µg kg^{-1}) (n^a=12) LOD^b 2.5 to 5 µg kg^{-1}	1.5–6.9 in 7/12	13.1 in 1/12	32.5 in 1/12	10.9–49.1 in 6/12	[26]
River water (ng L^{-1}) (n^a=5)	<1 4^c	 7^c	Median <1 <1–2	4-AMX median 3, 1–9 2-AMX median <0.5, <0.5–1	[15, 44]
Fish (µg kg^{-1} ww) (n^a=5)	Median 0.16 0.1–1.2	Median 0.05 <0.001–0.15	Median 0.3 0.1–0.5	4-AMX median 0.2 0.1–2.9 2-AMX median 0.06 <0.001–0.14	[15]
Sediment (µg kg^{-1} ww) (n^a=4)	0.135–0.189	–	0.185–0.296	2-AMX 0.003–0.054 4-AMX 0.080–0.127	[15]

[a] n=number of samples.
[b] LOD=limit of detection.
[c] Detected in the same sample.

The 4-AMX metabolite was detected in fish in concentrations varying around the parent concentration (from 0.2 to 9 times the musk xylene concentration). The concentrations of 2-AMX and 2-AMK were between 10 and 50% of their respective parent substances [15]. The presence in fish might be related to the direct uptake of amino metabolites from water, but is more likely related to metabolism by the fish of musk xylene or musk ketone taken up from the water. The bioaccumulation study on musk ketone showed the formation of three more polar transformation products in the fish [8].

The formation of amino metabolites has also been observed in toxicological studies with mice and rats. The amino metabolites were held responsible for the induction of cytochrome liver enzymes [45]. Musk xylene appeared a more potent inducer of the liver enzymes than musk ketone. From tests with and without antibiotics it was concluded that nitro reduction by the intestinal microflora is responsible for the induction [46].

Although the number of data is insufficient to make a complete mass balance or to estimate the environmental exposure to amino musks, it is clear that metabolism is partly responsible for the observed discrepancy between predicted and monitored data.

3.3
Aquatic Toxicity

The ecotoxicity data for musk ketone and musk xylene were reviewed by Tas et al. [8]. An updated summary with additional data of Chou and Dietrich [47] is presented in Table 5.

3.3.1
Acute Toxicity of Amino-Musk Xylene to Daphnia magna

The acute toxicity for *Daphnia magna* of the amino metabolites of musk xylene (AMX) was reported by Behechti et al. [34]; see Table 6. Whereas the toxicity of 2-AMX, 2,4-AMX and 2,4,6-AMX was of the same order of magnitude of musk xylene or less, the EC_{50} for the 4-aminometabolite was reported as 0.25 μg L^{-1}. It was highly remarkable that the sensitivity of *D. magna* for the 2-AMX and the 4-AMX differed by a factor of 400. On the basis of this single EC_{50} and an assessment factor of 1000, the PNEC for 4-AMX would be 0.25 ng L^{-1} and in view

Table 5 Aquatic and terrestrial toxicity data for musk ketone and musk xylene [8]

Test species	Musk ketone	Musk xylene
Algae		
Selenastrum capricornutum	72-h-NOEC 0.088 mg L^{-1}	5-day-NOEC >5.6 mg L^{-1}
Crustaceae		
Daphnia magna	21-day-NOEC 0.169 mg L^{-1}	48-h-EC50 >5.6 mg L^{-1}, 48 h-NOEC 0.32 mg L^{-1} 21-day-NOEC 0.056 mg L^{-1}
Fish		
Lepomis macrochirus (Bluegill sunfish)		96-h-LC50 1.2 mg L^{-1}
Onchorynchus mykiss (Trout)	21-day-NOEC 0.0625 mg L^{-1}	96-h-LC50 >1000 mg L^{-1}
Brachydanio rerio (Zebrafish)	96-h-NOEC >0.4 mg L^{-1} [a]	96-h-NOEC >0.4 mg L^{-1} [a] 14-day-NOEC <0.1 mg L^{-1}
Amphibians		
Xenopus laevis (Clawed frog)	4-day-NOEC >0.4 mg L^{-1} [a] 11-day-NOEC <0.4 mg L^{-1} [a]	4-day-NOEC >0.4 mg L^{-1} [a]
Worms		
Eisenia foetida (Earthworm)	56-day-NOEC 32 mg kg^{-1}	14-day NOEC >50 mg kg^{-1}
Insecta		
Folsomia candida (Springtail)	28-day-NOEC 100 mg kg^{-1}	

[a] Tested on embryos.

Table 6 Toxicity of the amino metabolites of musk xylene to *Daphnia magna*

Substance	48-h EC_{50}[a] ($mg\ L^{-1}$)	48-h EC_{50}[b] ($mg\ L^{-1}$)
2-Amino musk xylene	1.07	
4-Amino musk xylene	$0.25 \cdot 10^{-3}$	LD, S: 0.49 D, S: 0.51 LD, N: 0.37 D, N: 0.44
2,4-Diamino musk xylene	23.2	
2,4,6-Triamino musk xylene	58.8	

LD=Light and dark cycle.
D =In darkness.
N =Test medium natural water.
S =Synthetic water.
[a] Solvent ethanol, purity test material >98.5%, tested in darkness [34].
[b] Solvent dimethylformamide, purity test material >99.9% [50].

of the observed concentration of 9 ng L^{-1} in a sample of the Elbe, this was a cause for concern.

At the initiative of RIFM, the 48-h EC_{50} test with 4-AMX on *D. magna* was repeated by an independent laboratory according to the same OECD Test Guideline 202 and under the Principles of Good Laboratory Practice (GLP). The test substance was prepared by hydrogenation of musk xylene and further purification to yield 4-AMX (99.99%). Dimethylformamide was used as a solvent and the test solutions were prepared in reconstituted water (total hardness 160 mg $CaCO_3$ L^{-1}, pH 7.9–8.0 [48]) and in natural water from a pond (total hardness 12 mg L^{-1}, pH 6.5–6.9). Nominal concentrations ranged from 6.3 to 1000 µg L^{-1}. The concentration of 4-AMX in the test medium was analysed using HPLC and UV detection at initiation and after 48 h. Tests were carried out in the dark and under a light and dark regime.

The measured concentrations were 89–100% of nominal concentrations. The 48-h EC_{50}-values ranged from 370 to 510 µg L^{-1} under the four sets of test conditions; see Table 6. The confidence intervals were all between 400 and 600 µg L^{-1} [49, 50]. These EC_{50} values are above the earlier reported value [34] by a factor of 1500–2000 and they are of the same order of magnitude as reported by Behechti et al. for the 2-amino-derivative of musk xylene. After consultation with RIFM, Behechti et al. repeated their test. The extremely low EC_{50} for 4-AMX could not be reproduced and they now assume that the observed toxicity is not related to 4-AMX [51]. A difference of a factor of around 1000 in test results may also hint to an error in the units or in the dilution series.

In the meantime the results of Behechti et al. have been cited by several authors and caused concern. This one single irreproducible toxicity test result has turned out to be a poor basis for these concerns and subsequent evaluation in a risk characterisation.

3.3.2
Other Effects

3.3.2.1
Genotoxic Potential of Amino Metabolites

The effects of both 4-AMX and 2-AMX at the subcellular level were studied in microbial tests to detect genotoxic potential. The Salmonella microsome test detects mutagenic activity by reversion of an already existing mutation on the histidine operon (Ames test). The ara-test detects a mutation in the arabinose-operon or in other genes (L-arabinose resistance test). In the SOS *umu*-test, DNA damage induces the SOS repair system, including the *umu*C-gene. Induction of *umu*C is an indicator of genotoxicity.

The tests did not establish any mutagenic or genotoxic potential of the amino-musk xylenes in the Ames test and in the SOS *umu*-test, neither with nor without external metabolic activation by rat liver S9-mix. In the ara tests, no mutagenic effects were shown either [52].

3.3.2.2
Hormone Disrupting Potential

The competitive binding capability of nitro musks or their amino metabolites to the oestrogen receptors was investigated. Tests were conducted in vitro with preparations of oestrogen receptor from rainbow trout (*Oncorhynchus mykiss*) and the clawed frog *Xenopus laevis*. In these tests the affinity between the oestrogen receptor and 17β-oestradiol is measured and in the competition assay the inhibition of this affinity by competitive binding of the musks to the receptor is expressed as the IC_{50} (the concentration that causes 50% inhibition of affinity between the receptor and 17β-oestradiol). No competitive binding was observed for the parent compounds musk ketone or musk xylene. On the other hand all three amino musks (4-AMX, 2-AMX and 2-AMK) competitively inhibited the binding in both preparations. The IC_{50} for binding to the oestrogen receptor of the trout

Table 7 Binding capacity of nitro musks to the estrogen receptor (ER) of trout and frog [53]. Competitive binding to the ER is expressed as IC_{50} (inhibition concentration affecting 50% of the organisms)

Substance	IC_{50} for trout ER	IC_{50} for frog ER
Estradiol	5.3±1.2 nmol L^{-1}	187±76 nmol L^{-1}
Bisphenol A	8.8±1.8 µmol L^{-1}	441±76 nmol L^{-1}
Musk xylene	ncb[a]	ncb[a]
4-AMX	>1 mmol L^{-1} (>265 mg L^{-1})	30.8±28.5 µmol L^{-1} (8 mg L^{-1})
2-AMX	1.3±1.1 mmol L^{-1} (350 mg L^{-1})	12.9±10.3 µmol L^{-1} (3 mg L^{-1})
Musk ketone	ncb[a]	ncb[a]
2-AMK	>1 mmol L^{-1} (>265 mg L^{-1})	70.1±88.3 µmol L^{-1} (18 mg L^{-1})

[a] ncb=No competitive binding.

could only be obtained for 2-AMX, at a high concentration (1.3 mmol ~350 mg L^{-1}). In the preparations of xenopus the IC_{50} values were lower by two orders of magnitude, see Table 7. The authors state: "although competitive binding of nitro musk metabolites to the ER of rainbow trout and xenopus was demonstrated in vitro with this study, the question remains as to the relevance of these findings for the in vivo and specifically the environmental situation". While the authors point out that organisms are chronically exposed to these lipophilic metabolites, the environmental concentrations that fish and amphibians are chronically exposed to are on the order of 4 and 6 orders of magnitude, respectively, lower than the IC_{50} values [53].

4 Risk Evaluation

4.1 Polycyclic Musks

4.1.1 *Derivation of the PNEC*

For an environmental risk assessment, environmental concentrations are compared to a predicted no-effect concentration (PNEC). The PNEC in the aquatic environment for AHTN and HHCB was derived with an assessment factor of 10 to the lowest NOEC in the data set [13]; see Table 1. The $PNEC_{water}$ was 3.5 µg L^{-1} for AHTN and 6.8 µg L^{-1} for HHCB [11].

The PNEC for fish-eating predators was derived from a 90-day oral study on rats where the no adverse effect levels (NOAEL) were 15 mg AHTN kg^{-1} bodyweight (bw) day^{-1} and 150 mg HHCB kg^{-1} bw day^{-1}. The conversion factor for the daily dose for the rat to a level in the food is a factor of 20, giving 300 and 3000 mg kg^{-1} food for AHTN and HHCB, respectively. The application factor to derive a PNEC from a 90-day toxicity study is 30 [13]. Therefore, the $PNEC_{predator}$ is 10 mg kg^{-1} food (fish) for AHTN and 100 mg kg^{-1} food for HHCB. For the other polycyclic musks, in view of the close structural similarity, it was assumed that the substances have a toxicity similar to AHTN and HHCB. This was supported by the available toxicity data. Therefore, the existence of data for AHTN and HHCB was taken into account for the derivation of a $PNEC_{water}$ and $PNEC_{predator}$. In order to take the lack of data into account, the lowest PNEC is taken (AHTN) and divided by an additional safety factor of 3, resulting in an indicative $PNEC_{water}$ (0.0035/3=) 0.001 mg L^{-1} and an indicative $PNEC_{predator}$ (10/3=) 3.3 mg kg^{-1} food (ww) [19].

4.1.2 *Risks in the Aquatic Environment*

For an assessment of the environmental risks of the major polycyclic musks AHTN and HHCB, the exposure assessment was based on the 90th-percentile or the maximum of actually measured concentrations. Although the phenomenon

Table 8 Risk ratios for polycyclic musks and nitro musks

Substance	Surface water ($\mu g\ L^{-1}$)[a]	Fish ($mg\ kg^{-1}$ ww)[a]	PNEC water ($\mu g\ L^{-1}$)	PNEC fish predator ($mg\ kg^{-1}$ ww)	Risk ratio water	Risk ratio fish-predator
AHTN	0.3	0.10	3.5	10	0.09	0.01
HHCB	0.5	0.10	6.8	100	0.07	0.001
ADBI, ATII, AHDI	0.025	0.015	1	3.3	0.03	0.005
Musk ketone	0.04	0.01	6.3	2.5	0.006	0.004
Musk xylene	<0.04	≤0.01	1.1	7.5	<0.04	≤0.001
4-AMX	0.009	0.004	0.45	–	0.02	–

[a] 90th percentile of measured concentrations; for 4-AMX the maximum concentration.

of the missing tail fins in the fish early life stages related to AHTN exposure was confirmed, the effective concentration was just one step below the onset of mortality and growth inhibition. Therefore it seems to be just one of the phenomena related to a disturbed development and it does not show up at significantly lower concentrations than other effects. The PNEC for AHTN takes account of this effect. The risk ratios (PEC/PNEC) indicated that the risks of AHTN and HHCB for organisms in the aquatic and terrestrial compartment and for predators were low [11], see Table 8.

The initial concern about their persistence was related to the observed lack of mineralisation in standard biodegradation tests. More realistic laboratory studies have subsequently shown that HHCB and AHTN are biotransformed into polar metabolites. This process is a key for explaining the higher than predicted removal of these compounds in sewage treatment and the lower than predicted levels measured in aquatic environments.

For the other polycyclic musks, an indicative assessment was made where a conservative estimated level of exposure was based on the observed proportionality to the concentrations of AHTN and HHCB, and the no effect level was based on the structural similarity. The risk ratios given in Table 8 confirm that the risks of the other polycyclic musks for aquatic organisms and fish eating predators are also low.

4.2 Nitro Musks

4.2.1 *Derivation of the PNEC*

The PNEC in the aquatic environment for the nitro musks musk ketone and musk xylene was based on the lowest NOEC in the data set; see Table 5. The data set for musk ketone included long-term tests for species from three trophic levels, allowing an assessment factor of 10 [13]. The resulting PNEC is 6.3 μg L^{-1}. For musk xylene, however, only two of the NOECs were considered as long-term tests giving an assessment factor of 50. The PNEC for musk xylene is 1.1 μg L^{-1} [8]. For 4-AMX, an indication of the PNEC may be given with an assessment factor of 1000 on the only available EC_{50}, giving 0.45 μg L^{-1}.

For fish-eating predators, the lowest test results from studies on rat were used. The NOAEL (No observed adverse effects level) in a pre/post natal study where the parents were exposed to a dose in their food, was 2.5 mg musk ketone kg^{-1} bw day^{-1} and 7.5 mg musk xylene kg^{-1} bw day^{-1} [54, 55]. The daily dose is converted to a concentration in food (similar as for the polycyclic musks, factor of 20). With an assessment factor of 10 for the extrapolation from a test on reproduction toxicity [13], the PNEC for fish eating predators becomes $2.5\times20/10=5$ mg musk ketone kg^{-1} food (fish) and, likewise, 5 mg musk xylene kg^{-1}.

4.2.2
Risks in the Aquatic Environment

For the risk assessment of musk ketone and musk xylene, the 90th-percentiles of all freshwater and fish samples was used, whereas for 4-amino musk xylene, the maxima of the measured concentrations were taken. The amino metabolites of musk xylene are not far more toxic than the parent compound, as initially claimed. The amino metabolites proved to be neither mutagenic nor genotoxic. The PNEC is protective also for their competitive binding capacity to the oestrogen receptor (hormone disrupting potential).

As summarised in Table 8, the PEC/PNEC ratios for musk ketone, musk xylene for aquatic organisms and fish-eating predators and the risk ratio for aquatic organisms and 4-amino musk xylene are all well below 1, indicating that the environmental risks are low.

5
References

1. Eschke HD, Traud J, Dibowski HJ (1994) Vom Wasser 83:373
2. Eschke HD, Traud J, Dibowski HJ (1994) UWSF-Z Umweltchem Ökotox 6:183
3. Eschke HD, Dibowski HJ, Traud J (1995) UWSF-Z Umweltchem Ökotox 7:131
4. Eschke HD, Dibowski HJ, Traud J (1995) Dtsch Lebensm-Rundsch 91:375
5. Rimkus G, Wolf M (1995) Chemosphere 30:641
6. Rimkus G, Wolf M (1996) Chemosphere 33:2033
7. Tas JW, Van de Plassche EJ (1996) Initial environmental risk assessment of musk ketone and musk xylene in the Netherlands in accordance with the EU-TGD. RIVM report 601503 002, National Institute of Public Health and the Environment RIVM, Bilthoven, NL
8. Tas JW, Balk F, Ford RA, Van de Plassche EJ (1997) Chemosphere 35:2973
9. Van de Plassche EJ, Balk F (1997) Environmental risk assessment of polycyclic musks AHTN and HHCB according to the EU-TGD. RIVM report no. 601 503 008, National Institute of Public Health and the Environment RIVM, Bilthoven, NL
10. Balk F, Ford RA (1999) Toxicol Lett 111:57
11. Balk F, Ford RA (1999) Toxicol Lett 111:81

12. OSPAR Commission (2000) Background document on musk xylene and other musks, series point and diffuse sources no. 101, ISBN 0946956553
13. EU (1997) Technical Guidance Document in support of Directive 96/67/EEC on risk assessment of new notified substances and Regulation (EC) No. 1488/94 on risk assessment of existing substances. Office for Official Publications of the EC, Luxembourg, Lux, Part II
14. EU (1996) EUSES, the European Union System for the Evaluation of Substances. RIVM, NL (ed) European Chemicals Bureau (EC/JRC), Ispra, Italy
15. Rimkus G, Gatermann R, Hühnerfuss H (1999) Toxicol Lett 111:5
16. Geyer HJ, Rimkus G, Wolf M, Attar A, Steinberg C, Kettrup A (1994) UWSF-Z Umweltchem Ökotox 6:9
17. Rimkus G (1999) Toxicol Lett 111:37
18. Draisci R, Marchiafava C, Ferretti E, Palleschi L, Catellani G, Anastasio A (1998) J Chromatogr A 814:187
19. Balk F (1999) Indicative environmental risk assessment of the polycyclic musks ADBI, AHMI and AITI. Report to RIFM, HASKONING report H0903.D0/R001, Nijmegen, NL
20. Rijs GBJ (1998). Personal communication cited in [10]
21. Simonich SL, Begley WM, Debaere G, Eckhoff WS (2000) Environ Sci Technol 34:959
22. SAEFL (1998) Occurrence of polycyclic musk compounds in the aquatic environment in Switzerland. Swiss Agency for the Environment, Forests and Landscape SAEFL, Berne, CH
23. Paxéus N (1996) Water Res 30:1115
24. Sauer J, Antusch E, Ripp C (1997) Vom Wasser 88:49
25. Blok J (1997) Measurement of musk fragrances in sludges of sewage treatment plants in The Netherlands. Report to RIFM, BKH Consulting Engineers, Delft, NL
26. Herren D, Berset JD (2000) Chemosphere 40:565
27. Langworthy DE, Itrich NR, Federle TW (2000) Biotransformation of the polycyclic musk HHCB, in activated sludge and river water. Poster SETAC World Conference Brighton UK, May 2000
28. Langworthy DE, Federle TW, Salvito DT (2000). Current progress in evaluating the environmental fate of fragrance ingredients. Research Institute for Fragrance Materials, Inc, suppl I, vol. II, issue II, July 2000
29. Balk F, Blok J, Salvito DT (2000) Environmental risks of musk fragrance ingredients. In: Daughton C (ed) Pharmaceuticals and personal care products in the environment. American Chemical Society Symposium Series 791:168
30. Salvito DT (2003) Chronic effects of two polycyclic musks on *Pimephales promelas* and *Brachydanio rerio*. Water Res (submitted)
31. Balk F, Salvito DT (2003) Indicative risk assessment for the polycyclic musks ADBI, AHMI and AITI. Chemosphere (in preparation)
32. RIFM (1996). 13-week oral (dietary) toxicity study of AHTN in the rat with a 4 week treatment free period. Report 28069, May 1996
33. Api AM, Ford RA (1999) Toxicol Lett 111:143
34. Behechti A, Schramm KW, Attar A, Niederfellner J, Kettrup A (1998) Water Res 32:1704
35. Breukel RMA, Balk F (1996) Musken in Rijn en Maas. RIZA Werkdocument 96.197x. National Institute for Inland Water Management and Wastewater Treatment RIZA, Lelystad, NL
36. Heberer T, Gramer S, Stan HJ (1999) Acta Hydrochim Hydrobiol 27:150
37. Winkler M, Kopf G, Hauptvogel C, Neu T (1998) Chemosphere 37:1139
38. Rimkus GG, Wolf M, Attar A, Gatermann R, Hühnerfuss H (1998) Nitro musks metabolites in biota samples from the aquatic environment. Poster 20th International Symp on Cap Chrom, Riva del Garda, Italy, May 1998 (Proceedings Huethig Verlag Heidelberg)
39. Wiertz (1995) Technical Report:Nitromusks in fish samples. Handels- und Umweltschutzlaboratorium Hamburg. Report to the Netherlands Consumentenbond, The Hague, NL
40. De Boer J, Wester PG (1996) Het voorkomen van nitro musks in Nederlandse visserijprodukten. RIVO/DLO report C060/96, IJmuiden, NL
41. Schäfer A. RIZA Personal Communication in [10]

42. Ceschi M, De Rossa M, Jäggli M (1996) Trav Chim Aliment Hyg 87:189
43. SAEFL (1995) Occurrence of nitromusk compounds in the aquatic environment in Switzerland. Swiss Agency for the Environment, Forests and Landscape SAEFL, Berne, CH
44. Gatermann R, Hühnerfuss H, Rimkus G, Attar A, Kettrup A (1998) Chemosphere 36:2535
45. Stuard SB, Caudill D, Lehman-McKeeman LD (1997) Fund Appl Toxicol 40:264
46. Lehman-McKeeman LD, Johnson DR, Caudill D (1997) Toxicol Appl Pharmacol 142:169; Lehman-McKeeman LD, Johnson DR, Caudill D, Steward SB (1997) Drug Metab Dispos 25:384
47. Chou YJ, Dietrich DR (1999) Toxicol Lett 111:17
48. US EPA (1975) EPA-660/3-75-009
49. Salvito DT (2000) Water Res 34:2625
50. Giddings JM, Salvito DT, Putt AE (2000) Water Res 34:3686
51. Schramm KW (2000) Water Res 34:2625
52. Vahl HH, Biselli S, Gatermann R, Hühnerfuss H, Westendorf J (2000) Poster, SETAC World Conference Brighton UK, May 2000
53. Chou YJ, Dietrich DR (1999) Toxicol Lett 111:27
54. RIFM (1997) Musk ketone. Study for effects on peri- and post-natal development including maternal function in the rat (Gavage administration). Rep 29617, 10 March 1997
55. RIFM (1998) Musk xylene: study for effects on peri- and post-natal development including maternal function in the rat. Rep 32949, 5 August 1998

Subject Index

Printing: Saladruck, Berlin
Binding: Stein+Lehmann, Berlin